METHODS IN MOLECULAR BIOLOGY™

Series Editor
John M. Walker
School of Life Sciences
University of Hertfordshire
Hatfield, Hertfordshire, AL10 9AB, UK

For further volumes:
http://www.springer.com/series/7651

Array Comparative Genomic Hybridization

Protocols and Applications

Edited by

Diponkar Banerjee

Department of Pathology and Laboratory Medicine, The Ottawa Hospital, Ottawa, Canada

Sohrab P. Shah

Department of Molecular Oncology, BC Cancer Agency, Vancouver, Canada

Humana Press

Editors
Diponkar Banerjee
Department of Pathology
and Laboratory Medicine The Ottawa Hospital
Ottawa, Canada

Sohrab P. Shah
Department of Molecular Oncology
BC Cancer Agency
Vancouver, Canada

ISSN 1064-3745 ISSN 1940-6029 (electronic)
ISBN 978-1-4939-5955-6 ISBN 978-1-62703-281-0 (eBook)
DOI 10.1007/978-1-62703-281-0
Springer New York Heidelberg Dordrecht London

Printed on acid-free paper

Humana Press is a brand of Springer
Springer is part of Springer Science+Business Media (www.springer.com)

Preface

The earliest descriptions of human chromosomes in the late 1800s by Arnold and Flemming were the beginning of the genomics revolution now upon us. This book on "Array Comparative Genomic Hybridization: Protocols and Applications" explores the scope of what is now possible in documenting abnormalities associated with several types of human cancers. While the technology for interrogating the human genome continues to evolve, refinement of comparative genomic hybridization (CGH) using array CGH and related technologies have provided enormous insight into human cancers at an affordable scale in research and clinical laboratories.

As stated in Chapter 2 by Ewa Przybytkowski and colleagues, "Array CGH is a mature technology with low-cost competitive products, stable commercially produced software, and standardized protocols and therefore provides a low-cost, robust, and more easily accessible technology readily available to most wet lab scientists. In addition, aCGH allows the accurate characterization of gene copy number using as little as 0.5 μg of genomic DNA." Although costs of next generation sequencing is coming down rapidly, the true cost including bioinformatics support and computing power is daunting for most investigators and clinical laboratories. The challenge for diagnostic laboratories is also to produce results within a clinically meaningful time frame. Array CGH thus is not only cost effective as data analysis is simpler, it can produce clinically relevant data in a timely manner.

Some of the highlights include Chapter 3 in which Martin Hirst discusses the variety of sequence-based DNA methylation techniques to generate DNA methylation maps from normal and abnormal human tissue. Lisa Shaffer and colleagues discuss the two basic types of genomic microarrays in Chapter 4: Comparative genomic hybridization-based arrays (aCGH) and single nucleotide polymorphism-based (SNP) arrays. They describe benefits and challenges of each method in assessing Myelodysplastic Syndrome, and provide guidance in the interpretation of the results and reporting of the diagnostic or prognostic implications to physicians.

A variety of human neoplasia are covered in this book, including breast cancer, acute and chronic myeloid leukemia, chronic lymphocytic leukemia, diffuse large B cell lymphoma, mantle cell lymphoma, malt and marginal zone lymphomas, mycosis fungoides and Sezary syndrome, adult T-cell leukemia/lymphoma, cutaneous anaplastic large cell lymphoma, HIV-related B-cell lymphomas, pediatric osteosarcoma, Wilm's tumor, childhood myelodysplastic syndromes, adenocarcinoma of lung, and brain tumors.

Finally, two outstanding chapters have been contributed by Oscar Rueda and colleagues and Gavin Ha on computational aspects of assessing copy number alteration and variation.

We are indeed honored by the generous contributions of many of the leaders in the field who have provided overviews of the technologies and many have provided detailed step-by-step protocols to allow the readers to follow and learn these techniques for their own use in research or clinical diagnostic laboratories.

Ottawa, Canada | *Diponkar Banerjee*
Vancouver, Canada | *Sohrab P. Shah*

Contents

Contributors

ADRIANA AGUILAR-MAHECHA • *Department of Oncology, Lady Davis Institute, Sir Mortimer B. Davis, Jewish General Hospital, McGill University, Montreal, QC, Canada*
BLAKE C. BALLIF • *Signature Genomic Laboratories, PerkinElmer Inc., Spokane, WA, USA*
DIPONKAR BANERJEE • *Department of Pathology and Laboratory Medicine, The Ottawa Hospital, Ottawa, BC, Canada*
MARK BASIK • *Department of Oncology, Lady Davis Institute, Sir Mortimer B. Davis, Jewish General Hospital, McGill University, Montreal, QC, Canada*
FRANCESCO BERTONI • *Fondazione per la Ricerca e la Cura sui Linfomi, IOR Institute of Oncology Research, Lymphoma & Genomics Research Program, Bellinzona, Switzerland*
ESTEBAN BRAGGIO • *Mayo Clinic Arizona, Scottsdale, AZ, USA*
CARLOS CALDAS • *Department of Oncology, University of Cambridge, Cambridge, UK; Cancer Research UK Cambridge Research Institute, Cambridge, UK; Cambridge Breast Unit, Addenbrooke's Hospital, Cambridge University Hospital NHS Foundation Trust, Cambridge, UK; NIHR Cambridge Biomedical Research Centre, Cambridge, UK; Cambridge Experimental Cancer Medicine Centre, Cambridge, UK*
DANIELA CAPELLO • *Fondazione per la Ricerca e la Cura sui Linfomi, IOR Institute of Oncology Research, Lymphoma & Genomics Research Program, Bellinzona, Switzerland*
KYUSAM CHOI • *Samsung Biomedical Research Institute, Samsung Medical Center, Seoul, Korea*
YEUN-JUN CHUNG • *Department of Internal Medicine, Diagnostic DNA Chip Center, Seoul National University College of Medicine, Seoul, Korea*
KENNETH J. CRADDOCK • *Department of Pathology, Toronto General Hospital University Health Network, Toronto, ON, Canada*
RAMON DIAZ-URIARTE • *Departamento de Bioquímica, Instituto de Investigaciones Biomédicas "Alberto Sols," Universidad Autónoma de Madrid, Madrid, Spain*
BLANCA ESPINET • *Molecular Cytogenetics Laboratory, Pathology Department, Hospital del Mar, Barcelona, Spain*
RAFAEL FONSECA • *Department of Hematology/Oncology, Mayo Clinic Arizona, Scottsdale, AZ, USA*
GIANLUCA GAIDANO • *Fondazione per la Ricerca e la Cura sui Linfomi, IOR Institute of Oncology Research, Lymphoma & Genomics Research Program, Bellinzona, Switzerland*
JUAN LUIS GARCÍA • *Unidad de Diagnóstico Molecular y Celular del Cáncer, IBMCC, Centro de Investigación del Cáncer, Universidad de Salamanca-CSIC, Salamanca, Spain; Instituto de Estudios de Ciencias de la Salud de Castilla y León, León, Spain*
GAVIN HA • *Molecular Oncology, BC Cancer Agency, Vancouver, BC, Canada*
JESÚS M. HERNÁNDEZ • *Unidad de Diagnóstico Molecular y Celular del Cáncer, IBMCC, Centro de Investigación del Cáncer, Universidad de Salamanca-CSIC, Salamanca, Spain; Servicio de Hematología, Hospital Universitario de Salamanca, Salamanca, Spain*

Martin Hirst • *Department of Microbiology and Immunology, Centre for High-Throughput Biology, University of British Columbia, Vancouver, Canada; BC Cancer Agency, Canada's Michael Smith Genome Sciences Centre, Vancouver, Canada*
Göran Jönsson • *Department of Oncology, Clinical Sciences, and CREATE Health Strategic Center for Translational Cancer Research, Lund University, Lund, Sweden*
CREATE Health Strategic Center for Translational Cancer Research, Lund University, Lund, Sweden
Hyunjeong Ju • *Samsung Biomedical Research Institute, Samsung Medical Center, Seoul, Korea; Department of Pathology, Samsung Medical Center, Sungkyunkwan University School of Medicine, Seoul, South Korea*
Seung-Hyun Jung • *Department of Internal Medicine, Diagnostic DNA Chip Center, Seoul National University College of Medicine, Seoul, Korea*
Neil E. Kay • *Department of Hematology, Mayo Clinic, Arizona, Rochester, MN, USA*
Van Kester • *Department of Dermatology, Leiden University Medical Center, Leiden, The Netherlands*
Young Hyeh Ko • *Department of Pathology, Samsung Medical Center, Sungkyunkwan University School of Medicine, Seoul, South Korea*
Youngil Koh • *Department of Internal Medicine, Diagnostic DNA Chip Center, Seoul National University College of Medicine, Seoul, South Korea*
Wan L. Lam • *British Columbia Cancer Agency, Vancouver, BC, Canada; University of British Columbia, Vancouver, BC, Canada*
Cheng Li • *Department of Biostatistics, Harvard School of Public Health, Boston, MA, USA; Department of Computational Biology, Harvard School of Public Health, Boston, MA, USA; The Dana-Farber Cancer Institute, Boston, MA, USA*
Sami N. Malek • *Division of Hematology and Oncology, Department of Internal Medicine, University of Michigan, Comprehensive Cancer Center, Ann Arbor, MI, USA*
Jose A. Martinez-Climent • *Division of Oncology, Center for Applied Medical Research, University of Navarra, Pamplona, Spain*
Gayatry Mohapatra • *Department of Medicine, Massachusetts General Hospital, Boston, MA, USA*
Sheida Nabavi • *Department of Pathology, Beth Israel Deaconess Medical Center & Center for Biomedical Informatics, Harvard Medical School, Boston, MA, USA*
Peter D. Ouillette • *Division of Hematology and Oncology, Department of Internal Medicine, University of Michigan, Ann Arbor, MI, USA*
Paul C. Park • *Department of Pathology and Molecular Medicine, Queens University, Kingston, ON, Canada*
Seonyang Park • *Department of Internal Medicine, Diagnostic DNA Chip Center, Seoul National University College of Medicine, Seoul, South Korea*
Inka Praulich • *Institute of Cell and Molecular Pathology, Hannover Medical School, Hannover, Germany*
Ewa Przybytkowski • *Department of Oncology, Lady Davis Institute, Sir Mortimer B. Davis, Jewish General Hospital, McGill University, Montreal, QC, Canada*
Evica Rajcan-Separovic • *Child and Family Research Institute, and Cytogenetics, BC Children's Hospital, University of British Columbia, Vancouver, BC, Canada; Department of Pathology and Laboratory Medicine, University of British Columbia, Vancouver, BC, Canada*

SHAHRAD ROD RASSEKH • *Division of Hematology/Oncology/BMT, Department of Pediatrics, British Columbia's Children's Hospital, University of British Columbia, Vancouver, BC, Canada; Child and Family Research Institute, Vancouver, BC, Canada*

MELISSA RIEGER MENANTEAU • *Division of Oncology, Center for Applied Medical Research, University of Navarra, Pamplona, Spain*

ANDREA RINALDI • *Fondazione per la Ricerca e la Cura sui Linfomi, IOR Institute of Oncology Research, Lymphoma & Genomics Research Program, Bellinzona, Switzerland*

MARKUS RINGNÉR • *Department of Oncology, Clinical Sciences, Lund University, Lund, Sweden; CREATE Health Strategic Center for Translational Cancer Research, Lund University, Lund, Sweden*

CRISTINA ROBLEDO • *Unidad de Diagnóstico Molecular y Celular del Cáncer, IBMCC, Centro de Investigación del Cáncer, Universidad de Salamanca-CSIC, Salamanca, Spain*

OSCAR M. RUEDA • *Cancer Research UK Cambridge Research Institute, Li Ka Shing Centre, Cambridge, UK*

BEKIM SADIKOVIC • *Department of Molecular and Human Genetics, Baylor College of Medicine, Houston, TX, USA*

ROCÍO SALGADO • *Molecular Cytogenetics Laboratory, Pathology Department, Hospital del Mar, Barcelona, Spain*

ROGER A. SCHULTZ • *Signature Genomic Laboratories, PerkinElmer Inc., Spokane, WA, USA*

SHAMINI SELVARAJAH • *Department of Medical Oncology, Dana-Farber Cancer Institute, Harvard Medical School, Boston, MA, USA*

MASAO SETO • *Division of Molecular Medicine, Aichi Cancer Center Research Institute, Nagoya, Aichi, Japan; Department of Cancer Genetics, Nagoya University Graduate School of Medicine, Nagoya, Aichi, Japan*

LISA G. SHAFFER • *Signature Genomic Laboratories, PerkinElmer Inc., Spokane, WA, USA*

SOHRAB SHAH • *Molecular Oncology, BC Cancer Agency, Vancouver, BC, Canada*

JULIA SHARMA • *Division of Neurosurgery, Department of Surgery, Vancouver General Hospital, Vancouver, BC, Canada*

KERBY A. SHEDDEN • *Department of Statistics, University of Michigan, Ann Arbor, MI, USA*

JOHAN STAAF • *Department of Oncology, Clinical Sciences, Lund University, Lund, Sweden; CREATE Health Strategic Center for Translational Cancer Research, Lund University, Lund, Sweden*

DORIS STEINEMANN • *Institute of Cell and Molecular Pathology, Hannover Medical School, Hannover, Germany*

K. SZUHAI • *Department of Molecular Cell Biology, Leiden University Medical Center, Leiden, The Netherlands*

MARCEL TAUSCHER • *Institute of Cell and Molecular Pathology, Hannover Medical School, Hannover, Germany*

C.P. TENSEN • *Department of Dermatology, Leiden University Medical Center, Leiden, The Netherlands*

PETER J. TONELLATO • *Department of Pathology, Beth Israel Deaconess Medical Center & Center for Biomedical Informatics, Harvard Medical School, Boston, MA, USA*

MING-SOUND TSAO • *Department of Pathology, University Health Network, Toronto, ON, Canada; Department of Laboratory Medicine and Pathobiology, University of Toronto, Toronto, ON, Canada*

AKIRA UMINO • *Division of Molecular Medicine, Aichi Cancer Center Research Institute, Nagoya, Aichi, Japan; Department of Cancer Genetics, Nagoya University Graduate School of Medicine, Nagoya, Aichi, Japan*

R. VAN DOORN • *Department of Dermatology, Leiden University Medical Center, Leiden, The Netherlands*

STEPHEN YIP • *Department of Pathology and Laboratory Medicine, British Columbia Cancer Agency, Vancouver, BC, Canada; Centre for Translational and Applied Genomics, British Columbia Cancer Agency, Vancouver, BC, Canada*

MARIA ZIELENSKA • *Department of Laboratory Medicine and Pathobiology, The Hospital for Sick Children, Toronto, ON, Canada*

EMANUELE ZUCCA • *Fondazione per la Ricerca e la Cura sui Linfomi, IOR Institute of Oncology Research, Lymphoma & Genomics Research Program, Bellinzona, Switzerland*

Chapter 1

Array Comparative Genomic Hybridization: An Overview of Protocols, Applications, and Technology Trends

Diponkar Banerjee

Abstract

From the earliest observations of human chromosomes in the late 1800s to modern day next generation sequencing technologies, much has been learned about human cancers by the vigorous application of the techniques of the day. In general, resolution has improved tremendously, and correspondingly the size of the datasets generated has grown exponentially such that computational methods required to handle massive datasets have had to be devised. This chapter provides a brief synopsis of the evolution of such techniques as an introduction to the subsequent chapters that provide methods and applications, relevant to research, and clinical diagnostics.

Key words: Karyotyping, Comparative genomic hybridization, BAC arrays, Submegabase resolution tiling BAC array-CGH, cDNA and oligonucleotide microarrays, Single nucleotide polymorphism (SNP) arrays, Next generation sequencing (NGS), Copy number variation (CNV), Structural variants (SV)

1. Introduction

The first observations of human chromosomes were made in the late 1800s by Arnold (1) and Flemming (2), followed by the establishment of the correct complement of 46 human chromosomes per cell by Tjio and Levan in 1956 (3) and the development the first staining methods to demonstrate chromosomal banding in 1968 by Caspersson et al. (4). Comparative genomic hybridization (CGH) as a method to detect regions of loss or gain across the entire complement of chromosomes was introduced in the early 1990s by Kallioniemi et al. (5, 6). In the late 1990s, Solinas-Toldo et al. (7) and Pinkel et al. (8) introduced array-CGH using bacterial artificial chromosome (BAC) or P1-derived artificial

Diponkar Banerjee and Sohrab P. Shah (eds.), *Array Comparative Genomic Hybridization: Protocols and Applications*, Methods in Molecular Biology vol. 973, DOI 10.1007/978-1-62703-281-0_1, © Springer Science+Business Media, LLC 2013

chromosome (PAC) arrays, while genome-wide analysis of DNA copy number changes using cDNA microarrays was described by Pollack et al. in 1999 (9). Comprehensive reviews of current methods in molecular cytogenetics and the problem of structural genomic variants have been published recently (10–21). This chapter highlights some of these topics.

2. Conventional CGH

Whereas classical cytogenetics has played an important role in the assessment of nonrandom chromosomal abnormalities (loss, gain, deletion, duplication, rearrangements, translocations, and amplification) in hematopoietic and mesenchymal neoplasms (22), solid epithelial cancers have been difficult to study using this approach for a number of reasons. The growth of cancer cells from solid tumors in culture is often poor due to low mitotic rates. The specimens may be contaminated with microbial agents and may contain necrotic tissue. If there is successful growth of a major clone in culture, it may not reflect the in vivo heterogeneity, as the primary tumor may harbor multiple subclones which may exist with complex chromosomal alterations. In addition, solid tumor metaphase chromosomes often have poor morphology (23). Karyotyping/G-banding can detect deletions, duplications, insertions, unbalanced translocations, balanced translocations, inversions at low resolution (>3 Mb), but not loss of heterozygosity (LOH), uniparental disomy (UPD), copy number variants (CNV), and indels at a higher resolution (<50 kb) or small sequence variants that require a resolution of <1 kb (10, 13).

CGH was introduced as a way to identify segmental chromosomal alterations in solid tumor cells without the need for preparation of cancer cell metaphase spreads after culture (6, 24). The basic principle of the method is to extract DNA from the tumor for assessment and from normal reference tissue (preferably pooled DNA from 100 normal reference tissues to mitigate issues related to structural variants in the normal human genome which I will discuss later in this chapter) and to prepare biotinylated total tumor DNA and digoxigenin-labeled normal genomic reference DNA which are then simultaneously hybridized to normal cell metaphase spreads in the presence of unlabeled Cot-1 blocking DNA (6). Cot-1 DNA is required to block binding to repetitive DNA sequences (6). Repetitive sequences are now estimated to comprise over two-thirds of the human genome (25).

More recent studies, however, point to Cot-1 blocking as an imperfect solution to blocking repetitive sequences, as it may introduce its own biases. Cot-1 enhanced hybridization by up to threefold for genomic probes containing conserved repetitive elements and was also found to be enriched for linked single copy sequences (26).

Hybridization of tumor DNA is detected with fluorescein isothiocyanate (FITC)–avidin, and the reference DNA hybridization is detected with rhodamine antidigoxigenin. The relative amounts of tumor and reference DNA bound at a given chromosomal locus are influenced by the relative abundance of those sequences in the two DNA samples, respectively, and the ratio of green-to-red fluorescence is measured by image analysis. Software programs are used to plot the green and red fluorescence intensities in strips orthogonal to the chromosomal axis, subtracting background fluorescence. The reference DNA serves as a control for local variations in the ability to hybridize to target chromosomes; thus, pooled DNA from multiple normal sources is desirable to mitigate against artifacts caused by structural variants in the normal human genome. Gene amplification or chromosomal duplication in the tumor DNA is detected by an elevated green-to-red ratio, while deletions or chromosomal loss are detected by a reduced ratio. A limitation of conventional CGH is its resolution of about 20 Mb as it uses normal metaphase chromosomes as the hybridization target substrate (6).

3. BAC Arrays

Initially, BAC arrays were constructed using from several hundred BAC clones to study individual chromosomes (27) to about 2,500–3,500 BAC clones representing sequenced genomes at around 1 Mb resolution (28, 29). These were capable to detecting deletions, duplications, unbalanced translocations at 1 Mb resolution, but not balanced translocations, insertions, inversions, LOH, or UPD (10).

3.1. Submegabase Resolution Tiling BAC Array-CGH

The first submegabase resolution tiling (SMRT) DNA microarray consisting of 32,433 overlapping BAC clones covering the entire human genome was constructed by Ishkanian et al. in 2004 (30). This allowed the authors to detect microamplifications and deletions containing oncogenes, tumor-suppressor genes, and novel genes in cell lines (30). This platform has led to numerous publications regarding amplifications and deletions in various cancer cell lines and clinical samples (31–48). However, SMRT microarrays do not detect balanced translocations, insertions, inversions, LOH, or UPD (10).

4. cDNA and Oligonucleotide Microarrays

cDNA arrays (9, 49, 50) and oligonucleotide arrays (51, 52) have improved resolution even further, ranging from >400 bp for custom arrays to >1 kb for genome-wide arrays. These platforms have been used extensively to study copy number changes, small deletions, and insertions (53–72). However, they do not detect balanced translocations, insertions, inversions, LOH, or UPD (10).

4.1. Artifacts in Array-CGH Platforms

Artifacts associated with array-CGH include a phenomenon known as "waves" which is an undulating pattern of peaks in data across chromosomes, requiring smoothing algorithms (73–75), GC-bias (76), whole genome amplification bias (77), plate bias, background bias (78), and dye-bias (79–83).

5. SNP Arrays

Single nucleotide polymorphism (SNP) arrays are the highest resolution (5–10 kb) array-based platforms available. They not only detect amplifications and deletions but can also detect mosaicism, LOH, and UPD and provide highly accurate copy number assessment. Co-hybridization of test and reference DNA is not required as the test DNA is hybridized directly to the SNP array (10).

SNP arrays have been widely used to study cancer cell lines and clinical samples. LOH detection using low-density SNP arrays in small cell lung cancer was reported in 2000 by Lindblad-Toh et al. (84), and LOH detection in lung cancer cell lines was reported in 2004 by Janne et al. (85). Using a 10,000 SNP array, Janne et al. (85) were able to detect LOH missed by standard microsatellite marker panels. More recently, high-density SNP arrays have been used to study copy neutral LOH (CN-LOH) and copy number alterations (CNA) in various malignancies (86–94). To avoid false discovery and correct for naturally occurring structural variants, it is important to analyze patient-specific normal DNA in parallel (95).

SNP arrays have some disadvantages in comparison to custom array-CGH platforms for copy number variation detection due to sensitivity and resolution differences (10, 96–99).

Commercial platforms are available using high-resolution SNP arrays with resolutions around 3,000 bp but with variable performance (100). These have been used successfully with formalin-fixed paraffin-embedded (FFPE) tissue (101). Ultradense platforms, however, may detect artifacts inherent to whole genome amplification and have been recommended to be used only with non-amplified DNA samples (102).

6. Next Generation Sequencing

Next-generation sequencing (NGS) involves massively parallel sequencing of millions of fragments of DNA or RNA representing the entire genome and transcriptome of the test sample. Sequence reads are aligned to the reference genome (10). The resolution is at the base pair level. This platform has led to several recent discoveries, reviews, and commentaries (103–111).

Although the speed of sequencing has vastly improved and the sequencing costs have decreased drastically compared to Sanger sequencing, most recently to a mere $1000 per genome (Life Technologies' Benchtop Ion Proton™ Sequencer; Carlsbad, California) (112), the platform requires a significant computational and bioinformatics infrastructure that is not readily available to most clinical laboratories; thus, array-based platforms are likely to be utilized in clinical laboratories until this informatics bottleneck and associated costs have been mitigated.

7. Structural Variation in the Normal Human Genome

It had been estimated that humans share 99.9% of their DNA sequence (13, 113, 114). However, this is likely an overestimate as with higher resolution platforms, greater copy number variation (CNV) and structural variants (SV) are being discovered, which led to the prediction of finding at least 100 CNVs and SVs at >50 kb per individual genome, and a higher frequency of intermediate (8–40 kb) and small size (<8 kb) variants (13). A recent study detected 151 SVs per genome, based on 185 human genomes in which 22,025 deletions and 6,000 other SVs, including insertions and tandem duplications, were detected. More than half of these were mapped at single nucleotide resolution. SV hotspots formed by common mechanisms were also mapped (115). CNV in the human genome at >50 kb size has been reported recently (116). This study identified 952 CNVs >50 kbp across 159 individual genomes. It is likely higher frequencies will be found when over 1,000 genomes have been sequenced.

8. Chapters in the Book

The technologies summarized above and their applications to the study of several human neoplasms are provided in greater detail in the following chapters by experts in this field. Method details, discussion of platform choices, and the methods for data analysis are presented. I hope this will be a useful single source of information for years to come.

References

1. Arnold J (1879) Beobachtungen über Kerntheilungen in den Zellen der Geschwülste. Virchows Arch 78:279–301
2. Flemming W (1882) Beitrige zur Kenntniss der Zelle und ihrer Lebenserscheinungen. Part III. Arch mikr Anat 20:1–86
3. Tjio JH, Levan A (1956) The chromosome number of man. Hereditas 42:1–6
4. Caspersson T, Farber S, Foley GE, Kudynowski J, Modest EJ, Simonsson E, Wagh U, Zech L (1968) Chemical differentiation along metaphase chromosomes. Exp Cell Res 49(1):219–222. doi:10.1016/0014-4827(68)90538-7
5. Kallioniemi OP, Kallioniemi A, Piper J, Isola J, Waldman FM, Gray JW, Pinkel D (1994) Optimizing comparative genomic hybridization for analysis of DNA sequence copy number changes in solid tumors. Genes Chromosomes Cancer 10(4):231–243
6. Kallioniemi A, Kallioniemi OP, Sudar D, Rutovitz D, Gray JW, Waldman F, Pinkel D (1992) Comparative genomic hybridization for molecular cytogenetic analysis of solid tumors. Science 258(5083):818–821
7. Solinas-Toldo S, Lampel S, Stilgenbauer S, Nickolenko J, Benner A, Dohner H, Cremer T, Lichter P (1997) Matrix-based comparative genomic hybridization: biochips to screen for genomic imbalances. Genes Chromosomes Cancer 20(4):399–407. doi:10.1002/(SICI)1098-2264(199712)20:4<399::AID-GCC12>3.0.CO;2-I [pii]
8. Pinkel D, Segraves R, Sudar D, Clark S, Poole I, Kowbel D, Collins C, Kuo WL, Chen C, Zhai Y, Dairkee SH, Ljung BM, Gray JW, Albertson DG (1998) High resolution analysis of DNA copy number variation using comparative genomic hybridization to microarrays. Nat Genet 20(2):207–211. doi:10.1038/2524
9. Pollack JR, Perou CM, Alizadeh AA, Eisen MB, Pergamenschikov A, Williams CF, Jeffrey SS, Botstein D, Brown PO (1999) Genome-wide analysis of DNA copy-number changes using cDNA microarrays. Nat Genet 23(1):41–46. doi:10.1038/12640
10. Le Scouarnec S, Gribble SM (2012) Characterising chromosome rearrangements: recent technical advances in molecular cytogenetics. Heredity (Edinb) 108(1):75–85. doi:hdy2011100 [pii] 10.1038/hdy.2011.100
11. Church DM, Lappalainen I, Sneddon TP, Hinton J, Maguire M, Lopez J, Garner J, Paschall J, DiCuccio M, Yaschenko E, Scherer SW, Feuk L, Flicek P (2010) Public data archives for genomic structural variation. Nat Genet 42(10):813–814. doi:ng1010-813 [pii] 10.1038/ng1010-813
12. Conrad DF, Pinto D, Redon R, Feuk L, Gokcumen O, Zhang Y, Aerts J, Andrews TD, Barnes C, Campbell P, Fitzgerald T, Hu M, Ihm CH, Kristiansson K, Macarthur DG, Macdonald JR, Onyiah I, Pang AW, Robson S, Stirrups K, Valsesia A, Walter K, Wei J, Tyler-Smith C, Carter NP, Lee C, Scherer SW, Hurles ME (2010) Origins and functional impact of copy number variation in the human genome. Nature 464(7289):704–712. doi:nature08516 [pii] 10.1038/nature08516
13. Feuk L, Carson AR, Scherer SW (2006) Structural variation in the human genome. Nat Rev Genet 7(2):85–97. doi:nrg1767 [pii] 10.1038/nrg1767
14. Feuk L, Marshall CR, Wintle RF, Scherer SW (2006) Structural variants: changing the landscape of chromosomes and design of disease studies. Hum Mol Genet 15 Spec No 1:R57–66 doi: 15/suppl_1/R57 [pii] 10.1093/hmg/ddl057
15. Khaja R, Zhang J, MacDonald JR, He Y, Joseph-George AM, Wei J, Rafiq MA, Qian C, Shago M, Pantano L, Aburatani H, Jones K, Redon R, Hurles M, Armengol L, Estivill X, Mural RJ, Lee C, Scherer SW, Feuk L (2006) Genome assembly comparison identifies structural variants in the human genome. Nat Genet 38(12):1413–1418. doi:ng1921 [pii] 10.1038/ng1921
16. Komura D, Shen F, Ishikawa S, Fitch KR, Chen W, Zhang J, Liu G, Ihara S, Nakamura H, Hurles ME, Lee C, Scherer SW, Jones KW, Shapero MH, Huang J, Aburatani H (2006) Genome-wide detection of human copy number variations using high-density DNA oligonucleotide arrays. Genome Res 16(12):1575–1584. doi:gr.5629106 [pii] 10.1101/gr.5629106
17. Pang AW, MacDonald JR, Pinto D, Wei J, Rafiq MA, Conrad DF, Park H, Hurles ME, Lee C, Venter JC, Kirkness EF, Levy S, Feuk L,

Scherer SW (2010) Towards a comprehensive structural variation map of an individual human genome. Genome Biol 11(5):R52. doi:gb-2010-11-5-r52 [pii] 10.1186/gb-2010-11-5-r52

18. Pinto D, Marshall C, Feuk L, Scherer SW (2007) Copy-number variation in control population cohorts. Hum Mol Genet 16 Spec No. 2:R168–173. doi:16/R2/R168 [pii] 10.1093/hmg/ddm241
19. Scherer SW, Lee C, Birney E, Altshuler DM, Eichler EE, Carter NP, Hurles ME, Feuk L (2007) Challenges and standards in integrating surveys of structural variation. Nat Genet 39(7 Suppl):S7–S15. doi:ng2093 [pii] 10.1038/ng2093
20. Stranger BE, Forrest MS, Dunning M, Ingle CE, Beazley C, Thorne N, Redon R, Bird CP, de Grassi A, Lee C, Tyler-Smith C, Carter N, Scherer SW, Tavare S, Deloukas P, Hurles ME, Dermitzakis ET (2007) Relative impact of nucleotide and copy number variation on gene expression phenotypes. Science 315(5813):848–853. doi:315/5813/848 [pii] 10.1126/science.1136678
21. Zhang J, Feuk L, Duggan GE, Khaja R, Scherer SW (2006) Development of bioinformatics resources for display and analysis of copy number and other structural variants in the human genome. Cytogenet Genome Res 115(3–4):205–214. doi:95916 [pii] 10.1159/000095916
22. Solomon E, Borrow J, Goddard AD (1991) Chromosome aberrations and cancer. Science 254(5035):1153–1160
23. Houldsworth J, Chaganti RS (1994) Comparative genomic hybridization: an overview. Am J Pathol 145(6):1253–1260
24. Cremer T, Tesin D, Hopman AH, Manuelidis L (1988) Rapid interphase and metaphase assessment of specific chromosomal changes in neuroectodermal tumor cells by in situ hybridization with chemically modified DNA probes. Exp Cell Res 176(2):199–220
25. de Koning AP, Gu W, Castoe TA, Batzer MA, Pollock DD (2011) Repetitive elements may comprise over two-thirds of the human genome. PLoS Genet 7(12):e1002384. doi:10.1371/journal.pgen.1002384 PGENETICS-D-11-01686 [pii]
26. Newkirk HL, Knoll JH, Rogan PK (2005) Distortion of quantitative genomic and expression hybridization by Cot-1 DNA: mitigation of this effect. Nucleic Acids Res 33(22):e191. doi:33/22/e191 [pii] 10.1093/nar/gni190 [doi]
27. Chen X, Knauf JA, Gonsky R, Wang M, Lai EH, Chissoe S, Fagin JA, Korenberg JR (1998) From amplification to gene in thyroid cancer: a high-resolution mapped bacterial-artificial-chromosome resource for cancer chromosome aberrations guides gene discovery after comparative genome hybridization. Am J Hum Genet 63(2):625–637. doi: S0002-9297(07)61506-7 [pii]
28. Snijders AM, Nowak N, Segraves R, Blackwood S, Brown N, Conroy J, Hamilton G, Hindle AK, Huey B, Kimura K, Law S, Myambo K, Palmer J, Ylstra B, Yue JP, Gray JW, Jain AN, Pinkel D, Albertson DG (2001) Assembly of microarrays for genome-wide measurement of DNA copy number. Nat Genet 29(3):263–264. doi:10.1038/ng754 ng754 [pii]
29. Fiegler H, Carr P, Douglas EJ, Burford DC, Hunt S, Scott CE, Smith J, Vetrie D, Gorman P, Tomlinson IP, Carter NP (2003) DNA microarrays for comparative genomic hybridization based on DOP-PCR amplification of BAC and PAC clones. Genes Chromosomes Cancer 36(4):361–374. doi:10.1002/gcc.10155
30. Ishkanian AS, Malloff CA, Watson SK, DeLeeuw RJ, Chi B, Coe BP, Snijders A, Albertson DG, Pinkel D, Marra MA, Ling V, MacAulay C, Lam WL (2004) A tiling resolution DNA microarray with complete coverage of the human genome. Nat Genet 36(3):299–303. doi:10.1038/ng1307 ng1307 [pii]
31. Aarts M, Dannenberg H, deLeeuw RJ, van Nederveen FH, Verhofstad AA, Lenders JW, Dinjens WN, Speel EJ, Lam WL, de Krijger RR (2006) Microarray-based CGH of sporadic and syndrome-related pheochromocytomas using a 0.1–0.2 Mb bacterial artificial chromosome array spanning chromosome arm 1p. Genes Chromosomes Cancer 45(1):83–93. doi:10.1002/gcc.20268
32. Aviel-Ronen S, Coe BP, Lau SK, da Cunha SG, Zhu CQ, Strumpf D, Jurisica I, Lam WL, Tsao MS (2008) Genomic markers for malignant progression in pulmonary adenocarcinoma with bronchioloalveolar features. Proc Natl Acad Sci USA 105(29):10155–10160. doi:0709618105 [pii] 10.1073/pnas.0709618105
33. Baldwin C, Garnis C, Zhang L, Rosin MP, Lam WL (2005) Multiple microalterations detected at high frequency in oral cancer. Cancer Res 65(17):7561–7567. doi:65/17/7561 [pii] 10.1158/0008-5472.CAN-05-1513
34. Buys TP, Chari R, Lee EH, Zhang M, MacAulay C, Lam S, Lam WL, Ling V (2007) Genetic changes in the evolution of multidrug resistance for cultured human ovarian cancer cells. Genes Chromosomes Cancer 46(12):1069–1079. doi:10.1002/gcc.20492

35. Callagy G, Pharoah P, Chin SF, Sangan T, Daigo Y, Jackson L, Caldas C (2005) Identification and validation of prognostic markers in breast cancer with the complementary use of array-CGH and tissue microarrays. J Pathol 205(3):388–396. doi:10.1002/path.1694
36. Coe BP, Henderson LJ, Garnis C, Tsao MS, Gazdar AF, Minna J, Lam S, Macaulay C, Lam WL (2005) High-resolution chromosome arm 5p array CGH analysis of small cell lung carcinoma cell lines. Genes Chromosomes Cancer 42(3):308–313. doi:10.1002/gcc.20137
37. Coe BP, Lee EH, Chi B, Girard L, Minna JD, Gazdar AF, Lam S, MacAulay C, Lam WL (2006) Gain of a region on 7p22.3, containing MAD1L1, is the most frequent event in small-cell lung cancer cell lines. Genes Chromosomes Cancer 45(1):11–19. doi:10.1002/gcc.20260
38. de Leeuw RJ, Davies JJ, Rosenwald A, Bebb G, Gascoyne RD, Dyer MJ, Staudt LM, Martinez-Climent JA, Lam WL (2004) Comprehensive whole genome array CGH profiling of mantle cell lymphoma model genomes. Hum Mol Genet 13(17):1827–1837. doi:10.1093/hmg/ddh195 ddh195 [pii]
39. Espinosa AB, Mackintosh C, Maillo A, Gutierrez L, Sousa P, Merino M, Ortiz J, de Alava E, Orfao A, Tabernero MD (2008) Array-based comparative genomic hybridization of mapped BAC DNA clones to screen for chromosome 14 copy number abnormalities in meningiomas. Eur J Hum Genet 16(12):1450–1458. doi:ejhg2008128 [pii] 10.1038/ejhg.2008.128
40. Gao K, Lockwood WW, Li J, Lam W, Li G (2008) Genomic analyses identify gene candidates for acquired irinotecan resistance in melanoma cells. Int J Oncol 32(6):1343–1349
41. Garnis C, Lockwood WW, Vucic E, Ge Y, Girard L, Minna JD, Gazdar AF, Lam S, MacAulay C, Lam WL (2006) High resolution analysis of non-small cell lung cancer cell lines by whole genome tiling path array CGH. Int J Cancer 118(6):1556–1564. doi:10.1002/ijc.21491
42. Goldstein M, Meller I, Issakov J, Orr-Urtreger A (2006) Novel genes implicated in embryonal, alveolar, and pleomorphic rhabdomyosarcoma: a cytogenetic and molecular analysis of primary tumors. Neoplasia 8(5):332–343. doi:10.1593/neo.05829
43. Ishkanian AS, Mallof CA, Ho J, Meng A, Albert M, Syed A, van der Kwast T, Milosevic M, Yoshimoto M, Squire JA, Lam WL, Bristow RG (2009) High-resolution array CGH identifies novel regions of genomic alteration in intermediate-risk prostate cancer. Prostate 69(10):1091–1100. doi:10.1002/pros.20959
44. Lockwood WW, Coe BP, Williams AC, MacAulay C, Lam WL (2007) Whole genome tiling path array CGH analysis of segmental copy number alterations in cervical cancer cell lines. Int J Cancer 120(2):436–443. doi:10.1002/ijc.22335
45. O'Toole SA, Dunn E, Sheppard BL, Klocker H, Bektic J, Smyth P, Martin C, Sheils O, O'Leary JJ (2006) Genome-wide analysis of deoxyribonucleic acid in endometrial cancer using comparative genomic hybridization microarrays. Int J Gynecol Cancer 16(2):834–842. doi:IJG530 [pii] 10.1111/j.1525-1438.2006.00530.x
46. Savola S, Klami A, Tripathi A, Niini T, Serra M, Picci P, Kaski S, Zambelli D, Scotlandi K, Knuutila S (2009) Combined use of expression and CGH arrays pinpoints novel candidate genes in Ewing sarcoma family of tumors. BMC Cancer 9:17. doi:1471-2407-9-17 [pii] 10.1186/1471-2407-9-17
47. Murphy D, Parker J, Zhou M, Fadlelmola FM, Steidl C, Karsan A, Gascoyne RD, Chen H, Banerjee D (2010) Constitutively overexpressed 21 kDa protein in Hodgkin lymphoma and aggressive non-Hodgkin lymphomas identified as cytochrome B5b (CYB5B). Mol Cancer 9:14. doi:1476-4598-9-14 [pii] 10.1186/1476-4598-9-14
48. Fadlelmola FM, Zhou M, de Leeuw RJ, Dosanjh NS, Harmer K, Huntsman D, Lam WL, Banerjee D (2008) Sub-megabase resolution tiling (SMRT) array-based comparative genomic hybridization profiling reveals novel gains and losses of chromosomal regions in Hodgkin Lymphoma and Anaplastic Large Cell Lymphoma cell lines. Mol Cancer 7:2. doi:1476-4598-7-2 [pii] 10.1186/1476-4598-7-2
49. Heiskanen M, Kononen J, Barlund M, Torhorst J, Sauter G, Kallioniemi A, Kallioniemi O (2001) CGH, cDNA and tissue microarray analyses implicate FGFR2 amplification in a small subset of breast tumors. Anal Cell Pathol 22(4):229–234
50. Heiskanen MA, Bittner ML, Chen Y, Khan J, Adler KE, Trent JM, Meltzer PS (2000) Detection of gene amplification by genomic hybridization to cDNA microarrays. Cancer Res 60(4):799–802
51. Brennan C, Zhang Y, Leo C, Feng B, Cauwels C, Aguirre AJ, Kim M, Protopopov A, Chin L (2004) High-resolution global profiling of genomic alterations with long oligonucleotide microarray. Cancer Res 64(14):4744–4748. doi:10.1158/0008-5472.CAN-04-1241 64/14/4744 [pii]

52. Carvalho B, Ouwerkerk E, Meijer GA, Ylstra B (2004) High resolution microarray comparative genomic hybridisation analysis using spotted oligonucleotides. J Clin Pathol 57(6): 644–646
53. Smetana J, Frohlich J, Vranova V, Mikulasova A, Kuglik P, Hajek R (2011) Oligonucleotide-based array CGH as a diagnostic tool in multiple myeloma patients. Klin Onkol 24(Suppl):S43–S48
54. Waddell N, Arnold J, Cocciardi S, da Silva L, Marsh A, Riley J, Johnstone CN, Orloff M, Assie G, Eng C, Reid L, Keith P, Yan M, Fox S, Devilee P, Godwin AK, Hogervorst FB, Couch F, Grimmond S, Flanagan JM, Khanna K, Simpson PT, Lakhani SR, Chenevix-Trench G (2010) Subtypes of familial breast tumours revealed by expression and copy number profiling. Breast Cancer Res Treat 123(3): 661–677. doi:10.1007/s10549-009-0653-1
55. Toujani S, Dessen P, Ithzar N, Danglot G, Richon C, Vassetzky Y, Robert T, Lazar V, Bosq J, Da Costa L, Perot C, Ribrag V, Patte C, Wiels J, Bernheim A (2009) High resolution genome-wide analysis of chromosomal alterations in Burkitt's lymphoma. PLoS One 4(9):e7089. doi:10.1371/journal.pone.0007089
56. Tefferi A, Sirhan S, Sun Y, Lasho T, Finke CM, Weisberger J, Bale S, Compton J, LeDuc CA, Pardanani A, Thorland EC, Shevchenko Y, Grodman M, Chung WK (2009) Oligonucleotide array CGH studies in myeloproliferative neoplasms: comparison with JAK2V617F mutational status and conventional chromosome analysis. Leuk Res 33(5):662–664. doi:S0145-2126(08)00416-5 [pii] 10.1016/j.leukres.2008.09.009
57. Maciejewski JP, Tiu RV, O'Keefe C (2009) Application of array-based whole genome scanning technologies as a cytogenetic tool in haematological malignancies. Br J Haematol 146(5):479–488. doi:BJH7757 [pii] 10.1111/j.1365-2141.2009.07757.x
58. Legoffic A, Calvo EL, Barthet M, Delpero JR, Dagorn JC, Iovanna JL (2009) Identification of genomic alterations associated with the aggressiveness of pancreatic cancer using an ultra-high-resolution CGH array. Pancreatology 9(3):267–272. doi:000212092 [pii] 10.1159/000212092
59. Cooke SL, Pole JC, Chin SF, Ellis IO, Caldas C, Edwards PA (2008) High-resolution array CGH clarifies events occurring on 8p in carcinogenesis. BMC Cancer 8:288. doi:1471-2407-8-288 [pii] 10.1186/1471-2407-8-288
60. Lee JJ, Au AY, Foukakis T, Barbaro M, Kiss N, Clifton-Bligh R, Staaf J, Borg A, Delbridge L, Robinson BG, Wallin G, Hoog A, Larsson C (2008) Array-CGH identifies cyclin D1 and UBCH10 amplicons in anaplastic thyroid carcinoma. Endocr Relat Cancer 15(3):801–815. doi:15/3/801 [pii] 10.1677/ERC-08-0018
61. Giefing M, Arnemann J, Martin-Subero JI, Nielander I, Bug S, Hartmann S, Arnold N, Tiacci E, Frank M, Hansmann ML, Kuppers R, Siebert R (2008) Identification of candidate tumour suppressor gene loci for Hodgkin and Reed-Sternberg cells by characterisation of homozygous deletions in classical Hodgkin lymphoma cell lines. Br J Haematol 142(6): 916–924. doi:BJH7262 [pii] 10.1111/j.1365-2141.2008.07262.x
62. Chen HI, Hsu FH, Jiang Y, Tsai MH, Yang PC, Meltzer PS, Chuang EY, Chen Y (2008) A probe-density-based analysis method for array CGH data: simulation, normalization and centralization. Bioinformatics 24(16): 1749–1756. doi:btn321 [pii] 10.1093/bioinformatics/btn321
63. Fuhrmann C, Schmidt-Kittler O, Stoecklein NH, Petat-Dutter K, Vay C, Bockler K, Reinhardt R, Ragg T, Klein CA (2008) High-resolution array comparative genomic hybridization of single micrometastatic tumor cells. Nucleic Acids Res 36(7):e39. doi:gkn101 [pii] 10.1093/nar/gkn101
64. Bernheim A, Toujani S, Saulnier P, Robert T, Casiraghi O, Validire P, Temam S, Menard P, Dessen P, Fouret P (2008) High-resolution array comparative genomic hybridization analysis of human bronchial and salivary adenoid cystic carcinoma. Lab Invest 88(5):464–473. doi:labinvest200818 [pii] 10.1038/labinvest.2008.18
65. Costa JL, Meijer G, Ylstra B, Caldas C (2008) Array comparative genomic hybridization copy number profiling: a new tool for translational research in solid malignancies. Semin Radiat Oncol 18(2):98–104. doi:S1053-4296(07)00096-3 [pii] 10.1016/j.semradonc.2007.10.005
66. Steinemann D, Cario G, Stanulla M, Karawajew L, Tauscher M, Weigmann A, Gohring G, Ludwig WD, Harbott J, Radlwimmer B, Bartram C, Lichter P, Schrappe M, Schlegelberger B (2008) Copy number alterations in childhood acute lymphoblastic leukemia and their association with minimal residual disease. Genes Chromosomes Cancer 47(6):471–480. doi:10.1002/gcc.20557
67. Zafrakas M, Tarlatzis BC, Streichert T, Pournaropoulos F, Wolfle U, Smeets SJ, Wittek B, Grimbizis G, Brakenhoff RH, Pantel K, Bontis J, Gunes C (2008) Genome-wide microarray gene expression, array-CGH analysis, and telomerase activity in advanced

ovarian endometriosis: a high degree of differentiation rather than malignant potential. Int J Mol Med 21(3):335–344

68. Mantripragada KK, Spurlock G, Kluwe L, Chuzhanova N, Ferner RE, Frayling IM, Dumanski JP, Guha A, Mautner V, Upadhyaya M (2008) High-resolution DNA copy number profiling of malignant peripheral nerve sheath tumors using targeted microarray-based comparative genomic hybridization. Clin Cancer Res 14(4):1015–1024. doi:14/4/1015 [pii] 10.1158/1078-0432.CCR-07-1305
69. Patel A, Kang SH, Lennon PA, Li YF, Rao PN, Abruzzo L, Shaw C, Chinault AC, Cheung SW (2008) Validation of a targeted DNA microarray for the clinical evaluation of recurrent abnormalities in chronic lymphocytic leukemia. Am J Hematol 83(7):540–546. doi:10.1002/ajh.21145
70. Persson F, Winnes M, Andren Y, Wedell B, Dahlenfors R, Asp J, Mark J, Enlund F, Stenman G (2008) High-resolution array CGH analysis of salivary gland tumors reveals fusion and amplification of the FGFR1 and PLAG1 genes in ring chromosomes. Oncogene 27(21):3072–3080. doi:1210961 [pii] 10.1038/sj.onc.1210961
71. Ferreira BI, Alonso J, Carrillo J, Acquadro F, Largo C, Suela J, Teixeira MR, Cerveira N, Molares A, Gomez-Lopez G, Pestana A, Sastre A, Garcia-Miguel P, Cigudosa JC (2008) Array CGH and gene-expression profiling reveals distinct genomic instability patterns associated with DNA repair and cell-cycle checkpoint pathways in Ewing's sarcoma. Oncogene 27(14):2084–2090. doi:1210845 [pii] 10.1038/sj.onc.1210845
72. Flibotte S, Moerman DG (2008) Experimental analysis of oligonucleotide microarray design criteria to detect deletions by comparative genomic hybridization. BMC Genomics 9:497. doi:1471-2164-9-497 [pii] 10.1186/1471-2164-9-497
73. Lepretre F, Villenet C, Quief S, Nibourel O, Jacquemin C, Troussard X, Jardin F, Gibson F, Kerckaert JP, Roumier C, Figeac M (2010) Waved aCGH: to smooth or not to smooth. Nucleic Acids Res 38(7):e94. doi:gkp1215 [pii] 10.1093/nar/gkp1215
74. van de Wiel MA, Brosens R, Eilers PH, Kumps C, Meijer GA, Menten B, Sistermans E, Speleman F, Timmerman ME, Ylstra B (2009) Smoothing waves in array CGH tumor profiles. Bioinformatics 25(9):1099–1104. doi:btp132 [pii] 10.1093/bioinformatics/btp132
75. Marioni JC, Thorne NP, Valsesia A, Fitzgerald T, Redon R, Fiegler H, Andrews TD, Stranger BE, Lynch AG, Dermitzakis ET, Carter NP, Tavare S, Hurles ME (2007) Breaking the waves: improved detection of copy number variation from microarray-based comparative genomic hybridization. Genome Biol 8(10):R228. doi:gb-2007-8-10-r228 [pii] 10.1186/gb-2007-8-10-r228
76. Raiford DW, Krane DE, Doom TE, Raymer ML (2010) Automated isolation of translational efficiency bias that resists the confounding effect of GC(AT)-content. IEEE/ACM Trans Comput Biol Bioinform 7(2):238–250. doi:10.1109/TCBB.2008.65
77. Knijnenburg J, van der Burg M, Tanke HJ, Szuhai K (2007) Optimized amplification and fluorescent labeling of small cell samples for genomic array-CGH. Cytometry A 71(8):585–591. doi:10.1002/cyto.a. 20412
78. Khojasteh M, Lam WL, Ward RK, MacAulay C (2005) A stepwise framework for the normalization of array CGH data. BMC Bioinformatics 6:274. doi:1471-2105-6-274 [pii] 10.1186/1471-2105-6-274
79. Kelley R, Feizi H, Ideker T (2008) Correcting for gene-specific dye bias in DNA microarrays using the method of maximum likelihood. Bioinformatics 24(1):71–77. doi:btm347 [pii] 10.1093/bioinformatics/btm347
80. Dobbin KK, Kawasaki ES, Petersen DW, Simon RM (2005) Characterizing dye bias in microarray experiments. Bioinformatics 21(10):2430–2437. doi:bti378 [pii] 10.1093/bioinformatics/bti378
81. Martin-Magniette ML, Aubert J, Cabannes E, Daudin JJ (2005) Evaluation of the gene-specific dye bias in cDNA microarray experiments. Bioinformatics 21(9):1995–2000. doi:bti302 [pii] 10.1093/bioinformatics/bti302
82. Rosenzweig BA, Pine PS, Domon OE, Morris SM, Chen JJ, Sistare FD (2004) Dye bias correction in dual-labeled cDNA microarray gene expression measurements. Environ Health Perspect 112(4):480–487
83. Dombkowski AA, Thibodeau BJ, Starcevic SL, Novak RF (2004) Gene-specific dye bias in microarray reference designs. FEBS Lett 560 (1–3):120–124. doi:10.1016/S0014-5793(04)00083-3 S0014579304000833 [pii]
84. Lindblad-Toh K, Tanenbaum DM, Daly MJ, Winchester E, Lui WO, Villapakkam A, Stanton SE, Larsson C, Hudson TJ, Johnson BE, Lander ES, Meyerson M (2000) Loss-of-heterozygosity analysis of small-cell lung carcinomas using single-nucleotide polymorphism arrays. Nat Biotechnol 18(9):1001–1005. doi:10.1038/79269

85. Janne PA, Li C, Zhao X, Girard L, Chen TH, Minna J, Christiani DC, Johnson BE, Meyerson M (2004) High-resolution single-nucleotide polymorphism array and clustering analysis of loss of heterozygosity in human lung cancer cell lines. Oncogene 23(15):2716–2726. doi:10.1038/sj.onc.1207329 1207329 [pii]
86. Kawamata N, Ogawa S, Gueller S, Ross SH, Huynh T, Chen J, Chang A, Nabavi-Nouis S, Megrabian N, Siebert R, Martinez-Climent JA, Koeffler HP (2009) Identified hidden genomic changes in mantle cell lymphoma using high-resolution single nucleotide polymorphism genomic array. Exp Hematol 37(8):937–946. doi:S0301-472X(09)00175-1 [pii] 10.1016/j.exphem.2009.04.012
87. Shlien A, Malkin D (2009) Copy number variations and cancer. Genome Med 1(6):62. doi:gm62 [pii] 10.1186/gm62
88. Walter MJ, Payton JE, Ries RE, Shannon WD, Deshmukh H, Zhao Y, Baty J, Heath S, Westervelt P, Watson MA, Tomasson MH, Nagarajan R, O'Gara BP, Bloomfield CD, Mrozek K, Selzer RR, Richmond TA, Kitzman J, Geoghegan J, Eis PS, Maupin R, Fulton RS, McLellan M, Wilson RK, Mardis ER, Link DC, Graubert TA, DiPersio JF, Ley TJ (2009) Acquired copy number alterations in adult acute myeloid leukemia genomes. Proc Natl Acad Sci USA 106(31):12950–12955. doi:0903091106 [pii] 10.1073/pnas.0903091106
89. Barresi V, Romano A, Musso N, Capizzi C, Consoli C, Martelli MP, Palumbo G, Di Raimondo F, Condorelli DF (2010) Broad copy neutral-loss of heterozygosity regions and rare recurring copy number abnormalities in normal karyotype-acute myeloid leukemia genomes. Genes Chromosomes Cancer 49(11):1014–1023. doi:10.1002/gcc.20810
90. Cheung KJ, Delaney A, Ben-Neriah S, Schein J, Lee T, Shah SP, Cheung D, Johnson NA, Mungall AJ, Telenius A, Lai B, Boyle M, Connors JM, Gascoyne RD, Marra MA, Horsman DE (2010) High resolution analysis of follicular lymphoma genomes reveals somatic recurrent sites of copy-neutral loss of heterozygosity and copy number alterations that target single genes. Genes Chromosomes Cancer 49(8):669–681. doi:10.1002/gcc.20780
91. Hagenkord JM, Monzon FA, Kash SF, Lilleberg S, Xie Q, Kant JA (2010) Array-based karyotyping for prognostic assessment in chronic lymphocytic leukemia: performance comparison of Affymetrix 10 K2.0, 250 K Nsp, and SNP6.0 arrays. J Mol Diagn: JMD 12(2):184–196. doi:S1525-1578(10)60047-5 [pii] 10.2353/jmoldx.2010.090118
92. Hartmann S, Gesk S, Scholtysik R, Kreuz M, Bug S, Vater I, Doring C, Cogliatti S, Parrens M, Merlio JP, Kwiecinska A, Porwit A, Piccaluga PP, Pileri S, Hoefler G, Kuppers R, Siebert R, Hansmann ML (2010) High resolution SNP array genomic profiling of peripheral T cell lymphomas, not otherwise specified, identifies a subgroup with chromosomal aberrations affecting the REL locus. Br J Haematol 148(3):402–412. doi:BJH7956 [pii] 10.1111/j.1365-2141.2009.07956.x
93. Zarghooni M, Bartels U, Lee E, Buczkowicz P, Morrison A, Huang A, Bouffet E, Hawkins C (2010) Whole-genome profiling of pediatric diffuse intrinsic pontine gliomas highlights platelet-derived growth factor receptor alpha and poly (ADP-ribose) polymerase as potential therapeutic targets. J Clin Oncol 28(8):1337–1344. doi:JCO.2009.25.5463 [pii] 10.1200/JCO.2009.25.5463
94. Tiu RV, Gondek LP, O'Keefe CL, Elson P, Huh J, Mohamedali A, Kulasekararaj A, Advani AS, Paquette R, List AF, Sekeres MA, McDevitt MA, Mufti GJ, Maciejewski JP (2011) Prognostic impact of SNP array karyotyping in myelodysplastic syndromes and related myeloid malignancies. Blood 117(17):4552–4560. doi:blood-2010-07-295857 [pii] 10.1182/blood-2010-07-295857
95. Heinrichs S, Li C, Look AT (2010) SNP array analysis in hematologic malignancies: avoiding false discoveries. Blood 115(21):4157–4161. doi:blood-2009-11-203182 [pii] 10.1182/blood-2009-11-203182
96. Curtis C, Lynch AG, Dunning MJ, Spiteri I, Marioni JC, Hadfield J, Chin SF, Brenton JD, Tavare S, Caldas C (2009) The pitfalls of platform comparison: DNA copy number array technologies assessed. BMC Genomics 10:588. doi:1471-2164-10-588 [pii] 10.1186/1471-2164-10-588
97. Alkan C, Coe BP, Eichler EE (2011) Genome structural variation discovery and genotyping. Nat Rev Genet 12(5):363–376. doi:nrg2958 [pii] 10.1038/nrg2958 [doi]
98. Cooper GM, Zerr T, Kidd JM, Eichler EE, Nickerson DA (2008) Systematic assessment of copy number variant detection via genome-wide SNP genotyping. Nat Genet 40(10):1199–1203. doi:ng.236 [pii] 10.1038/ng.236
99. Pinto D, Darvishi K, Shi X, Rajan D, Rigler D, Fitzgerald T, Lionel AC, Thiruvahindrapuram B, Macdonald JR, Mills R, Prasad A, Noonan K, Gribble S, Prigmore E, Donahoe PK, Smith RS, Park JH, Hurles ME, Carter NP, Lee C, Scherer SW, Feuk L (2011) Comprehensive assessment of array-based platforms and calling

algorithms for detection of copy number variants. Nat Biotechnol 29(6):512–520. doi:nbt.1852 [pii] 10.1038/nbt.1852

100. Halper-Stromberg E, Frelin L, Ruczinski I, Scharpf R, Jie C, Carvalho B, Hao H, Hetrick K, Jedlicka A, Dziedzic A, Doheny K, Scott AF, Baylin S, Pevsner J, Spencer F, Irizarry RA (2011) Performance assessment of copy number microarray platforms using a spike-in experiment. Bioinformatics 27(8):1052–1060. doi:10.1093/bioinformatics/btr106
101. Krijgsman O, Israeli D, Haan JC, van Essen HF, Smeets SJ, Eijk PP, Steenbergen RD, Kok K, Tejpar S, Meijer GA, Ylstra B (2012) CGH arrays compared for DNA isolated from formalin-fixed, paraffin-embedded material. Genes Chromosomes Cancer 51(4):344–352. doi:10.1002/gcc.21920
102. Przybytkowski E, Ferrario C, Basik M (2011) The use of ultra-dense array CGH analysis for the discovery of micro-copy number alterations and gene fusions in the cancer genome. BMC Med Genomics 4:16. doi:1755-8794-4-16 [pii] 10.1186/1755-8794-4-16
103. Keller A, Backes C, Leidinger P, Kefer N, Boisguerin V, Barbacioru C, Vogel B, Matzas M, Huwer H, Katus HA, Stahler C, Meder B, Meese E (2011) Next-generation sequencing identifies novel microRNAs in peripheral blood oflung cancer patients. Mol Biosyst 7(12):3187–3199. doi:10.1039/c1mb05353a
104. Watahiki A, Wang Y, Morris J, Dennis K, O'Dwyer HM, Gleave M, Gout PW (2011) MicroRNAs associated with metastatic prostate cancer. PLoS One 6(9):e24950. doi:10.1371/journal.pone.0024950 PONE-D-11-04350 [pii]
105. Edgren H, Murumagi A, Kangaspeska S, Nicorici D, Hongisto V, Kleivi K, Rye IH, Nyberg S, Wolf M, Borresen-Dale AL, Kallioniemi O (2011) Identification of fusion genes in breast cancer by paired-end RNA-sequencing. Genome Biol 12(1):R6. doi:gb-2011-12-1-r6 [pii] 10.1186/gb-2011-12-1-r6
106. Robbins CM, Tembe WA, Baker A, Sinari S, Moses TY, Beckstrom-Sternberg S, Beckstrom-Sternberg J, Barrett M, Long J, Chinnaiyan A, Lowey J, Suh E, Pearson JV, Craig DW, Agus DB, Pienta KJ, Carpten JD (2011) Copy number and targeted mutational analysis reveals novel somatic events in metastatic prostate tumors. Genome Res 21(1):47–55. doi:gr.107961.110 [pii] 10.1101/gr.107961.110
107. Walsh T, Lee MK, Casadei S, Thornton AM, Stray SM, Pennil C, Nord AS, Mandell JB, Swisher EM, King MC (2010) Detection of inherited mutations for breast and ovarian cancer using genomic capture and massively parallel sequencing. Proc Natl Acad Sci USA 107(28):12629–12633. doi:1007983107 [pii] 10.1073/pnas.1007983107
108. Wood HM, Belvedere O, Conway C, Daly C, Chalkley R, Bickerdike M, McKinley C, Egan P, Ross L, Hayward B, Morgan J, Davidson L, MacLennan K, Ong TK, Papagiannopoulos K, Cook I, Adams DJ, Taylor GR, Rabbitts P (2010) Using next-generation sequencing for high resolution multiplex analysis of copy number variation from nanogram quantities of DNA from formalin-fixed paraffin-embedded specimens. Nucleic Acids Res 38(14):e151. doi:gkq510 [pii] 10.1093/nar/gkq510
109. Walter MJ, Graubert TA, Dipersio JF, Mardis ER, Wilson RK, Ley TJ (2009) Next-generation sequencing of cancer genomes: back to the future. Per Med 6(6):653. doi:10.2217/pme.09.52
110. Aparicio SA, Huntsman DG (2010) Does massively parallel DNA resequencing signify the end of histopathology as we know it? J Pathol 220(2):307–315. doi:10.1002/path.2636
111. Shah SP, Morin RD, Khattra J, Prentice L, Pugh T, Burleigh A, Delaney A, Gelmon K, Guliany R, Senz J, Steidl C, Holt RA, Jones S, Sun M, Leung G, Moore R, Severson T, Taylor GA, Teschendorff AE, Tse K, Turashvili G, Varhol R, Warren RL, Watson P, Zhao Y, Caldas C, Huntsman D, Hirst M, Marra MA, Aparicio S (2009) Mutational evolution in a lobular breast tumour profiled at single nucleotide resolution. Nature 461(7265):809–813. doi:nature08489 [pii] 10.1038/nature08489
112. Rothberg JM (2012) Life Technologies Introduces the Benchtop Ion Proton™ Sequencer; Designed to Decode a Human Genome in One Day for $1,000. http://www.lifetechnologies.com/us/en/home/about-us/news-gallery/press-releases/2012/life-techologies-itroduces-the-bechtop-io-proto.html.
113. Przeworski M, Hudson RR, Di Rienzo A (2000) Adjusting the focus on human variation. Trends Genet: TIG 16(7):296–302. doi:S0168-9525(00)02030-8 [pii]
114. Reich DE, Schaffner SF, Daly MJ, McVean G, Mullikin JC, Higgins JM, Richter DJ, Lander ES, Altshuler D (2002) Human genome sequence variation and the influence of gene history, mutation and recombination. Nat Genet 32(1):135–142. doi:10.1038/ng947 ng947 [pii]

115. Mills RE, Walter K, Stewart C, Handsaker RE, Chen K, Alkan C, Abyzov A, Yoon SC, Ye K, Cheetham RK, Chinwalla A, Conrad DF, Fu Y, Grubert F, Hajirasouliha I, Hormozdiari F, Iakoucheva LM, Iqbal Z, Kang S, Kidd JM, Konkel MK, Korn J, Khurana E, Kural D, Lam HY, Leng J, Li R, Li Y, Lin CY, Luo R, Mu XJ, Nemesh J, Peckham HE, Rausch T, Scally A, Shi X, Stromberg MP, Stutz AM, Urban AE, Walker JA, Wu J, Zhang Y, Zhang ZD, Batzer MA, Ding L, Marth GT, McVean G, Sebat J, Snyder M, Wang J, Eichler EE, Gerstein MB, Hurles ME, Lee C, McCarroll SA, Korbel JO (2011) Mapping copy number variation by population-scale genome sequencing. Nature 470(7332):59–65. doi:nature09708 [pii] 10.1038/nature09708

116. Sudmant PH, Kitzman JO, Antonacci F, Alkan C, Malig M, Tsalenko A, Sampas N, Bruhn L, Shendure J, Eichler EE (2010) Diversity of human copy number variation and multicopy genes. Science 330(6004): 641–646. doi:330/6004/641 [pii] 10.1126/science.1197005

Chapter 2

Ultradense Array CGH and Discovery of Micro-Copy Number Alterations and Gene Fusions in the Cancer Genome

Ewa Przybytkowski, Adriana Aguilar-Mahecha, Sheida Nabavi, Peter J. Tonellato, and Mark Basik

Abstract

The characterization of molecular alterations specific to cancer facilitates the discovery of predictive and prognostic biomarkers important to targeted therapeutics. Alterations critical to cancer therapeutics include copy number alterations (CNAs) such as gene amplifications and deletions as well as genomic rearrangements resulting in gene fusions. There are two genome-wide technologies used to detect CNAs: next generation sequencing (NGS) and dense microarray based comparative genomic hybridization, termed array CGH (aCGH). aCGH is a mature robust technology of lower cost and more accessible than NGS. This chapter describes the protocol steps and analysis required to obtain reliable aCGH results from clinical samples. Technical options and various necessary compromises related to the nature of clinical material are considered and the consequences of these choices for data analysis and interpretation are discussed. The chapter includes brief description of the data analysis, even though analysis is often performed by bioinformaticians. Today's cancer research requires collaboration of clinicians, molecular biologists, and mathematicians. Acquaintance with the basic principles related to the extraction of the data from arrays, its normalization and the algorithms available for analysis provides a baseline for mutual understanding and communication.

Key words: Breast cancer, Needle biopsies, Aneuploidy, Whole genome amplification, Array CGH, Copy number alterations, Gene fusions, Micro-aberration, Feature extraction, ADM-2

1. Introduction

1.1. Application of Array CGH in Cancer Research

Technological advances have dramatically increased our capacity to analyze and reveal a multitude of changes in cancer cells. The characterization of molecular alterations specific to cancer has enabled the discovery of novel predictive and prognostic biomarkers now an integral part of the development of novel targeted therapeutics. Molecular alterations critical to cancer therapeutics

Diponkar Banerjee and Sohrab P. Shah (eds.), *Array Comparative Genomic Hybridization: Protocols and Applications*, Methods in Molecular Biology vol. 973, DOI 10.1007/978-1-62703-281-0_2,

include copy number alterations (CNAs) such as gene amplifications and deletions as well as genomic rearrangements resulting in gene fusions. DNA amplifications have been shown to contain important druggable oncogenes, such as the genes encoding for the HER2 and EGF receptors (1, 2). The discovery of chromosomal translocations in solid tumors, such as that involving the *ALK* gene resulting in a novel oncogenic fusion protein in lung adenocarcinoma, have also led to promising novel therapies directed against these changes (3, 4).

There are two genome-wide technologies used to detect CNAs associated with cancer: massively parallel next generation sequencing (NGS) (5) and microarray based comparative genomic hybridization (aCGH) (6). NGS requires access to expensive NGS platforms or services, extensive novel software and bioinformatics support, large computer storage and computational resources, uses considerable quantities of genomic DNA (>5 μg), and is therefore not easily accessible or of low cost. Even when the sequencing cost decreases, the computational resource and bioinformatics needs will remain. Array CGH (aCGH) is a mature technology with low cost competitive products, stable commercially produced software and standardized protocols and therefore provides a low cost, robust, and more easily accessible technology readily available to most wet lab scientists. In addition, aCGH allows the accurate characterization of gene copy number using as little as 0.5 μg of genomic DNA (7). Such sensitivity becomes important when one considers that genomic technologies are increasingly applied to minute tumor samples such as those obtained from clinical biopsies.

aCGH is a microarray-based technique detecting the relative efficacy with which two differentially fluorescently labeled DNAs (test and reference DNA) hybridize to the array of oligonucleotide probes immobilized on a slide. The probe sequences are generally designed to cover the entire genome with precision of coverage defined by the probe density across the genome. aCGH technology and probe density continue to rapidly improve allowing accurate detection of very small copy number variations (CNVs). Currently, ultradense arrays (also called high-resolution arrays) contain over 200,000 probes and interrogate the genome with an average resolution of 10–50 kb (8). Ultrahigh resolution allows better delineation of DNA breakpoints at regions of CNAs as well as the identification of very small, focal CNAs in addition to the large chromosomal regions involving several Mb thus bridging the gap between conventional cytogenetic technologies, which address gross, microscopic chromosomal aberrations (>1 Mb) and sequencing technologies, which detect variations at the level of a single nucleotide.

There are many commercially available arrays with varying genome coverage and resolution, but as yet, there are no clear standards for reporting CNAs in the cancer genome and only a few studies that discuss the comparison between platforms and the reproducibility of the data (7, 9).

Obtaining reproducible, high-resolution copy number data with high sensitivity and few false positives is the primary objective for any high throughput experimental study. According to the most comprehensive analysis published so far, reproducibility of replicate experiments of aCGH for finding CNV in DNA from normal healthy individuals is <70% (9) and may decrease even further when cancer biopsy samples are analyzed or when test DNA is subjected to whole genome amplification (WGA) prior to labeling and hybridization (8) . Thus, the processing and analysis of the data must be carefully performed (see Subheading 3.8) and interpretation of results must take into consideration the quality of the biological material, the nature of the experimental protocols as well as analytical methods (see Subheading 3.9).

1.2. Methods for Array CGH Testing of Clinical Cancer Samples

High throughput techniques such as ultradense aCGH are not only susceptible to the "garbage in-garbage out" principle but the vast amount of data produced in a semi-automated fashion amplifies the potential complications of producing poor quality data resulting in flawed interpretation. This is especially true when dealing with clinical samples which are often very small, sometimes partially degraded, and generally of heterogeneous composition. Several steps can be taken to avoid the "garbage-in" problem and thus vastly improve the likelihood of producing high quality data and corresponding interpretation: (1) Careful assessment of sample quality, (2) Rigorous quality control for each step in sample processing, (3) meticulous experimental procedure, and (4) well informed and statistically accurate data processing and analysis. These steps are reviewed herein and will lead to confidence in the data, results, and interpretation.

Clinical cancer samples are either formalin-fixed paraffin embedded (FFPE) specimens or snap-frozen tissues of tumor pieces or tumor biopsies. FFPE specimens provide relatively large numbers of archived samples with known clinical outcomes, but the quality of DNA obtained from these specimens is usually very poor (10). In this chapter we describe the use of snap-frozen tissues from core needle biopsies (see Subheading 3.1). Snap-frozen biopsies are an excellent source of high quality DNA, usually in sufficient amounts for aCGH.

An important cancer sample issue is tumor heterogeneity. Tumorigenesis is a micro-evolution in which variants of cells clonally expand within a tissue. This dynamic process results in genetic diversity within tumors and marked changes in tissue architecture. Thus tumor samples typically contain various proportions of tumor cells and normal cells as well as components of stroma and infiltrating lymphocytes (11). Consequently, DNA from tumor samples consists of a pool of heterogeneous DNA. To overcome the pooling problem and improve DNA homogeneity tumor tissue slices can be microdissected and only tumor cells selected for further analysis (12). This strategy is often applied to FFPE samples. When fresh frozen biopsies are used, the pathologist may first screen the biopsies, and retain only those, which contain a high percentage of tumor cells (see Subheading 3.1). In the protocol shown in this chapter we describe an additional strategy to overcome tumor heterogeneity, using flow sorting of fluorescently labeled nuclei according to ploidy (see Subheadings 3.2 and 3.3). This procedure allows the selection of relatively pure tumor cell subpopulations especially for aneuploid tumors (13, 14). Diploid fractions, however, are less reliable since they may represent diploid tumor cells as well as normal contaminants.

While flow sorting improves tumor sample homogeneity and therefore the quality of DNA samples for further analysis, the post-cell sorted quantities of DNA are often insufficient for aCGH protocols (see Subheading 3.4 and Note 7) so WGA may be required (see Subheading 3.5). There are excellent WGA protocols (15), but all can introduce bias since various genome sequences have different amplification efficiencies (16, 17). The consequences of WGA seem to be more relevant when denser arrays are used and when analysis focuses on micro-aberrations (8).

aCGH requires a reference genomic DNA. The best reference for studies of cancer genome would be matched normal DNA from the same individual. However, this may be difficult to obtain for every clinical sample. Thus, in a protocol described here we used commercially available normal human genomic DNA (see Subheadings 3.5, 3.6, and 3.7). This compromise is not without consequences for data interpretation. We briefly discuss this problem in a section on data interpretation (see Subheading 3.9).

Finally, considering the potentially confounding factors of aCGH analysis of clinical cancer samples, validation of copy number changes with other molecular biology techniques such as polymerase chain reaction (PCR) is very important. Methods and strategies applied for validation of the aCGH results are not covered in this chapter.

2. Materials

2.1. General Lab Equipment

Mini-Sub DNA Cell Bio-Rad and power supply (Bio-Rad Power Pac 3000).

Gel-Doc EZ imaging system with UV tray (Bio-Rad).

DNA 120 SpeedVac Concentrator (Thermo Electron Corp.)

NanoDrop Spectrophotometer ND-1000. FLUOstar OPTIMA microplate reader (BMG LABTECH) equipped with 485 nm excitation filter and 520 nm emission filter (for PicoGreen Assay).

Microcentrifuge.

Vortex mixer.

Microwave oven.

Magnetic stir plate (2×) and stir bars (2×).

Single-channel pipettes (P10, P20, P200, P1000).

Sterile, low binding, barrier tips 100, 200, 20, and 10 μl.

Ice bucket.

Timer.

Circulating water bath.

Heat block.

Nuclease-free Eppendorf tubes (autoclaved). MICROTEST Flat Bottom Polystyrene, non-tissue culture treated 96 well plates (Falcon #351172) for PicoGreen Assay.

Powder-free gloves (latex) (see Note 1).

Sterile tweezers.

2.2. General Lab Reagents

Molecular biology grade water from Milli-Q Synthesis system (Millipore) (autoclaved).

1× PBS, prepared from 10× stock (Fisher BioReagents #BP399-20).

0.5 M EDTA (Fisher BioReagents BP2483-500).

1× TE (pH 8.0), Molecular grade (Promega #V6231).

EtOH (100/96%).

UltraPure Agarose (Invitrogen #15515-027).

Nucleic Acid Stain SYBR safe (Invitrogen #S33102).

Gene Ruler 1 kb Ladder (Fermentas #SM0311).

2.3. For Biospecimen Collection and Processing

Portable liquid nitrogen canister filled with liquid nitrogen.

Powder-free gloves (latex).

Bar-coded cryovials.

Sterile tweezers.

Cryogenic labels and permanent marker.

Log sheets.

Dry ice.

Standard pathology equipment and materials: cryomolds, optimal cutting temperature (O.C.T) compound, isopentane (2-methyl butane), cryostat, blades, glass slides, hematoxylin and eosin (H&E), etc.

2.4. If Sorting of Nuclei is Performed

Sterile 35 × 10 mm petri dishes and sterile tweezers.

Extraction buffer: 10 mM Tris HCl pH 7.4, 146 mM NaCl, 2 mM $CaCl_2$, 22 mM $MgCl_2$, BSA 0.005%, Igepal CA-630 0.1%.

Sterile single use stainless surgical blades (Lance Paragon LTD, Sheffield, England).

20G × 1 needles and 1 ml syringes.

DAPI stock solution, 10 mg/ml in DMSO (1,000×; final concentration of DAPI is 10 μg/ml).

DAPI working solution (100×), DAPI stock diluted in extraction buffer.

FACS tubes: BD Falcon 352235 12 × 75 mm, 5 ml polystyrene round bottom test tubes with a 35 μm nylon mesh cell-strainer cap.

BD FACSAria flow cytometer sorter with blue (488 nm) and violet (408 nm) lasers and 450/40 emission filter.

2.5. For Isolation of DNA from Nuclei

QIAmp DNA MicroKit (Qiagen #56304).

RNase A (100 mg/ml) (Qiagen #19101). Quant-iT PicoGreen dsDNA Assay Kit (Invitrogen/Molecular Probes #P11496)

2.6. For DNA Amplification

Human Genomic DNA (*Female*): Promega #G1521.

Illustra GenomiPhi V2 DNA Amplification Kit GE Healthcare #256600-31.

2.7. For DNA Labeling and Purification

Bioprime Array CGH genomic labeling kit (Invitrogen #45-0048).

Fluorescent nucleotides Amersham; CY3-dUTP #CA95040-146L.

CY5-dUTP #CA95040-158L.

Microcon Centrifugal Filters Ultracel YM-30 (Millipore #42410).

2.8. For Hybridization, Washing, and Scanning of the Image

SurePrint HD CGH Microarray Kits: G4411B Human Genome CGH Microarray Kit 1 × 244 K.

SurePrint G3 CGH Microarray Kits: G4447A SurePrint G3 Human CGH Microarray Kit 1 × 1 M (see Note 2).

Agilent High-Resolution Microarray Scanner Bundle G2565CA (include: Scanner, PC and LCD Monitor, Barcode reader and Feature extraction software). We used upgraded B scanner.

Hybridization chamber gasket slides, 5-pack (Agilent #G2534-60003).

Hybridization oven (Agilent #G2545A) and Hybridization oven rotator for Agilent Microarray hybridization chambers (Agilent #G2530-60029).

Hybridization Chambers (Agilent #G2534A).

Ozone-barrier slide covers (Agilent #G2505-60550).

Cot-1 DNA 1 mg/ml (Invitrogen #15279-101).

Agilent Oligo aCGH Hybridization Kit (Agilent #5188-5380).

Agilent Oligo aCGH Wash Buffer 1 and 2 set (Agilent #5188-522).

250 ml rectangular slide staining dishes (Wheaton).

1.5 l flat Pyrex baking dish for warming up the dish with washing buffer #2.

3. Methods

3.1. Biopsy Collection and Processing

Ideally, the clinical specimen should be preserved as soon as it is removed from the patient. The setting where it is relatively easy to do this is in the context of tissue biopsies. Biopsies for research purposes can be collected at the time of diagnosis or during the course of the disease.

1. The day the procedure is scheduled prepare all the material necessary for biospecimen collection (see Subheading 2.3), mark cryogenic labels with patient identifier and bring all to the room where the biopsy will take place.
2. When the specimen has been removed from the patient, take the core of tissue from the needle using sterile tweezers and put it into a pre-labeled bar-coded cryovial. Tightly close the cryovial and immediately submerge it into liquid nitrogen (see Note 3).
3. The sample can be stored for future analysis (−80 °C freezer or vapor phase of liquid nitrogen tank) or processed immediately.
4. Use a log sheet to write down the time of collection, preservation, long-term storage, or any deviation to the standard operation procedure (SOP).
5. The percent tumor cells present in the biospecimen has to be evaluated by a pathologist, thus the following steps need to be performed in a standard pathology lab.
6. Place on dry ice a beaker containing precooled isopentane (2-methyl butane) and the cryomold, let both chill.
7. Carefully remove the frozen biopsy from the cryovial using sterile tweezers, place on the chilled cryomold and cover with O.C.T compound avoiding bubbles.
8. Take the cryomold with forceps and submerge into the ice-cold isopentane for approximately 30 s (until the O.C.T has solidified).
9. Mount the O.C.T block containing the biopsy on to the cryostat (set between −25 and −30 °C) and proceed with cryosectioning of 5 μm sections (standard pathology procedure).
10. Mount 5 μm sections on glass slides and proceed with H&E staining (standard pathology protocol).
11. Have the H&E slides evaluated by a pathologist. Specimens containing at least 70% tumor cells are accepted for sorting experiments (see Note 4 and 5).

3.2. Preparation of Biopsies for Sorting Nuclei (See Note 6)

1. Place frozen O.C.T. block containing biopsy on a small sterile petri dish.
2. Thaw O.C.T. block at room temperature until O.C.T. becomes soft and transparent.
3. Using sterile tweezers pull out biopsy from O.C.T. and place it on another sterile petri dish.
4. Add 100 μl of extraction buffer to the biopsy.
5. Mince the tissue using a surgical scalpel blade.
6. Add 100 μl of extraction buffer and continue mincing the tissue (see Note 7).
7. Transfer the buffer and debris to a sterile Eppendorf tube and put it on ice.
8. Wash petri dish with another 100 μl of extraction buffer and add it to the Eppendorf tube.
9. Pass the suspension in the Eppendorf tube through a 20G needle five times using a syringe to disaggregate nuclei.
10. Add 3 μl of a DAPI working solution (100×) per 300 μl of extraction buffer and incubate on ice for 3 min.
11. Filter sample through the mesh (35 μm) into the FACS tube.
12. Wash the filter with 200 μl of extraction buffer containing 1× DAPI.

3.3. Flow Sorting of Nuclei

1. Nuclei are sorted according to DAPI intensity using BD FACSAria flow cytometer/sorter. Excitation is achieved with violet laser (408 nm) and emission of DAPI is detected with a 450/40 filter (see Note 8).
2. For optimal DNA yield, extraction from at least 20,000 nuclei is recommended (see Note 9). Figure 1 shows two examples of sorting performed with different specimens.
3. After sorting keep the nuclear suspension on ice. Do not freeze and proceed with the isolation of DNA as soon as possible.

3.4. Isolation of DNA from Nuclei and Determination of DNA Quality

It would be useful at this point to read carefully the manual published by Agilent, which describes and discusses in depth the experimental protocols and various options (18).

The following sections describe experimental choices used in our laboratory.

DNA from nuclei is isolated using QIAmp DNA MicroKit (Qiagen #56304) and a protocol designed by the manufacturer for genomic DNA isolation from tissues with some modifications.

1. Measure the volume of the nuclear suspension; if larger than 80 μl, the sample needs to be concentrated. This can be achieved by centrifugation (microcentrifuge at maximum speed, 13,000 rpm (16000×*g*) for 25 min at 4 °C) (see Note 10).

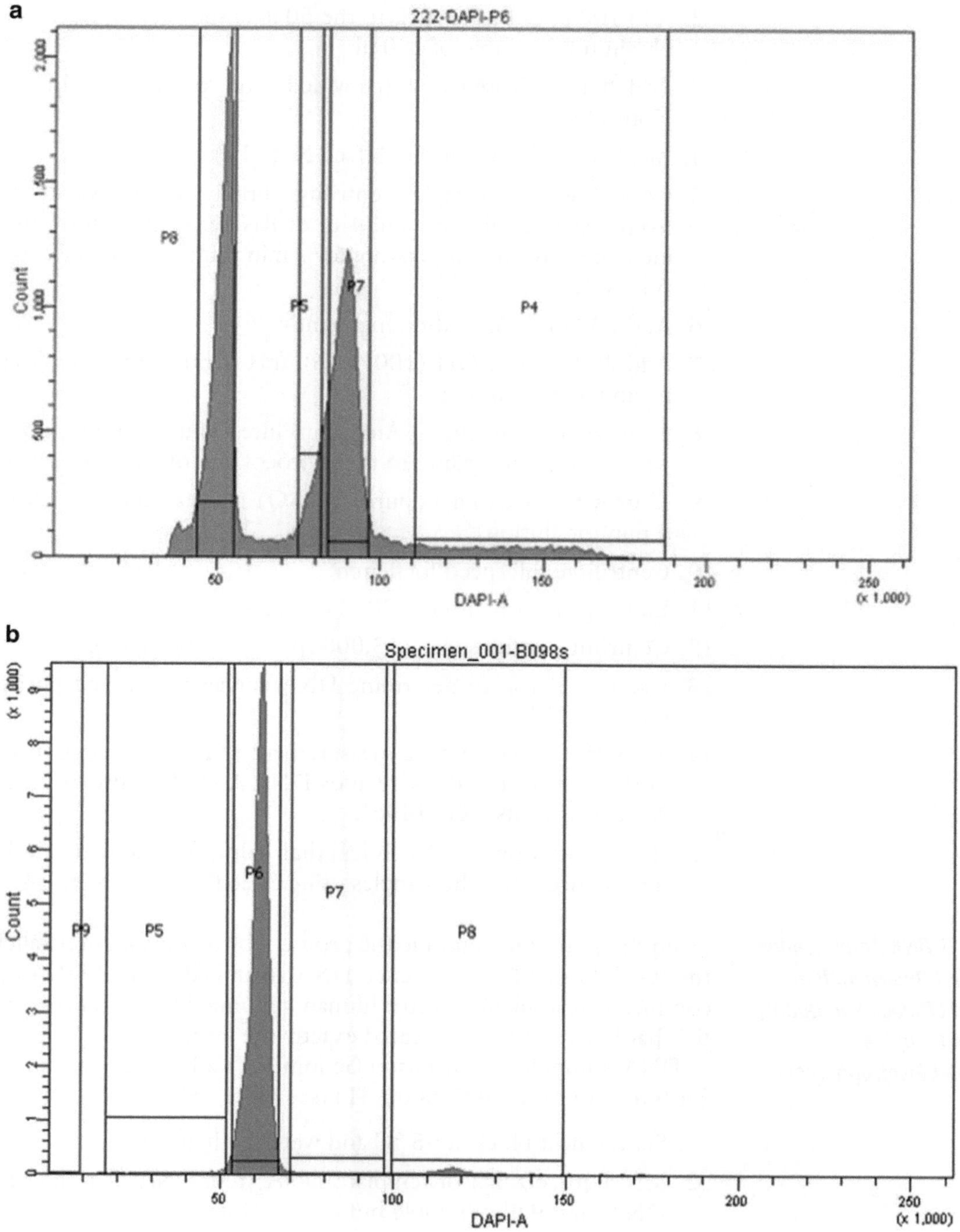

Fig. 1. FACS profiles of specimens. The numbers of nuclei are plotted against intensity of fluorescence from DAPI (**a**) Tumor specimen T-222 had two clearly defined subpopulations of nuclei, diploid P8 (105,246 nuclei were collected) and aneuploid P7 (96,166 nuclei were collected). We collected also two other, less defined subpopulations, P4 and P5. The number of nuclei in these two subpopulations was sufficient for DNA extraction (19,770 and 20,043 respectively). However, often this may not be the case and sufficient numbers of nuclei can be obtained only from one or two main subpopulations (**b**) Needle biopsy B-098 contained only one main diploid tumor subpopulation P6, which yielded 11,343,376 nuclei. The number of nuclei collected from P7 and P8 subpopulations was not sufficient to obtain enough DNA for array CGH analysis (9,600 and 3,891 respectively).

2. Add 100 µl of buffer ATL to the 50 µl suspension of nuclei to obtain total volume of 180 µl.
3. Add 20 µl of Proteinase K (provided in the kit), mix gently (see Note 11).
4. Incubate at 56 °C for 2–3 h (see Note 12).
5. Cool down the sample, centrifuge briefly to remove drops from inside the lid and add 4 µl of RNase A, mix gently and incubate at room temperature for 2 min (digestion with RNase A is optional).
6. Add 200 µl of AL buffer, mix gently.
7. Add 200 µl of EtOH (100/96%), mix gently and let it sit at room temp for 5 min.
8. Transfer lysate to the QIAmp minElute Column and proceed with washing according to the protocol described in the kit.
9. Elute with 30 µl of nanopure (Milli-Q) autoclaved water (allow 5 min for elution).
10. Centrifuge full speed for 1 min.
11. Elute again with another 30 µl of water.
12. Centrifuge at full speed (13,000 rpm (16,000×*g*)) for 1 min.
13. Use NanoDrop to determine DNA concentration and purity (see Note 13).
14. To obtain a more precise measurement of DNA concentration in the samples use PicoGreen ds DNA Assay Kit (Invitrogen/Molecular Probes #P11496).
15. If concentration of DNA is less than 10ng/µl (PicoGreen values), concentrate the samples using SpeedVac, (see Note 14).

3.5. DNA Amplification and Determination of Efficacy and Quality of Samples (Gel Electrophoresis)

From this point on, continue the protocol by processing in parallel the test DNA and the reference DNA (matched normal DNA or commercially available normal human genomic DNA). It is essential that both samples be treated exactly the same.

DNA is amplified with illustra Genomiphi V2 DNA amplification Kit (GE Healthcare #25-6600-31) (see Note 15).

1. Set the heat block at 95 °C and water bath at 30 °C.
2. Mix 1 µl (60 ng) of template DNA (test DNA or reference DNA) and 9 µl of sample buffer (see Note 16).
3. Heat samples at 95 °C for exactly 3 min (not longer!) to denature DNA then cool on ice for 3 min.
4. Prepare amplification mix (9 µl of reaction buffer + 1 µl of enzyme for each reaction), add 10 µl of the mix to each tube and incubate samples for 2 h at 30 °C.
5. Set the temperature of heat block to 65 °C, which will be needed for the next step.

6. After 2 h at 30 °C, inactivate the enzyme by heating the reaction mix at 65 °C for 10 min.
7. Cool samples on ice.
8. Verify if amplification was successful by running 1 μl of each sample on 0.8% agarose gel (see Note 17).

3.6. DNA Labeling Purification and Determination of Efficiency of Labeling and DNA Quality

Samples are fluorescently labeled using Invitrogen Array CGH genomic labeling kit (#45-0048), fluorescent nucleotides from Amersham: CY3-dUTP (#CA95040146L) and CY5-dUTP (#CA95040-158L) and manufacturer protocol:

1. Use total amount of each amplified sample, adjust their volume to 21 μl and add 20 μl of 2.5× Random Primers Solution.
2. Incubate at 95 °C for 5 min and then immediately cool on ice for 5 min.
3. On ice, add 5 μl of 10× dUTP Nucleotide Mix, 3 μl of CY5 dUTP (test DNA) or 3 μl CY3 dUTP (reference DNA) and 1 μl of Exo-Klenow Fragment.
4. Mix gently and incubate at 37 °C for 2 h.
5. Add 5 μl of Stop Buffer to each tube and place on ice. The reaction can be stored at −20 °C overnight.
6. Purify sample with Microcon Centrifugal Filters Ultracel YM-30 (Millipore #42410);

Add 430 μl of TE buffer to each tube, load on a column, and centrifuge in microcentrifuge for 10 min at 10,000 rpm (6000×*g*). Proceed with washing and recovery as described by manufacturer.

7. Adjust volume of each sample to 80.5 μl with TE buffer.
8. Check the efficiency of labeling and the yield of DNA using NanoDrop. Typical yield is 7–10 μg of DNA and samples contain 50–80 pmol of labeled nucleotides per μg of DNA. The ratio of absorbance 260 nm/280 nm should be >1.8.

3.7. Hybridization, Washing and Scanning of the Image

Hybridization is set according to the protocol described in detail in the manual from Agilent (Agilent Oligonucleotide Array-Based CGH for Genomic DNA Analysis, available on web).

1. Prepare hybridization mix: 50 μl of Cot-1 DNA (1 mg/ml, Invitrogen #15279-101), 52 μl of 10× blocking agent and 260 μl of 2× Hi-RPM Buffer (Agilent Oligo aCGH Hybridization Kit #5188-5380), and labeled DNAs (test and reference).
2. Incubate at 95 °C for 3 min following immediately by incubation at 37 °C for 30 min.
3. Use entire hybridization mixture for hybridization assembly and hybridize at 65 °C for 40 h.

4. Washing is performed with Washing Buffers from Agilent and according to their procedure, except the second wash at 37 °C is done for 3 min. We do not use stabilization or drying solutions (see Note 18).
5. Immediately after washing, slides are covered with Agilent Ozone-Barrier Slide Covers (#G2505-60550) and scanned using the Agilent scanner with 5 μm (244 K array) or 3 μm (1 × 1 M arrays) resolution.

3.8. Image Processing and Data Analysis

Ultradense aCGH images generated by the Agilent scanner require processing performed by two highly specialized software packages. First, Agilent's *Feature Extraction* software is used to review assay quality and then to process the array spot intensities to extract the raw data. To conduct subsequent array data analysis we use Agilent's *Genomic Workbench* (GW). Although platform-independent analysis algorithms exist (both commercial and open source) GW tends to provide higher quality analysis with Agilent data (9). Each step in image processing and subsequent data analysis is highly dependent on the previous step and clear understanding of the implications and ultimately the functional interpretation and hypothesis development conditioned on the results requires significant insight into the potential Type I and Type II errors at each step. The potential origin of those errors is complicated. Generally, experimentalists without experience in the analysis would be wise to develop collaborations with bioinformaticians or others with the background, experience, and insight to navigate the iterative analysis described below and produce the highest possible quality results with accurate and informed interpretation.

3.8.1. Image Processing and Feature Extraction

The Agilent *Feature Extraction* (FE) software for aCGH image processing is highly complex and requires 97 parameters (Agilent Feature Extraction Software Reference Guide, pp. 119–137) (19). Default settings should be used unless an expert in microarray image processing is available. Feature extraction software is use to:

1. View images of arrays obtained with the data generated by the Agilent scanner for preliminary inspection (.tiff files).
2. Extract the data and perform preliminary processing (.text files).
3. Create a quality control (QC) report on arrays (.pdf files).

To extract and process the data, the FE software first creates and aligns a grid to the image, identifies image spots, recognizes and rejects outlier pixels and then flags outlier features. Finally, signal and background biases are computed, calculation of signal log ratios is performed, and errors are estimated. The log ratio of fluorescence is the final readout of each spot on the array.

Explicit steps to run FE are found in the user guide (20) and briefly described in Agilent's Feature Extraction software tutorial (21).

3.8.2. aCGH Quality

Quality Control (QC) metrics integrated into the Agilent FE software facilitate the visual inspection of the overall quality of Agilent arrays. The QC Report generated by the software is an important review of the likelihood of success of the assay and the quality of the image analysis (see FE User Guide) (20).

Basic QC metrics provided by FE include the following:

1. DLRSpread (derivative log ratio spread): A measure of the log ratio noise calculated as the standard deviation (spread) of the log ratio differences between consecutive probes.
2. BGNoiseRed/BGNoiseGreen (background noise): Calculated as the standard deviations of negative control probes after rejecting nonuniform outliers (outliers significantly different from background), saturated features and feature population outliers (outliers the same as background), it is a measure of background fluorescence for each channel (Red and Green). High background fluorescence can result from inefficient washing procedure, contamination of washing dishes with detergent, or use of colored gloves when manipulating arrays (see Subheading 3.7 and Note 1).
3. SignalIntensityRed/SignalIntensityGreen (signal intensity): For each Red and Green channel, this metric is calculated as the median background-subtracted signal after rejecting nonuniform outliers and saturated features. For high quality arrays this metric should be more than 150. Low values (<50) are indicative of a failed array and marginal values (between 50 and 150), are indicative of noisy assays. Low values can result from poor quality DNA or loss during labeling and washing (see Notes 13 and 17 and Subheadings 3.6 and 3.7).
4. SignalToNoiseGreen/SignalToNoiseRed (signal to noise): For each channel this metric is calculated as SignalIntensity divided by BGNoise. The higher this ratio the better the quality of the array. If this ratio is low, there is significant noise in the data. A ratio over 100 indicates good DNA quantity and no significant problems with hybridization, washing, or scanning (see Subheading 3.7).
5. ReproducibilityRed/ReproducibilityGreen (reproducibility): For each channel, this metric calculates the median percent coefficient of variation (% CV) of background-subtracted signal for probes replicated on the chip, called noncontrol probes, after outlier rejection. It is a measure of the reproducibility of features. A large value of this metric (>0.2) indicates a problem with the hybridization procedure such as slide leakage.

3.8.3. Data Analysis

The primary goal of aCGH in cancer research is to identify regions of altered copy number such as amplifications and deletions in the cancer genome relative to the reference DNA. The aCGH data should also permit calculation of precise copy number for aberrant regions relative to the reference DNA. Analysis of aCGH assay results consists of a three-step process: (1) Loading the data, (2) Preprocessing, and (3) Identification of CNAs. Further analysis of multiple arrays or grouped assays as well as the analysis of correlations with phenotypes or medical outcomes may follow the three-step analysis of each individual array, but is outside the scope of this chapter.

1. Loading the data
   Step by step importation of FE files to GW is found in the GW User guide p. 50 (22). In addition, the correct microarray design files containing probe annotation and details can be downloaded (see Workbench user guide, pp. 51–54) (22). It is critical that the correct design files are loaded with the FE data files to ensure accurate data analysis.
2. Preprocessing
   Most aCGH-based aberration detection algorithms assume that the log ratio intensity values are approximately zero when no "true" aberration exists and adjacent probe noise signals are independent. However, this may not always be the case and GW has two preprocessing options to correct for biases before aberration detection algorithms are applied: centralization and GC correction. The centralization algorithm "zeros" the total data by adding or subtracting the same constant to all log ratio measurements thus setting the average log ratio for all array elements to zero. However, centralization should be applied with caution when highly aberrant cancer genomes are analyzed to avoid elimination of real aberration before a CNA detection algorithm is applied. This normalization can be used prior analysis with Aberration Detection Method 1 (ADM-1) and 2 (ADM-2) as well as circular binary segmentation (CBS) aberration detection algorithms.

   GC correction compensates for local GC content of the genome, which tends to cause wavy artifacts in aCGH signal intensities. Such artifacts tend to interfere with detection algorithms and therefore lead to inaccurate aberration calls. GC correction algorithms are designed to remove this correlated signal. The effects of the algorithm are often negligible, but in the case when the baseline visually shows these artifacts we recommend using GC Correction when you perform CGH + SNP analysis.

Another GW normalization algorithm is Fuzzy Zero that applies a "global error model" to all aberrant intervals. In many samples, the log ratio errors of successive probes are correlated over wide genomic intervals and aberration detection methods which are based on estimating independent errors from probe to probe, (e.g., ADM algorithms) will underestimate the error for long intervals. Consequently, for such algorithms, long aberrations with low average log ratios are often incorrectly considered significant. The global error model is a more realistic approach to avoid such erroneous aberration. We recommend using the Fuzzy Zero algorithm with GW's ADM-1 and ADM-2 aberration detection algorithms.

3. Identification of CNAs

Aberration detection algorithms typically interrogate genomic stretches or "windows" of log ratio intensities and corresponding probe quality measures to accurately identify genomic regions of altered copy number of the test genome relative to a reference genome. Agilent GW includes five such algorithms: Z-score, Aberration Detection Methods (ADM-1, and ADM-2), Hidden Markov Model (HMM), and CBS.

The Z-score method is a relatively simple method for detecting aberrant regions that are statistically different from the reference genome and tends to work well when genomic intervals have a high density of probes each with relatively high log ratios within a window defined by the user. The Z-Score method is sensitive to window size and Z-score "threshold." Changing the window size tends to influence the length of detected aberration regions and changing the Z-score threshold affects the number of probes that are rejected and thus influence the aberration calls.

The ADM algorithms avoid the fixed window size limitation and thus may be more suitable to those users with limited aberration detection analysis experience. Both ADM-1 and ADM-2 sample adjacent probe's log ratio of intensity to produce a robust estimation of the range of each aberrant segment. ADM-1 searches for intervals in which a statistical score exceeds a user specified threshold. The score is proportional to the absolute average log ratio of the genomic interval and to the square root of the number of probes in the interval. Intervals that have only a small number of probes with high log ratio values or those intervals that include relatively large number of probes with only slightly elevated log ratios will generate high score and therefore will be "called" aberrant. The statistical score represents the deviation of the average log ratios from the expected value of zero, in units of standard deviation. The ADM-2 algorithm follows the same principle but includes a "quality" estimate based on the probe log ratio errors to produce a more robust aberration "call" when the assay is of generally lower quality. Consequently, we recommend the ADM-2 method for noisy arrays.

Both ADM-1 and ADM-2 depend on a judicious selection of the statistical score threshold. Although a threshold of six is recommended in the instruction manual, we recommend the use of a more conservative threshold (8–10). The threshold's effect can be verified by visual inspection. Usually, when the array plots suggest the rejection of "true" aberrations, the threshold is too high. In addition, the aberrant regions defined by ADM-2 can be filtered (Aberration filters in a Discovery menu of GW). We have considered as aberrant only those regions, which contained minimum of five probes per region and/or those with minimum absolute $\log_2$ ratio for region >0.3. The validity of aberration detection with ADM-1 and -2 is based on the assumption that errors from probe to probe are independent. However, this assumption is often incorrect and ADM-like algorithms tend to underestimate the error for long intervals. Consequently, long aberrations with low average log ratios are often incorrectly considered significant. To solve this problem we recommend application of the Fuzzy Zero algorithm (see section on data preprocessing, above).

The HMM aberration detection algorithm is based on a stochastic process which estimates the likelihood of observing a probe with aberrant signal in the presence of uncertainty in the signal measurement. In this algorithm, probes are sequentially binned according to whether they are estimated to likely measure an amplification, a deletion, or a region in which no copy number change occurred. The state with maximum probability (deletion, amplification, or no change) is assigned to each probe. HMM's probabilistic approach assigns confidence measures or p-value to the estimate of each probe state. However, this algorithm tends to require significant computational time.

The CBS algorithm is similar to ADM-like algorithm in that it samples adjacent probes to arrive at a robust estimation of the true range of the aberrant segment. However, unlike ADM, CBS iterates within any given initial segment and measures the difference of log ratio means between two adjacent candidate segments to find a breakpoint. End points of the segment are then joined to test the combined mean log ratio. Consequently, this method identifies those break points that yield the greatest discrimination between segmental means. The output of the CBS algorithm does not include any score, and therefore, other methods, especially ADM-1 and ADM-2, are better suited for aberration classification. In addition, because of searching iteratively for the best breakpoints the CBS algorithm is not as fast as ADM and Z-score algorithms.

3.9. Data Interpretation

3.9.1. Impact of Experimental Procedures on Data Generation, Quality, Analysis, and Interpretation

The general design of high throughput experiments using clinical samples does not allow multiple measurements. To gain confidence about the data analysis and interpretation careful review and testing of the experimental protocols, data processing and analysis as well as setting standards of reproducibility with cell lines should be performed before clinical samples are tested. Protocols described in this chapter should aid your workbench preparation, but we suggest that in house testing and review be done before clinical specimens are used.

Data analysis and interpretation of aCGH experiments is a complex multi step process. Quality control at every step cannot be overemphasized, but even when done carefully, experimental complications related to clinical samples (such as sample heterogeneity, genome amplification, and choice of reference) may have consequences for data interpretation and thus should be accounted for accordingly.

1. DNA Heterogeneity: We apply two independent steps to avoid problems with DNA heterogeneity; screening of specimens by the pathologist for percentage of tumor cells and flow sorting of tumor subpopulations according to DNA ploidy. However, some diploid samples may be more difficult to interpret than others and the conclusions about precise copy number for such samples have to be interpreted with caution.
2. WGA: The consequences of WGA for ultradense aCGH are discussed in detail in our paper (8), where we concluded that the artifacts related to WGA are more pertinent to the densest 1×1 M Agilent arrays. We recommend using 244 K arrays when WGA is applied.
3. Reference Selection: The best reference for studies of cancer genome would be matched normal DNA from the same individual. However, this may be difficult to obtain for every clinical sample. The problems, which may arise due to the use of commercial normal human genomic DNA as a reference, are discussed in the following section.

3.9.2. Impact of Scientific Questions on the Analytical Approach and the Interpretation

aCGH is a hypothesis-free, genome-wide approach to investigate structural genomic rearrangements. The instability of a cancer genome is a well-accepted phenomenon that produces CNAs varying greatly in size and location within genomes (23). Analytical tools such as the algorithms described in Subheading 3.8 are meant to objectively evaluate the data to identify regions of CNAs within a cancer genome relative to the chosen reference. The next challenge is to interpret this overwhelming amount of information and translate it into scientifically useful information and clinically important knowledge.

One way to do this is to focus the attention on patterns, which can be assigned to a particular phenotype determining its molecular signature for diagnostic or prognostic purposes. However, there is no standard way of describing complex cancer aberrations. In the literature, the regions of copy number changes are often described using cytoband nomenclature taken from cytogenetics. Few terms such as "focal amplifications/deletions" or "amplicons" are commonly used, but they are not precisely defined (Fig. 2a). Focal amplification/deletion are usually considered small (there is no indication about their size), while amplicons are usually considered large (again, no indication about their size) and focused around one area (e.g., ERBB2 amplicon on chromosome 17). However, the amplified areas within large amplicons may not be adjacent to each other, even though aCGH viewing tools display them like this. Parts of chromosomes may be translocated and rearranged, creating fusions between sequences, which in a normal genome are never close to each other. The complex structure of copy number changes within amplicons suggests that many events likely contributed to their existence (Fig. 2b).

DNA amplifications have been shown to contain important druggable oncogenes, such as the genes encoding for the HER2 and EGF receptors (1, 2). Thus, another common approach is search for such genes. The potential complication here is that some large amplifications or deletions contain many genes and finding those, which are pertinent to the phenotype (so-called drivers, as opposed to passengers, which are altered, but not important for tumor development) is not a trivial task.

In our approach we focused our attention on micro-CNAs, which we defined as ranging from 100 kb (the limit of detection for 244 K Agilent platform) or 20 kb (the limit for 1 M platform) up to 1 Mb in genomic length (submicroscopic limit). In searching for genes (biomarkers and/or drug targets), such small aberrations likely affect only a few genes, (in the MCF7 genome, most of such aberrations contain only one gene) (8). However, many smaller CNAs map to regions of known CNVs in the human genome. CNVs are structural genomic variants, which are responsible for the diversity between healthy individuals, but they also play a role in determining predisposition to some diseases such as autism or HIV infection (24). Some of these variants may be inherited and benign, but others could potentially contribute to tumor development. CNVs in the size range of micro-CNAs are very common in the human genome (25). Thus, when addressing micro-CNAs in the cancer genome we need to deal with the complex issues related to natural variation between individuals. The best way to avoid these issues is to use normal matched DNA as a reference for hybridization.

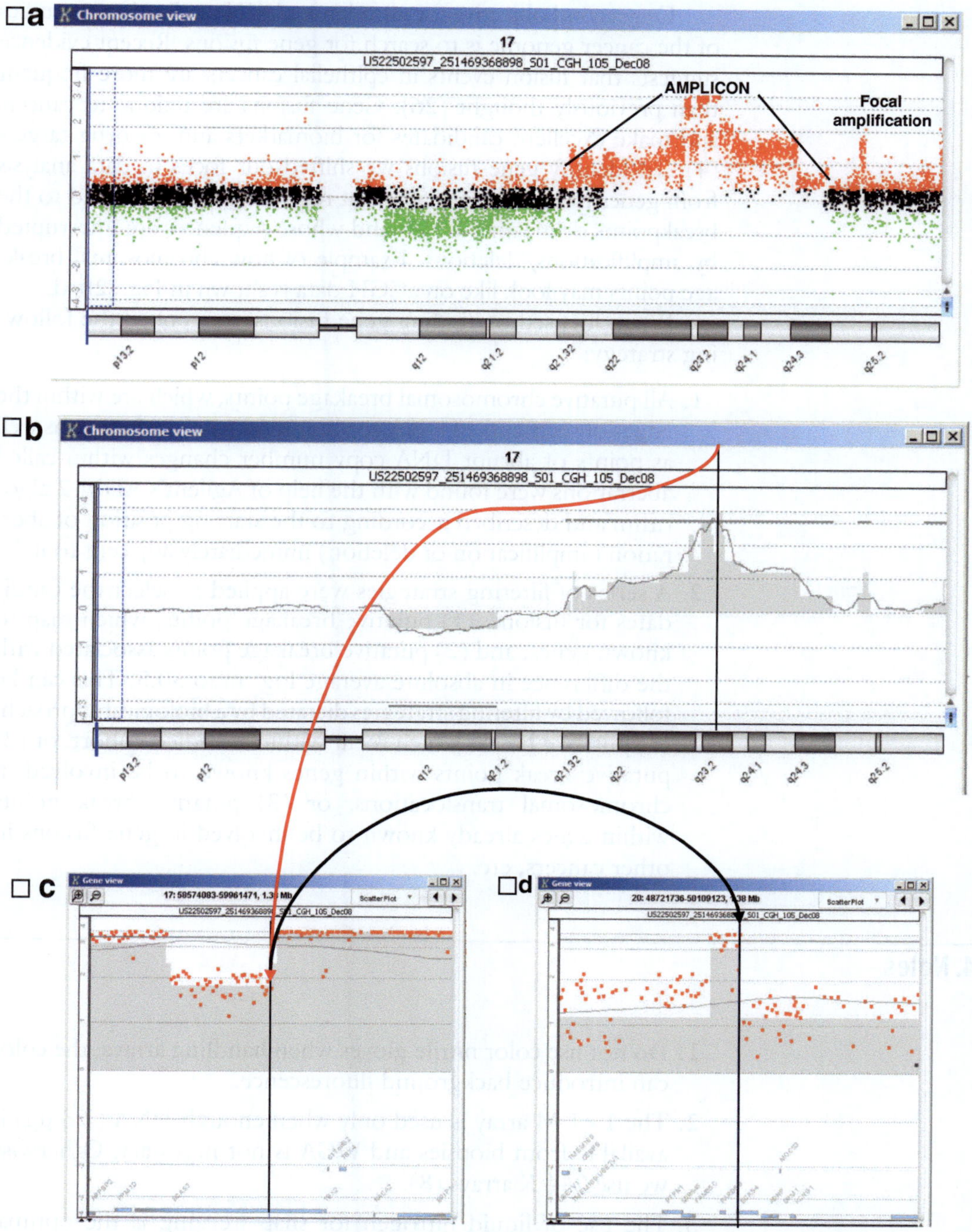

Fig. 2. Copy number alterations (CNAs) found on chromosome 17 in a genome of MCF-7 cells with ultradense aCGH platform (Agilent 244 K). (**a**) Chromosome view of a scatter plot showing data points for $\log_2$ ratios (*green*: values below $\log_2 = -0.3$; *red*: values above $\log_2 = 0.3$; *black*: values above $\log_2 = -0.3$ and below $\log_2 = 0.3$) (**b**) Chromosome view of overlaid plots showing moving average (2 Mb window) for $\log_2$ ratios of fluorescence between labeled MCF-7 DNA and the differentially labeled normal human reference, showing aberrations found with the ADM-2 algorithm (*shaded areas*) (**c**) Zoom-in on 1.38 Mb fragment of chromosome 17 showing close-up of the putative chromosomal breakage point which cut through the sequence of BCAS3 gene (genes are indicated by *blue boxes*) (**d**) Zoom-in 1.38 Mb fragment of chromosome 20 showing close-up of a putative chromosomal breakage point, which cut through the sequence of BCAS4 gene. The fusion of the BCAS3 gene on chromosome 17 with BCAS4 gene located on chromosome 20 was found by sequencing in the MCF 7 genome and validated with PCR (27).

Our second objective for ultradense aCGH application in studies of the cancer genome is to search for gene fusions. Recent evidence suggests that fusion events in epithelial cancers are more frequent than previously thought (26). Gene fusions are unique to cancers and make excellent candidates for biomarkers and/or drug targets (4). To identify gene fusions we shifted our focus in data analysis from genes mapped to the aberrant regions to those mapped to the breakpoints delineating CNAs and whose sequences are interrupted by amplifications/deletions. Example of how chromosomal breakage points may look like on aCGH data is shown in Fig. 2b–d.

When focused on finding gene fusions, we applied the following strategy:

1. All putative chromosomal breakage points, which are within the edges of segments of DNA copy number gains and losses as well as points of abrupt DNA copy number changes within called aberrations were found with the help of Agilent's ADM-2 algorithm and described according to the starting position of aberration (amplification or deletion) immediately adjacent to it.
2. A series of filtering strategies were applied to select the candidates for fusions: (1) putative breakage points, which map to known genes, and (2) putative breakage points associated with the difference in absolute average $\log_2$ ratio >0.5. This can be followed by filtering strategies dictated by a biological approach: (1) putative breaks which recur within a studied cohort, or (2) putative break points within genes known to be involved in chromosomal translocations, or (3) putative break points within genes already known to be involved in gene fusions in other cancers, etc.

4. Notes

1. Do not use color nitrile gloves when handling arrays; the color can introduce background fluorescence.
2. The 1 × 1 M array is used only when enough DNA (2.5 μg) is available from biopsies and WGA is not necessary. Otherwise we use 244 K arrays (8).
3. The use of liquid nitrogen for snap freezing is the optimal method to preserve the molecular integrity of the sample. However, liquid nitrogen is often not accessible in the clinical setting making the adoption of this method challenging. To circumvent this limitation we have attempted to use RNAlater (Ambion), a solution that allows the collection of tissue specimens at room temperature, preserving the integrity of the genetic material. Although, useful for many downstream

applications, RNAlater caused disintegration of nuclei and thus is not compatible with flow sorting prior aCGH.

4. It is possible to obtain enough nuclei from needle core biopsies with tumor content less than 70% as long as the core biopsy is at least 1mm × 5mm in size.
5. It is important to make sure that blood is collected from the patients before or after the biopsy procedure. Blood lymphocytes could be the source of matched normal DNA.
6. Flow sorting of nuclei is an excellent way to obtain pure tumor DNA from the dominant subpopulation in the specimen. However, this step can be skipped and DNA can be extracted directly from the frozen biopsy sample at the risk of having nonhomogenous DNA. The procedure for isolation of DNA in this case will be similar (Subheading 3.4) except that the digestion with Proteinase K should be carried overnight. However, flow sorting of nuclei from a biopsy may not only assure a more homogenous DNA sample, but it may also permit the isolation of different subpopulations from a single specimen, thus addressing the interesting issue of tumor evolution (11).
7. The minimum volume is 100 μl. More extraction buffer can be added if needed or if the biopsy is bigger, but keep in mind that increasing the volume of extraction buffer will dilute the sample. This will result in longer sorting time and a more dilute nuclear suspension.
8. To accurately determine the diploid reference position in a FACS profile, a diploid control sample (lymphoblastoid cell line of a normal person) is first run and then a small amount of nuclear suspension from the specimen is mixed into this control sample in order to establish FACS collecting gates.
9. The number of nuclei varies from biopsy to biopsy. We have always sorted the entire sample to get as many nuclei as possible. In some cases we obtain as many as a few millions of nuclei from one major fraction (Fig. 1). The amount of DNA obtained from this high number of nuclei would be sufficient to avoid WGA. However, the majority of specimens are not cellular and/or large enough to allow this. If WGA is not performed, DNA is digested with restriction enzymes exactly as described in the Agilent manual (Agilent Oligonucleotide Array-Based CGH for Genomic DNA Analysis). In this case the reference normal DNA will have to be digested as well.
10. The nuclear pellet is not visible in the Eppendorf tube. For better recovery of nuclei, samples with volume greater than 200 μl should be split and re-aliquoted in smaller volumes before centrifugation. After centrifugation, gently aspirate the supernatant using a pipette, do not use the vacuum. Leave a small volume of supernatant in the tube (about 80 μl).

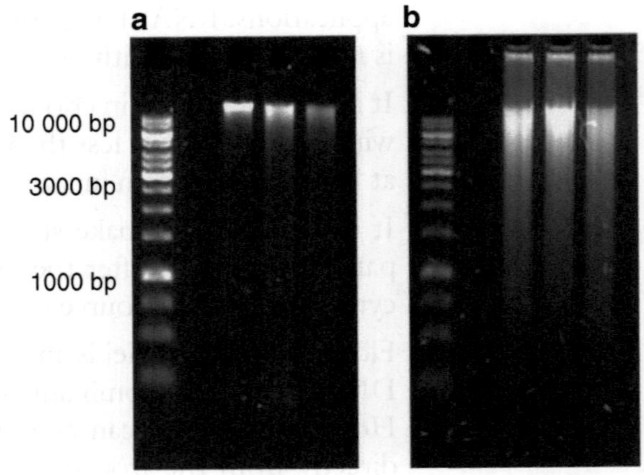

Fig. 3. Gel image of DNA samples. Samples were analyzed on an 0.8% agarose gel (**a**) DNA ladder and a clear single band of high molecular weight DNA, indicating a good quality sample (**b**) DNA Ladder and fragments of various molecular weights obtained after WGA.

11. Do not vortex, mix the solutions by gently tapping the tube with your finger.
12. We have never tested shorter times of digestion for this protocol. However, it is likely that digestion may be completed in less than an hour. This is true for samples containing nuclei. If the whole biopsy is used without sorting, samples should be digested overnight.
13. An excellent way to verify if the sample contains high molecular weight DNA is agarose gel electrophoresis. We could easily visualize quantities of about 30–50 ng of genomic DNA. A single band of high molecular weight and no smear indicates high quality (Fig. 3a). However, when the number of nuclei is small, the amount of DNA obtained might be too small to run on a gel. Our yield was about 100–200 ng of DNA from 20,000 nuclei and the following step requires a minimum of 60 ng. Thus, we most often relied on the NanoDrop and PicoGreen measurements for DNA quantitation and quality control at this step.
14. Bring down the volume to about 20 μl. Frequently check the sample during drying to make sure that it will not dry completely. Complete drying and reconstitution may cause degradation of high molecular weight DNA.
15. If you encounter problems during this or the following steps with one of the samples, the test DNA or the reference DNA, it is preferable that you restart again from the amplification step with both samples in parallel. Do not adjust volumes or put aside any aliquots.
16. Although smaller amounts of DNA can be successfully amplified (we have amplified as little as 5 ng of DNA), we do not recommend decreasing this amount; the fewer the templates, the

greater the risk of amplifying nonspecific sequences. If your sample has a concentration below 60 ng/μl, you will need to concentrate it; pipette the volume corresponding to 60 ng into the tube, which will be used for the amplification reaction and place it in a SpeedVac for 1–2 min. Make sure that your sample will never dry. Add 9 μl of sample buffer directly to this tube and continue the protocol.

While pipetting very small volumes of DNA for amplification, make sure that you see the liquid in the tip after aspiration and then deposit it to the bottom of the tube, making sure that the tip touches the bottom. The tiny drop needs to be drawn to the tube by adhesion, since micro liter drops will not fall by themselves. Note that even without any DNA in your tube you may still end up with amplified DNA, which will be visible on an agarose gel. However, this will not be your sample, but a contaminant. Unfortunately, you may realize this only after you see the final result of aCGH. Note: Cleanliness is extremely important when doing WGA.

17. Gel electrophoresis. Prepare 0.8% agarose gel (Ultrapure agarose) in TBE buffer, add Syber safe Nucleic Acid Staining solution and pour into Bio-Rad minigel apparatus. To prepare the samples use 1 μl of amplified DNA, 9 μl of water and 2 μl of loading buffer. Run a gel at stable voltage (70 V) for 2 h. A typical image of amplified DNA is shown in Figs. 3b.
18. To obtain 37 °C temperature of washing buffer #2 in an open staining dish, we proceeded as follows: a 1.5 l flat Pyrex baking dish filled with water was placed on the top of heating stir plate. After the water in the baking dish reached 50 °C, the staining dish was placed in it. The temperature in the dish reached 37 °C 15–20 min later. At that moment the washing with buffer #1 (5 min room temp.) was started and immediately followed with wash #2 at 37 °C (3 min).

References

1. Ross JS, Fletcher JA, Bloom KJ et al (2004) Targeted therapy in breast cancer: the HER-2/neu gene and protein. Mol Cell Proteomics 3:379–398
2. Shigematsu H, Gazdar AF (2006) Somatic mutations of epidermal growth factor receptor signaling pathway in lung cancers. Int J Cancer 118:257–262
3. Mano H (2008) Non-solid oncogenes in solid tumors: EML4-ALK fusion genes in lung cancer. Cancer Sci 99:2349–2355
4. Neal JW, Sequist LV (2010) Exciting new targets in lung cancer therapy: ALK, IGF-1R, HDAC, and Hh. Curr Treat Options Oncol 11:36–44
5. Meyerson M, Gabriel S, Getz G (2010) Advances in understanding cancer genomes through second-generation sequencing. Nat Rev Genet 11:685–696
6. Pinkel D, Albertson DG (2005) Array comparative genomic hybridization and its applications in cancer. Nat Genet 37(Suppl):S11–S17
7. Curtis C, Lynch AG, Dunning MJ et al (2009) The pitfalls of platform comparison: DNA copy number array technologies assessed. BMC Genomics 10:588

8. Przybytkowski E, Ferrario C, Basik M (2011) The use of ultra-dense array CGH analysis for the discovery of micro-copy number alterations and gene fusions in the cancer genome. BMC Med Genomics 4:16
9. Pinto D, Darvishi K, Shi X et al (2011) Comprehensive assessment of array-based platforms and calling algorithms for detection of copy number variants. Nat Biotechnol 29:512–520
10. Klopfleisch R, Weiss AT, Gruber AD (2011) Excavation of a buried treasure–DNA, mRNA, miRNA and protein analysis in formalin fixed, paraffin embedded tissues. Histol Histopathol 26:797–810
11. Navin N, Kendall J, Troge J et al (2011) Tumour evolution inferred by single-cell sequencing. Nature 472:90–94
12. Callagy G, Jackson L, Caldas C (2005) Comparative genomic hybridization using DNA from laser capture microdissected tissue. Methods Mol Biol 293:39–55
13. Ruiz C, Lenkiewicz E, Evers L et al (2011) Advancing a clinically relevant perspective of the clonal nature of cancer. Proc Natl Acad Sci USA 108:12054–12059
14. Navin N, Krasnitz A, Rodgers L et al (2010) Inferring tumor progression from genomic heterogeneity. Genome Res 20:68–80
15. Hughes S, Lasken R (2005) Whole genome amplification. Scion Publishing Ltd, Bloxham
16. Hughes S, Lim G, Beheshti B et al (2004) Use of whole genome amplification and comparative genomic hybridisation to detect chromosomal copy number alterations in cell line material and tumour tissue. Cytogenet Genome Res 105:18–24
17. Pugh TJ, Delaney AD, Farnoud N et al (2008) Impact of whole genome amplification on analysis of copy number variants. Nucleic Acids Res 36:e80
18. Agilent Technology, Inc. Agilent Oligonucleotide Array-Based CGH for Genomic DNA Analysis Enzymatic Labeling for Blood, Cells, or Tissues (with a High Throughput option) Protocol (February 2010) #G4410-90010_CGH_Enzymatic_Protocol_v6.2.1.pdf
19. Agilent Technology, Inc. Agilent Feature Extraction Software (v10.7) Reference Guide (October 2009) #G4460-90026_FE_Reference.pdf
20. Agilent Technology, Inc. Agilent Feature Extraction Software (v10.5) User Guide (december 2008) #G4460-90019_FE_10.5_User.pdf
21. Agilent Technology, Inc. http://www.genomics.agilent.com/GenericA.aspx?pagetype=Custom&subpagetype=Custom&pageid=2151, resource on the web.
22. Agilent Technology, Inc. Agilent Genomic Workbench 6.5 CGH Interactive Analysis User Guide (September 2010) #g3800-90028_cgh_interactive.pdf
23. Hicks J, Krasnitz A, Lakshmi B et al (2006) Novel patterns of genome rearrangement and their association with survival in breast cancer. Genome Res 16:1465–1479
24. Stankiewicz P, Lupski JR (2010) Structural variation in the human genome and its role in disease. Annu Rev Med 61:437–455
25. Speleman F, Kumps C, Buysse K et al (2008) Copy number alterations and copy number variation in cancer: close encounters of the bad kind. Cytogenet Genome Res 123:176–182
26. Prensner JR, Chinnaiyan AM (2009) Oncogenic gene fusions in epithelial carcinomas. Curr Opin Genet Dev 19:82–91
27. Hampton OA, Den Hollander P, Miller CA et al (2009) A sequence-level map of chromosomal breakpoints in the MCF-7 breast cancer cell line yields insights into the evolution of a cancer genome. Genome Res 19:167–177

Chapter 3

Epigenomics: Sequencing the Methylome

Martin Hirst

Abstract

DNA methylation patterns are increasingly surveyed through methods that utilize massively parallel sequencing. Sequence-based assays developed to detect DNA methylation can be broadly divided into those that depend on affinity enrichment, chemical conversion, or enzymatic restriction. The DNA fragments resulting from these methods are uniformly subjected to library construction and massively parallel sequencing. The sequence reads are subsequently aligned to a reference genome and subjected to specialized analytical tools to extract the underlying methylation signature. This chapter will outline these emerging techniques.

Key words: Epigenetics, Epigenomics, DNA methylation, Massively parallel sequencing

1. Introduction

Chemical modifications to both DNA and histone proteins, the building blocks of nucleosomes, provide a partial explanation for how organisms with identical DNA sequences can differ in disease susceptibility (1–3). The term "epigenetics," first coined by Conrad Waddington (4, 5), is now used to describe a field of research that seeks to understand the function of these chromatin modifications which can be both heritable and transient in nature. Recent technological advancements have catalyzed great strides forward in our understanding of how chromatin modifications are patterned genome wide. Normal cell development is accompanied by marked changes in the epigenome and specific epigenome signatures distinguish pluripotent, developing, and terminally differentiated cells. Anomalies in epigenetic programming characterize many human diseases including cancer, cardiovascular disease, neuropsychiatric disorders, imprinting disorders, inflammation, and autoimmune disease (6). In contrast to pathogenic somatic and germ line

Diponkar Banerjee and Sohrab P. Shah (eds.), *Array Comparative Genomic Hybridization: Protocols and Applications*, Methods in Molecular Biology vol. 973, DOI 10.1007/978-1-62703-281-0_3,

mutations, epimutations are potentially reversible, and inhibitors of chromatin-modifying enzymes, including histone deacetylase and DNA methyltransferase inhibitors, have demonstrated clinical anticancer activity (7).

An exponential growth in sequencing throughput (8) coupled with innovative molecular biology techniques (9–14) has enabled the reproducible assessment of epigenetic marks across the entire genome of mammalian cells, and large-scale international efforts are now underway to generate high-resolution reference epigenome maps to accelerate scientific exploitation of human epigenomic information (15, 16). These epigenome maps integrate detailed DNA methylation, histone modification, nucleosome occupancy, and coding and noncoding RNA expression in different normal and disease cell types, with the goal of providing new insights into many diseases, and the discovery of new means to control them.

DNA methylation occurs on the C5 position of cytosines (5mC) that precede guanines (CpG dinucleotides) and at non-CpG cytosines in plants and embryonic stem cells in mammals (17–19). Additional modifications of cytosine, including 5-hydroxymethylation (5hmC) and its oxidized derivatives, 5-formylcytosine (5fC) and 5-carboxylcytosine (5caC), catalyzed by the ten-eleven translocation (TET) (20) family of proteins acting on methylated cytosine, have been shown to exist in mammalian cells(21–29). 5hmC, either directly or through 5caC, is thought to lead to the demethylation of 5mC through both passive and active mechanisms, although direct evidence of an active 5mC to cytosine conversion has remained elusive (21, 29–31).

Approximately 5% of cytosines in a somatic human genome are methylcytosines, and 2% are hydroxymethylcytosines (23), and the majority of the CpG dinucleotides are found to be methylated in a somatic mammalian genome (32). CpG methylation occurs predominantly in repetitive genomic regions, including satellite repeats, long and short interspersed transposable elements, and endogenous retroviruses (33). Cytosines within regions of increased CpG density, known as CpG islands (CGIs), are predominately unmethylated, and these regions, along with their flanking sequences (CpG Shores), are believed to play a regulatory role in the genome (34, 35).

In general, cytosine methylation is associated with gene silencing and methylated cytosines have been shown to recruit histone deacetylases and other factors involved in transcriptional silencing (36–38). However, methylated cytosines are also found enriched within transcriptionally active gene bodies where they appear to influence aspects of RNA polymerase II activity including elongation (39), splicing (40), and initiation from intragenic alternative promoters (41). At least one methyl-CpG-binding protein, MeCP2, can act as a transcriptional activator in the context of sparse CpG methylation (42, 43). In addition many CpG islands, located at genes that have a tissue-restricted expression pattern, remain unmethylated at all stages of development and in all tissue types (44). Thus, other mechanisms must be in play to keep these genes silent. A small but significant proportion of CpG islands become methylated during development, and when this happens, the associated promoter is stably repressed (17).

Genome-wide methylcytosine patterns are profoundly altered in cancer, and localized alterations methylcytosine are being identified in an increasing number of inherited and sporadic human diseases (7, 45). Cancer genomes show overall loss of methylcytosine primarily due to hypomethylation of repetitive DNA sequences and demethylation of intronic sequences. During the development of a neoplasm, the degree of hypomethylation of genomic DNA increases as the lesion progresses from a benign proliferation of cells to a metastatic cell (46). Three mechanisms have been proposed to explain the contribution of DNA hypomethylation to the development of cancer: (1) increase in genomic instability, (2) reactivation of transposable elements, and (3) loss of imprinting (47).

Paradoxically, along with global hypomethylation, the genomes of cancer cells are also characterized with localized regions of de novo hypermethylation, predominantly in CGIs of tumor suppressor genes and microRNA (miRNA) genes. Recent evidence linking recurrent inactivating mutations to TET2 in a myeloid malignancies (48–50) and neomorphic mutations to *IDH1/2* in glioblastoma (51, 52) and acute myeloid leukemia (53) to disease progression provides a tantalizing mechanism for a localized gain in 5mC at CGIs in these malignancies.

A paradigm shift in DNA sequencing has made it possible to survey DNA methylation at a scale and resolution unthinkable 5 years ago. Numerous sequence-based methodologies have been developed to exploit an ever-increasing sequence output afforded by massively parallel sequencing platforms. This review will provide an overview of sequence-based methodologies developed to profile cytosine methylation and provide examples of how these methodologies are being applied to address fundamental questions on the role of DNA methylation in normal development and human disease.

2. Methods

Sequence-based assays developed to detect cytosine methylation can be broadly divided into those that depend on affinity enrichment, chemical conversion, or enzymatic restriction. A theme common to all methodologies is the requirement to convert the enriched or chemically converted DNA fragments into DNA structures that are capable of being sequenced on a massively parallel sequencing platform. In the simplest form this involves the addition of defined sequences on the termini of the DNA fragments to be sequenced to allow for clonal amplification and to provide a priming site to initiate the sequence-by-synthesis process. The resulting pool of sequence-ready fragments is referred to as a "library" in this review.

2.1. Enrichment-Based Methods

2.1.1. Methylated DNA Immunoprecipitation

Methylated DNA immunoprecipitation sequencing (MeDIP-Seq and hMeDIP-seq) is an immunoprecipitation-based technique where fragmented DNA is enriched based on either 5mC or 5hmC content (41, 54–61). Commercially available antibodies used in this technique (diagenode, eurogentec, active motif, and others) are raised against conjugated 5-methylcytidine or 5-hydroxymethylcytidine, and thus, the immunoprecipitation is performed in a denatured state. To facilitate sequencing of the enriched nucleic acid fragments, genomic DNA is subjected to library construction prior to enrichment. Following enrichment, primers that hybridize to the adapters are used to amplify the product directly.

In addition to direct 5hmC immunoprecipitation, methodologies that rely on specific chemical conversion of 5hmC have been developed (24, 62, 63). In the first (24), the hydroxyl group of 5hmC in duplex DNA is glucosylated by β-glucosyltransferase to form N_3-β-glucosyl-5-hydroxymethylcytosine using UDP-6-N3-Glu as a cofactor. The resulting product is subsequently labeled with a biotin moiety using click chemistry (dibenzocyclooctyne-modified biotin) followed by affinity purification (avidin), library construction, and deep sequencing. The second method (63) takes advantage of the fact that sodium bisulfite treatment converts 5hmC into cytosine 5-methylene-sulphonate (CMS) and uses specific antisera against CMS (64). As with MeDIP-seq, libraries are constructed prior to bisulfite conversion and amplified for sequencing following affinity enrichment.

At sufficient sequencing depths, on the order of two Gigabases aligned (30 million 75 nt reads; ~25% of a HiSeq2000 lane), MeDIP-seq is capable of identifying 70–80% of the 28 million CpGs in the human haploid genome at a resolution of 100–300 bases (65). This is near to the expected frequency of methylation in the human genome (17, 23). A recent comparison of published 5hmC datasets showed striking differences in enrichment profiles suggesting that additional standardization in methodologies and/or reagents is required (66).

2.1.2. Methylated DNA Binding Domain Sequencing

Methylated DNA binding domain sequencing (MBD-seq) is similar in concept to MeDIP-seq where genomic fragments are enriched based on their 5mC content (67). In this technique bead immobilized recombinant methylated-CpG binding proteins MECP2 (68) (MethylCap-Seq) or MBD2 (67) (MiGS) are used to enrich for methylated DNA fragments from a pool of fragmented duplex

DNA. Following enrichment of methylated duplex DNA fragments, standard library construction techniques are utilized for library construction. In theory, it would also be possible to construct a library prior to enrichment, and this may be particularly useful in the context of limited cellular quantities. Following enrichment the resulting fragments are amplified following a protocol similar to that for MeDIP-seq. An analogous hMBD-seq protocol has not been developed.

MBD-seq differs from MeDIP-seq in the ability to perform multiple salt cuts during elution of the 5mC containing DNA fragments bound to the immobilized methyl-binding domain. In this technique, weakly methylated DNA fragments are eluted at lower salt concentrations compared with moderately or densely methylated DNA fragments (i.e., methylated CpG islands). Thus, it is possible to selectively enrich for weakly or densely methylated DNA fragments during immunoprecipitation, potentially reducing downstream sequencing requirements. As might be expected, the use of salt cuts will also skew the underlying CG content (69).

In the absence of specific sequence or methylation content enrichment (~1 M NaCl), MBD-seq requires a similar degree of sequencing as MeDIP-seq, and at this depth (30 million aligned reads), MBD-seq is capable of identifying 70–80% of the 28 million CpGs in the human genome at a resolution of 100–300 bases (65).

A comprehensive comparison of MeDIP-seq and MBD-seq (65) (1 M NaCl cut) methylation calls demonstrated an overall high concordance but revealed differences in CpG content of regions interrogated and in sensitivity to non-CpG methylation. Consistent with a similar comparison performed using promoter arrays (69), the rate of enrichment was found to differ slightly with respect to local CpG density, with MeDIP-seq enriching more at regions with relatively low CpG density and MBD-seq enriching more at regions with slightly higher CpG density. In this study, MeDIP-seq density was found to increase with increasing non-CpG cytosine methylation (as measured by MethylC-seq), whereas MBD-seq did not, suggesting a differential sensitivity in these two enrichment methods. However, it was noted that the power to distinguish non-CpG cytosine methylation from CpG methylation was poor as non-CpG cytosine methylation was often embedded within regions of high CpG methylation (65).

2.2. Bisulfite Conversion-Based Methods

2.2.1. Whole-Genome Bisulfite Sequencing

The "gold standard" for profiling methylated cytosine is bisulfite-mediated deamination of cytosine followed by PCR amplification and sequencing. Bisulfite-mediated deamination, discovered simultaneously by the Shapiro and Hayatsu groups in the early 1970s, relies on the selectivity of the bisulfite reaction to deaminate cytosine, but not 5-methylcytosine, to uracil which is subsequently read as thymidine during sequencing (70, 71). In the original methodology, genomic DNA is bisulfite treated on mass, and genomic regions are amplified using carefully designed PCR primers, cloned and subjected to Sanger sequencing (71). The resulting sequence reads are assessed individually and visualized as a matrix with the CpG content of each clone represented as a row (41). While this approach has been extremely valuable in the analysis of select regulatory regions, and remains the first choice for validating findings from whole-genome datasets, it cannot be feasibly applied to whole-genome studies.

With the advent of next-generation sequencing, it is now possible to directly shotgun sequence bisulfite-treated genomic DNA. In this method, genomic DNA is fragmented, and sequencing adapters that have 5-methylcytosines substituted for cytosines are added to generate a pool of fragments with known sequences on the termini (72, 73). Following one or two rounds of bisulfite treatment (e.g., Qiagen EpiTect bisulfite kit), a process which is performed under denaturing conditions, the library is PCR amplified using oligonucleotides which extend the adapter to allow for clonal amplification and sequencing. This technique, termed MethylC-seq and BS-seq (a related method which does not preserve strand specificity), was first performed genome wide on the genome of the flowering plant *Arabidopsis thaliana* (72, 74) but more recently has been applied to both human and mouse genomes (73, 75–78).

In contrast to largely qualitative methylation measurements obtained from enrichment-based techniques, bisulfite sequencing provides the opportunity for quantitative methylation measurement. Analogous to variant detection in whole-genome shotgun (WGS) sequencing, redundant genome sequence coverage is required to adequately sample a bisulfite-converted library and allow for single cytosine methylation measurements. The current standard for coverage, recommended by the ENCODE (79, 80) and NIH Reference Epigenome Mapping (15) consortiums, is 30-fold redundant coverage or roughly 90 billion bases aligned. As with WGS, coverage requirements for clinical and diagnostic purposes are likely to be significantly higher (81). However, ongoing advancements in throughput and efficiency of sequencing platforms have reduced the direct sequencing costs dramatically (from >$200,000 in 2009 to ~$5,000 in 2012) to the point where it is now possible to begin to consider large-scale application of bisulfite whole-genome sequencing in a research setting.

Even with its adoption as the de facto standard in the detection of 5mC genome wide, there are limitations to a bisulfite-mediated approach. First among these is the fact that 5hmC protects cytosine from deamination to at least the same degree as 5mC (82). Thus, while bisulfite-based methods can detect 5hmC, they cannot distinguish this modification from 5mC. Second, accurate 5mC detection requires complete conversion during the bisulfite reaction, as under converted cytosine will appear as 5mC in any subsequent analyses. To estimate conversion frequency, DNA completely devoid of 5mC is spiked into the reaction prior to bisulfite conversion (73). However there is some evidence that bisulfite conversion rates of spike-ins do not accurately reflect the conversion rate of the genome as a whole. Finally, the computational requirements for whole-genome bisulfite sequencing are significantly greater, and the methodologies are less developed than WGS sequencing, primarily due to the increased complexity in the sequence alignment step.

2.2.2. Targeted Bisulfite Sequencing

The high cost and complexity of performing sodium bisulfite sequencing on a complete mammalian genome has spurred the development of strategies for enriching genomic regions of interest followed by bisulfite sequencing (78, 83–90). Two general strategies for genomic partitioning have emerged. In the first, termed reduced representation bisulfite sequencing (RRBS) (87–89, 91), the genome (formalin-fixed and paraffin-embedded samples have been successfully profiled by RRBS (91)) is digested by a methylation-insensitive restriction enzyme (Msp1) to generate short fragments that contain CpG dinucleotides at the ends. The resulting pool of DNA fragments is subjected to standard library construction using methylated adapters and size selected to generate a library of CpG rich fragments. Following sodium bisulfite conversion, the library is PCR amplified and subjected to sequencing. Due to the reduced content of RRBS libraries, multiple libraries can be sequenced in a single lane utilizing a sequence barcode that is introduced during PCR (92) substantially reducing per sample sequencing requirements to allow for larger sample sizes (90). Due to reduced diversity of bases at the 5′ end of the RRBS reads (CG), these libraries require specialized sequencing protocols that offset cluster/bead recognition. RRBS libraries cover 12% of CpGs genome wide, and these CpGs are highly enriched within CpG islands (65, 93). As with MethylC-seq and BS-seq, non-CpG methylation is captured by RRBS, and 5hmC is detected but cannot be distinguished from 5mC.

The second general strategy involves oligonucleotide-mediated genome partitioning combined with sodium bisulfite conversion and deep sequencing. Published strategies include the use of gene-specific primers (85), solid- and solution-phase capture (83, 84), and molecular inversion probes (86, 94). Methods that rely on the amplification of specific genomic segments (85, 86, 94) require carefully designed oligonucleotides that either include all possible combinations of cytosines and uracils contained within the hybridized region or avoid CpGs to mitigate specificity loss associated with the cytosine to uracil conversion. Solid- and solution-phase enrichment can be performed before (83) or after (84) bisulfite conversion. Once amplified, the targeted regions can be directly sequenced on a next-generation sequencing platform following standard techniques. As with RRBS, the reduced content of these libraries allows for highly multiplexed sequencing (10s–100s of samples). Collectively bisulfite enrichment methods are suitable for exploring methylation patterns in genomic regions that have been defined by orthologous experiments (e.g., validation (95) or frequency assessment) or functional classifications (e.g., promoter sequences (94)).

2.3. Methyl-Sensitive Restriction-Based Methods

A number of strategies have been developed that leverage the sensitivity of restriction enzymes to the presence of 5mC (96). These profile the CpG fraction of the genome by digesting the genome with restriction enzymes that are blocked by the presence of 5mCpG within their sequence recognition domain. Protocols involving a single methyl-sensitive restriction enzyme digestion (HpaII; HELP-seq, Methyl-seq, and MSCC) as well as multiple digestions (HpaII, AciI, Hinc6I; MRE-seq; and AciI, HpaII, HinP1I, and HpyCH4IV expanded MSCC) have been developed (41, 86, 97–99). Collectively, these protocols involve the digestion of genomic DNA by one, or more, methyl-sensitive enzyme followed by size selection, pooling where appropriate and library construction. Minor modifications to the standard library construction procedures are used to account nature of the overhangs generated by the restriction digests. As with RRBS the enzymes used enrich for CpG-containing fragments, and thus, survey CGIs preferentially and custom sequencing protocols are required to account for the reduced sequence diversity in the 5′ end of the digested fragments. With a four enzyme cocktail (expanded MSCC), up to 30% of CpGs genome can be assessed (97), and the libraries can be highly multiplexed prior to deep sequencing.

2.4. Direct Detection

Innovative methodologies for nucleic acid detection have raised the possibility of sequencing platforms with the capability to detect nucleotide modifications directly. In the forefront of these efforts is Pacific Biosciences who have recently demonstrated an ability to directly detect DNA methylation during single-molecule, real-time (SMRT) DNA sequencing, a technique for studying nucleic acid sequence and structure (100–102). Similarly, Oxford Nanopore Technologies has published proof of concept data for the direct detection of 5-methylcytosine (103). At the appropriate scale, these techniques offer the exciting possibility of the direct, "de novo" detection of 5mC, 5hmC, 5caC, 5fC, and 5fM genome wide and may lead to the discovery of as yet undiscovered modifications.

3. Data Analysis and Interpretation

Significant progress has been made in the stability, length, and accuracy of sequence reads generated from massively parallel sequencing platforms (NGS) since their inception nearly a decade ago. Large-scale genomics projects, such as the 1,000 genomes project (104) and The Cancer Genome Atlas, have spurred the development of well-tested "production scale" open-source informatic tools for NGS data. However, the field of NGS is relatively immature, and a myriad of tools have been, and continue to be, produced for analyzing NGS datasets, each claiming superiority over predecessors. Proposals for standardization of bioinformatic infrastructure have met with limited success (notable exception being the SAM/BAM file format). It is beyond the scope of this review to provide a detailed review of existing NGS bioinformatic tools (see http://seqanswers.com/wiki/Software/list for an updated list). Instead a high-level overview of the analysis process highlighting select tools with direct application to sequence-based cytosine methylation analysis will be provided.

Analysis of the sequence data generated from the sequence-based methylcytosine assays detailed above can be broken into two main steps: sequence alignment and segmentation. Data generated from a next-generation sequence platform consists of strings of bases and associated quality scores. The first step in analysis is to align this primary data to a reference genome assembly to generate a file containing the genomic coordinates of the alignments and orientation on the reference genome. A number of specialized aligners have been developed to map the 100s of millions reads generated in a single experiment to a mammalian-sized reference genome (105). The majority of widely adopted aligners use a "seed- and extend"-based algorithm where a sub-string contained within the read is matched to an indexed reference genome. Reads that cannot be placed uniquely in the genome are discarded. BWA (106) and Bowtie (107) are two widely adopted open-source aligners capable of generating alignments in the standardized SAM/BAM file format (108). Bisulfite-treated DNA requires specialized alignment to account for the cytosine to thymidine conversion. Several short-read alignment algorithms are available that can be configured for bisulfite-converted DNA alignment (107, 109–112), and more recently integrated packages have been made available which both align and segment (call methylation states) bisulfite datasets (113). Following read alignment, enrichment- and restriction-based datasets (meDIP, hmeDIP, MBD-seq, and MRE-seq) may be viewed directly by converting the read alignments into read density maps and displaying the result on a genome browser (114) (e.g., http://genome.ucsc.edu) or further processed through segmentation.

Segmentation methods transform raw sequence alignments into regions of signal and background. Bisulfite- and restriction-based methods provide single-nucleotide resolution while enrichment-based strategies are limited to the average length of the enriched genomic fragments sequenced (~200 bp). Segmentation of bisulfite data involves a genome-wide enumeration of C > T conversions at each cytosine in the genome; the results of which are displayed as a percentage for each base (73, 78). Alternatively the raw percentages can be transformed into defined states (113) (high, low, mixed, etc.) Restriction-based datasets are generally displayed as normalized read counts for each CpG; however, a wide distribution of read counts (65) (typically 1–1,000s) suggests that the application of a background threshold or state-based analysis would be more appropriate. Enrichment-based methods can be analyzed using a standard "peak-calling" algorithm (e.g., FindPeaks (115)) and displayed directly, or transformed into CpG calls. Alternatively normalized density measures corrected for local CpG content can be generated. In contrast to ChIP-seq datasets, input controls are generally not required as the DNA is naked during fragmentation.

There a number of stand-alone and web-based options available for visualization of aligned or segmented epigenomic datasets (116). The most mature and widely used are the genome browsers maintained by the University of California Santa Cruz (114) and Ensemble (117). These "first-generation" genome browsers enable visualization of genome-wide datasets as linear tracks provided in the context of tunable genome annotations. While extremely powerful for manual genome "browsing" and focused gene-by-gene visualizing, linear browsers become unwieldy when large numbers of individual tracks are visualized at once. To meet this challenge, "second-generation" genome browsers are being developed that allow for the integration of sample metadata (e.g., disease vs. normal) with DNA methylation tracks and other epigenetic datasets and provide user-friendly genome-wide analysis tools (118).

4. Discussion

Exponential reductions in sequencing costs have brought sequence-based DNA methylation techniques to the fore and large-scale initiatives to generate reference DNA methylation maps from normal and diseased human tissue have begun (15, 16). Genome-wide DNA methylation maps have been generated for human and murine embryonic stem cells, induced pluripotent cells, and differentiated cell types providing fundamental insights into how DNA methylation contributes to normal development. Many of these techniques are now being applied to delineate disease-specific methylation signatures to provide diagnostic and prognostic molecular biomarkers. A key challenge to the field remains the lack of an integrated, single-nucleotide resolution, assay capable of quantifying 5mC and all its oxidized derivatives. Until this is developed, a complete methylome will require the integration of multiple methods (e.g., MethylC-seq and hmeDIP-seq). This limitation should not be used as a reason to defer application of existing sequence-based methodologies as great strides in our understanding have been, and continue to be, made through these techniques. Indeed, it is the iterative improvements in molecular, sequencing, and informatics methodologies that have allowed for ever-increasing specificity and sensitivity in whole-genome sequencing. The same will be true for whole-genome methylation analysis, and in the process, our understanding of its role in normal development and disease will continue to refine.

Acknowledgments

I would like to thank all of my colleagues in the field who have contributed the data discussed in this article and apologize to all those whose work has not been included because of space constraints. This work was supported by the US National Institutes of Health (NIH) Roadmap Epigenomics Program, NIH grant 5U01ES017154-02, and the Canadian Institutes of Health Research Grant 92093.

References

1. Fraga MF, Ballestar E, Paz MF et al (2005) Epigenetic differences arise during the lifetime of monozygotic twins. Proc Natl Acad Sci U S A 102(30):10604–10609
2. Humpherys D, Eggan K, Akutsu H et al (2001) Epigenetic instability in ES cells and cloned mice. Science 293(5527):95–97
3. Esteller M (2008) Epigenetics in cancer. N Engl J Med 358(11):1148–1159
4. Waddington C (1942) The pupil contraction as an epigenetic crisis in drosophila. Proc Zool Soc Lond A11(3–4):181–188
5. Waddington CH (1942) The epigenotype. Endeavour 1(1):18–20
6. Hirst M, Marra MA (2009) Epigenetics and human disease. Int J Biochem Cell Biol 41(1):136–146. doi:10.1016/j.biocel.2008.09.011
7. Baylin SB, Jones PA (2011) A decade of exploring the cancer epigenome—biological and translational implications. Nat Rev Cancer 11(10):726–734. doi:10.1038/nrc3130
8. Metzker ML (2010) Sequencing technologies—the next generation. Nat Rev Genet 11(1):31–46. doi:10.1038/nrg2626
9. Bamshad MJ, Ng SB, Bigham AW et al (2011) Exome sequencing as a tool for mendelian disease gene discovery. Nat Rev Genet 12(11):745–755. doi:10.1038/nrg3031
10. Meyerson M, Gabriel S, Getz G (2010) Advances in understanding cancer genomes through second-generation sequencing. Nat Rev Genet 11(10):685–696. doi:10.1038/nrg2841
11. Hirst M, Marra MA (2011) Next generation sequencing based approaches to epigenomics. Brief Funct Genomic Proteomic 9(5–6):455–465. doi:10.1093/bfgp/elq035
12. Morozova O, Hirst M, Marra MA (2009) Applications of new sequencing technologies for transcriptome analysis. Annu Rev Genomics Hum Genet 10:135–151. doi:10.1146/annurev-genom-082908-145957

13. Barski A, Cuddapah S, Cui K et al (2007) High-resolution profiling of histone methylations in the human genome. Cell 129(4):823–837
14. Mortazavi A, Williams BA, Mccue K et al (2008) Mapping and quantifying mammalian transcriptomes by RNA-Seq. Nat Methods 5(7):621–628. doi:10.1038/nmeth.1226
15. Bernstein BE, Stamatoyannopoulos JA, Costello JF et al (2010) The nih roadmap epigenomics mapping consortium. Nat Biotechnol 28(10):1045–1048. doi:10.1038/nbt1010-1045
16. Abbott A (2011) Europe to map the human epigenome. Nature 477(7366):518. doi:10.1038/477518a
17. Bird A (2002) DNA methylation patterns and epigenetic memory. Genes Dev 16(1):6–21
18. Feinberg AP, Vogelstein B (1983) Hypomethylation distinguishes genes of some human cancers from their normal counterparts. Nature 301(5895):89–92
19. Gama-Sosa MA, Slagel VA, Trewyn RW et al (1983) The 5-methylcytosine content of DNA from human tumors. Nucleic Acids Res 11(19):6883–6894
20. Lorsbach RB, Moore J, Mathew S et al (2003) TET1, a member of a novel protein family, is fused to MLL in acute myeloid leukemia containing the t(10;11)(q22;q23). Leukemia 17(3):637–641. doi:10.1038/sj.leu.2402834
21. Ito S, D'alessio AC, Taranova OV et al (2010) Role of tet proteins in 5mC to 5hmC conversion, ES-cell self-renewal and inner cell mass specification. Nature 466(7310):1129–1133. doi:10.1038/nature09303
22. Koh KP, Yabuuchi A, Rao S et al (2011) Tet1 and Tet2 regulate 5-hydroxymethylcytosine production and cell lineage specification in mouse embryonic stem cells. Cell Stem Cell 8(2):200–213. doi:10.1016/j.stem.2011.01.008
23. Kriaucionis S, Heintz N (2009) The nuclear DNA base 5-hydroxymethylcytosine is present in purkinje neurons and the brain. Science 324(5929):929–930. doi:10.1126/science.1169786
24. Song CX, Szulwach KE, Fu Y et al (2011) Selective chemical labeling reveals the genome-wide distribution of 5-hydroxymethylcytosine. Nat Biotechnol 29(1):68–72. doi:10.1038/nbt.1732
25. Szwagierczak A, Bultmann S, Schmidt CS et al (2010) Sensitive enzymatic quantification of 5-hydroxymethylcytosine in genomic DNA. Nucleic Acids Res 38(19):e181. doi:10.1093/nar/gkq684
26. Tahiliani M, Koh KP, Shen Y et al (2009) Conversion of 5-methylcytosine to 5-hydroxymethylcytosine in mammalian DNA by mll partner TET1. Science 324(5929):930–935. doi:10.1126/science.1170116
27. Penn NW, Suwalski R, O'riley C et al (1972) The presence of 5-hydroxymethylcytosine in animal deoxyribonucleic acid. Biochem J 126(4):781–790
28. Ito S, Shen L, Dai Q et al (2011) Tet proteins can convert 5-methylcytosine to 5-formylcytosine and 5-carboxylcytosine. Science 333(6047):1300–1303. doi:10.1126/science.1210597
29. He YF, Li BZ, Li Z et al (2011) Tet-mediated formation of 5-carboxylcytosine and its excision by TDG in mammalian DNA. Science 333(6047):1303–1307. doi:10.1126/science.1210944
30. Wu H, Zhang Y (2011) Mechanisms and functions of tet protein-mediated 5-methylcytosine oxidation. Genes Dev 25(23):2436–2452. doi:10.1101/gad.179184.111
31. Cimmino L, Abdel-Wahab O, Levine RL et al (2011) Tet family proteins and their role in stem cell differentiation and transformation. Cell Stem Cell 9(3):193–204. doi:10.1016/j.stem.2011.08.007
32. Ehrlich M, Gama-Sosa MA, Huang LH et al (1982) Amount and distribution of 5-methylcytosine in human DNA from different types of tissues of cells. Nucleic Acids Res 10(8):2709–2721
33. Yoder JA, Walsh CP, Bestor TH (1997) Cytosine methylation and the ecology of intragenomic parasites. Trends Genet 13(8):335–340
34. Irizarry RA, Ladd-Acosta C, Wen B et al (2009) The human colon cancer methylome shows similar hypo- and hypermethylation at conserved tissue-specific CpG island shores. Nat Genet 41(2):178–186. doi:10.1038/ng.298
35. Gardiner-Garden M, Frommer M (1987) CpG islands in vertebrate genomes. J Mol Biol 196(2):261–282
36. Meehan R, Lewis J, Cross S et al (1992) Transcriptional repression by methylation of CpG. J Cell Sci Suppl 16:9–14
37. Lewis JD, Meehan RR, Henzel WJ et al (1992) Purification, sequence, and cellular localization of a novel chromosomal protein that binds to methylated DNA. Cell 69(6):905–914
38. Jones PL, Veenstra GJ, Wade PA et al (1998) Methylated DNA and MeCP2 recruit histone deacetylase to repress transcription. Nat Genet 19(2):187–191

39. Lorincz MC, Dickerson DR, Schmitt M et al (2004) Intragenic DNA methylation alters chromatin structure and elongation efficiency in mammalian cells. Nat Struct Mol Biol 11(11):1068–1075. doi:10.1038/nsmb840
40. Chodavarapu RK, Feng S, Bernatavichute YV et al (2010) Relationship between nucleosome positioning and DNA methylation. Nature 466(7304):388–392. doi:10.1038/nature09147
41. Maunakea AK, Nagarajan RP, Bilenky M et al (2010) Conserved role of intragenic DNA methylation in regulating alternative promoters. Nature 466(7303):253–257. doi:10.1038/nature09165
42. Yasui DH, Peddada S, Bieda MC et al (2007) Integrated epigenomic analyses of neuronal MeCP2 reveal a role for long-range interaction with active genes. Proc Natl Acad Sci U S A 104(49):19416–19421
43. Chahrour M, Jung SY, Shaw C et al (2008) MeCP2, a key contributor to neurological disease, activates and represses transcription. Science 320(5880):1224–1229
44. Antequera F, Bird A (1993) Number of CpG islands and genes in human and mouse. Proc Natl Acad Sci U S A 90(24):11995–11999
45. Robertson KD (2005) DNA methylation and human disease. Nat Rev Genet 6(8):597–610. doi:10.1038/nrg1655
46. Fraga MF, Herranz M, Espada J et al (2004) A mouse skin multistage carcinogenesis model reflects the aberrant DNA methylation patterns of human tumors. Cancer Res 64(16):5527–5534. doi:10.1158/0008-5472.CAN-03-4061
47. Esteller M (2008) Epigenetics in cancer. N Engl J Med 358(11):1148–1159. doi:10.1056/NEJMra072067
48. Weissmann S, Alpermann T, Grossmann V et al (2011) Landscape of TET2 mutations in acute myeloid leukemia. Leukemia. doi:10.1038/leu.2011.326
49. Abdel-Wahab O, Mullally A, Hedvat C et al (2009) Genetic characterization of TET1, TET2, and TET3 alterations in myeloid malignancies. Blood 114(1):144–147. doi:10.1182/blood-2009-03-210039
50. Jankowska AM, Szpurka H, Tiu RV et al (2009) Loss of heterozygosity 4q24 and TET2 mutations associated with myelodysplastic/myeloproliferative neoplasms. Blood 113(25):6403–6410. doi:10.1182/blood-2009-02-205690
51. Yan H, Parsons DW, Jin G et al (2009) IDH1 and IDH2 mutations in gliomas. N Engl J Med 360(8):765–773. doi:10.1056/NEJMoa0808710
52. Parsons DW, Jones S, Zhang X et al (2008) An integrated genomic analysis of human glioblastoma multiforme. Science 321(5897):1807–1812. doi:10.1126/science.1164382
53. Mardis ER, Ding L, Dooling DJ et al (2009) Recurring mutations found by sequencing an acute myeloid leukemia genome. N Engl J Med 361(11):1058–1066. doi:10.1056/NEJMoa0903840
54. Jacinto FV, Ballestar E, Ropero S et al (2007) Discovery of epigenetically silenced genes by methylated DNA immunoprecipitation in colon cancer cells. Cancer Res 67(24):11481–11486. doi:10.1158/0008-5472.CAN-07-2687
55. Weber M, Hellmann I, Stadler MB et al (2007) Distribution, silencing potential and evolutionary impact of promoter DNA methylation in the human genome. Nat Genet 39(4):457–466. doi:10.1038/ng1990
56. Weber M, Davies JJ, Wittig D et al (2005) Chromosome-wide and promoter-specific analyses identify sites of differential DNA methylation in normal and transformed human cells. Nat Genet 37(8):853–862. doi:10.1038/ng1598
57. Ruike Y, Imanaka Y, Sato F et al (2010) Genome-wide analysis of aberrant methylation in human breast cancer cells using methyl-DNA immunoprecipitation combined with high-throughput sequencing. BMC Genomics 11:137. doi:10.1186/1471-2164-11-137
58. Stroud H, Feng S, Morey Kinney S et al (2011) 5-Hydroxymethylcytosine is associated with enhancers and gene bodies in human embryonic stem cells. Genome Biol 12(6):R54. doi:10.1186/gb-2011-12-6-r54
59. Ficz G, Branco MR, Seisenberger S et al (2011) Dynamic regulation of 5-hydroxymethylcytosine in mouse ES cells and during differentiation. Nature 473(7347):398–402. doi:10.1038/nature10008
60. Wu H, D'alessio AC, Ito S et al (2011) Genome-wide analysis of 5-hydroxymethylcytosine distribution reveals its dual function in transcriptional regulation in mouse embryonic stem cells. Genes Dev 25(7):679–684. doi:10.1101/gad.2036011
61. Xu Y, Wu F, Tan L et al (2011) Genome-wide regulation of 5hmC, 5mC, and gene expression by Tet1 hydroxylase in mouse embryonic stem cells. Mol Cell 42(4):451–464. doi:10.1016/j.molcel.2011.04.005
62. Szulwach KE, Li X, Li Y et al (2011) 5hmC-mediated epigenetic dynamics during postnatal neurodevelopment and aging. Nat Neurosci 14(12):1607–1616. doi:10.1038/nn.2959

63. Pastor WA, Pape UJ, Huang Y et al (2011) Genome-wide mapping of 5-hydroxymethylcytosine in embryonic stem cells. Nature 473(7347):394–397. doi:10.1038/nature10102

64. Ko M, Huang Y, Jankowska AM et al (2010) Impaired hydroxylation of 5-methylcytosine in myeloid cancers with mutant TET2. Nature 468(7325):839–843. doi:10.1038/nature09586

65. Harris RA, Wang T, Coarfa C et al (2010) Comparison of sequencing-based methods to profile DNA methylation and identification of monoallelic epigenetic modifications. Nat Biotechnol 28(10):1097–1105. doi:10.1038/nbt.1682

66. Matarese F, Carrillo-De Santa Pau E, Stunnenberg HG (2011) 5-Hydroxymethylcytosine: a new kid on the epigenetic block? Mol Syst Biol 7:562. doi:10.1038/msb.2011.95

67. Serre D, Lee BH, Ting AH (2010) Mbd-isolated genome sequencing provides a high-throughput and comprehensive survey of DNA methylation in the human genome. Nucleic Acids Res 38(2):391–399. doi:10.1093/nar/gkp992

68. Brinkman AB, Simmer F, Ma K et al (2010) Whole-genome DNA methylation profiling using MethylCap-seq. Methods 52(3):232–236. doi:10.1016/j.ymeth.2010.06.012

69. Nair SS, Coolen MW, Stirzaker C et al (2011) Comparison of methyl-DNA immunoprecipitation (MeDIP) and methyl-CpG binding domain (MBD) protein capture for genome-wide DNA methylation analysis reveal CpG sequence coverage bias. Epigenetics 6(1):34–44. doi:10.4161/epi.6.1.13313

70. Hayatsu H (2008) Discovery of bisulfite-mediated cytosine conversion to uracil, the key reaction for DNA methylation analysis–a personal account. Proc Jpn Acad Ser B Phys Biol Sci 84(8):321–330

71. Frommer M, Mcdonald LE, Millar DS et al (1992) A genomic sequencing protocol that yields a positive display of 5-methylcytosine residues in individual DNA strands. Proc Natl Acad Sci U S A 89(5):1827–1831

72. Lister R, O'malley RC, Tonti-Filippini J et al (2008) Highly integrated single-base resolution maps of the epigenome in arabidopsis. Cell 133(3):523–536. doi:10.1016/j.cell.2008.03.029

73. Lister R, Pelizzola M, Dowen RH et al (2009) Human DNA methylomes at base resolution show widespread epigenomic differences. Nature 462(7271):315–322. doi:10.1038/nature08514

74. Cokus SJ, Feng S, Zhang X et al (2008) Shotgun bisulphite sequencing of the arabidopsis genome reveals DNA methylation patterning. Nature 452(7184):215–219

75. Li Y, Zhu J, Tian G et al (2010) The DNA methylome of human peripheral blood mononuclear cells. PLoS Biol 8(11):e1000533. doi:10.1371/journal.pbio.1000533

76. Bormann Chung CA, Boyd VL, Mckernan KJ et al (2010) Whole methylome analysis by ultra-deep sequencing using two-base encoding. PLoS One 5(2):e9320. doi:10.1371/journal.pone.0009320

77. Laurent L, Wong E, Li G et al (2010) Dynamic changes in the human methylome during differentiation. Genome Res 20(3):320–331. doi:10.1101/gr.101907.109

78. Stadler MB, Murr R, Burger L et al (2011) DNA-binding factors shape the mouse methylome at distal regulatory regions. Nature 480(7378):490–495. doi:10.1038/nature10716

79. Birney E, Stamatoyannopoulos JA, Dutta A et al (2007) Identification and analysis of functional elements in 1% of the human genome by the encode pilot project. Nature 447(7146):799–816. doi:10.1038/nature05874

80. Myers RM, Stamatoyannopoulos J, Snyder M et al (2011) A user's guide to the encyclopedia of DNA elements (encode). PLoS Biol 9(4):e1001046. doi:10.1371/journal.pbio.1001046

81. Ajay SS, Parker SC, Abaan HO et al (2011) Accurate and comprehensive sequencing of personal genomes. Genome Res 21(9):1498–1505. doi:10.1101/gr.123638.111

82. Huang Y, Pastor WA, Shen Y et al (2010) The behaviour of 5-hydroxymethylcytosine in bisulfite sequencing. PLoS One 5(1):e8888. doi:10.1371/journal.pone.0008888

83. Hodges E, Smith AD, Kendall J et al (2009) High definition profiling of mammalian DNA methylation by array capture and single molecule bisulfite sequencing. Genome Res 19(9):1593–1605. doi:10.1101/gr.095190.109

84. Lee EJ, Pei L, Srivastava G et al (2011) Targeted bisulfite sequencing by solution hybrid selection and massively parallel sequencing. Nucleic Acids Res 39(19):e127. doi:10.1093/nar/gkr598

85. Taylor KH, Kramer RS, Davis JW et al (2007) Ultradeep bisulfite sequencing analysis of DNA methylation patterns in multiple gene promoters by 454 sequencing. Cancer Res 67(18):8511–8518. doi:10.1158/0008-5472.CAN-07-1016

86. Ball MP, Li JB, Gao Y et al (2009) Targeted and genome-scale strategies reveal gene-body methylation signatures in human cells. Nat Biotechnol 27(4):361–368. doi:10.1038/nbt.1533
87. Meissner A, Gnirke A, Bell GW et al (2005) Reduced representation bisulfite sequencing for comparative high-resolution DNA methylation analysis. Nucleic Acids Res 33(18):5868–5877. doi:10.1093/nar/gki901
88. Smith ZD, Gu H, Bock C et al (2009) High-throughput bisulfite sequencing in mammalian genomes. Methods 48(3):226–232. doi:10.1016/j.ymeth.2009.05.003
89. Gu H, Smith ZD, Bock C et al (2011) Preparation of reduced representation bisulfite sequencing libraries for genome-scale DNA methylation profiling. Nat Protoc 6(4):468–481. doi:10.1038/nprot.2010.190
90. Wang L, Sun J, Wu H et al (2012) Systematic assessment of reduced representation bisulfite sequencing to human blood samples: a promising method for large-sample-scale epigenomic studies. J Biotechnol 157(1):1–6. doi:10.1016/j.jbiotec.2011.06.034
91. Gu H, Bock C, Mikkelsen TS et al (2010) Genome-scale DNA methylation mapping of clinical samples at single-nucleotide resolution. Nat Methods 7(2):133–136. doi:10.1038/nmeth.1414
92. Wiegand KC, Shah SP, Al-Agha OM et al (2010) ARID1a mutations in endometriosis-associated ovarian carcinomas. N Engl J Med 363(16):1532–1543. doi:10.1056/NEJMoa1008433
93. Bock C, Tomazou EM, Brinkman AB et al (2010) Quantitative comparison of genome-wide DNA methylation mapping technologies. Nat Biotechnol 28(10):1106–1114. doi:10.1038/nbt.1681
94. Deng J, Shoemaker R, Xie B et al (2009) Targeted bisulfite sequencing reveals changes in DNA methylation associated with nuclear reprogramming. Nat Biotechnol 27(4):353–360. doi:10.1038/nbt.1530
95. Hansen KD, Timp W, Bravo HC et al (2011) Increased methylation variation in epigenetic domains across cancer types. Nat Genet 43(8):768–775. doi:10.1038/ng.865
96. Bird AP, Taggart MH, Smith BA (1979) Methylated and unmethylated DNA compartments in the sea urchin genome. Cell 17(4):889–901
97. Colaneri A, Staffa N, Fargo DC et al (2011) Expanded methyl-sensitive cut counting reveals hypomethylation as an epigenetic state that highlights functional sequences of the genome. Proc Natl Acad Sci U S A 108(23):9715–9720. doi:10.1073/pnas.1105713108
98. Brunner AL, Johnson DS, Kim SW et al (2009) Distinct DNA methylation patterns characterize differentiated human embryonic stem cells and developing human fetal liver. Genome Res 19(6):1044–1056. doi:10.1101/gr.088773.108
99. Oda M, Glass JL, Thompson RF et al (2009) High-resolution genome-wide cytosine methylation profiling with simultaneous copy number analysis and optimization for limited cell numbers. Nucleic Acids Res 37(12):3829–3839. doi:10.1093/nar/gkp260
100. Flusberg BA, Webster DR, Lee JH et al (2010) Direct detection of DNA methylation during single-molecule, real-time sequencing. Nat Methods 7(6):461–465. doi:10.1038/nmeth.1459
101. Korlach J, Bjornson KP, Chaudhuri BP et al (2010) Real-time DNA sequencing from single polymerase molecules. Methods Enzymol 472:431–455. doi:10.1016/S0076-6879(10)72001-2
102. Eid J, Fehr A, Gray J et al (2009) Real-time DNA sequencing from single polymerase molecules. Science 323(5910):133–138. doi:10.1126/science.1162986
103. Clarke J, Wu H-C, Jayasinghe L et al (2009) Continuous base identification for single-molecule nanopore DNA sequencing. Nat Nanotechnol 4(4):265–270. doi:10.1038/nnano.2009.12
104. The 1000 Genomes Project Consortium (2010) A map of human genome variation from population-scale sequencing. Nature 467(7319):1061–1073. doi:10.1038/nature09534
105. Flicek P, Birney E (2009) Sense from sequence reads: methods for alignment and assembly. Nat Methods 6(11 Suppl):S6–S12. doi:10.1038/nmeth.1376
106. Li H, Durbin R (2009) Fast and accurate short read alignment with burrows-wheeler transform. Bioinformatics 25(14):1754–1760. doi:10.1093/bioinformatics/btp324
107. Langmead B, Trapnell C, Pop M et al (2009) Ultrafast and memory-efficient alignment of short DNA sequences to the human genome. Genome Biol 10(3):R25. doi:10.1186/gb-2009-10-3-r25
108. Li H, Handsaker B, Wysoker A et al (2009) The sequence alignment/map format and samtools. Bioinformatics 25(16):2078–2079. doi:10.1093/bioinformatics/btp352

109. Xi Y, Li W (2009) BSMAP: whole genome bisulfite sequence mapping program. BMC Bioinformatics 10:232. doi:10.1186/1471-2105-10-232
110. Coarfa C, Yu F, Miller CA et al (2010) Pash 3.0: a versatile software package for read mapping and integrative analysis of genomic and epigenomic variation using massively parallel DNA sequencing. BMC Bioinformatics 11:572. doi:10.1186/1471-2105-11-572
111. Chen PY, Cokus SJ, Pellegrini M (2010) BS seeker: precise mapping for bisulfite sequencing. BMC Bioinformatics 11:203. doi:10.1186/1471-2105-11-203
112. Xi Y, Bock C, Muller F et al (2011) RRBSMAP: a fast, accurate and user-friendly alignment tool for reduced representation bisulfite sequencing. Bioinformatics. doi:10.1093/bioinformatics/btr668
113. Krueger F, Andrews SR (2011) Bismark: a flexible aligner and methylation caller for bisulfite-Seq applications. Bioinformatics 27(11):1571–1572. doi:10.1093/bioinformatics/btr167
114. Dreszer TR, Karolchik D, Zweig AS et al (2012) The UCSC genome browser database: extensions and updates 2011. Nucleic Acids Res 40(Database issue):D918–D923. doi:10.1093/nar/gkr1055
115. Fejes AP, Robertson G, Bilenky M et al (2008) FindPeaks 3.1: a tool for identifying areas of enrichment from massively parallel short-read sequencing technology. Bioinformatics 24(15):1729–1730. doi:10.1093/bioinformatics/btn305
116. Nielsen CB, Cantor M, Dubchak I et al (2010) Visualizing genomes: techniques and challenges. Nat Methods 7(3 Suppl):S5–S15. doi:10.1038/nmeth.1422
117. Flicek P, Amode MR, Barrell D et al (2012) Ensemble 2012. Nucleic Acids Res 40(Database issue):D84–D90. doi:10.1093/nar/gkr991
118. Zhou X, Maricque B, Xie M et al (2011) The human epigenome browser at Washington University. Nat Methods 8(12):989–990. doi:10.1038/nmeth.1772

Chapter 4

Application of Array Comparative Genomic Hybridization in Chronic Myeloid Leukemia

Seonyang Park, Youngil Koh, Seung-Hyun Jung, and Yeun-Jun Chung

Abstract

Chromosomal alteration is one of the hallmarks of chronic myeloid leukemia (CML), and the Philadelphia chromosome is the most important and key example of the chromosomal changes in this disease. Indeed, the *BCR–ABL1* fusion product is a target against which many tyrosine kinase inhibitors (TKIs) have been proven to be effective in the treatment of CML. However, the reality is that CML patients show resistance to TKIs both in an acquired and de novo manner, and the mechanism of TKI resistance is still largely unknown. This phenomenon suggests that in addition to the *BCR–ABL* mutation, further genetic alterations such as copy number aberration may be involved in unexplained TKI resistance. Although the recent array comparative genomic hybridization analyses (array-CGH) across the whole genome have detected multiple genetic aberrations in CML, the detailed feature of chromosomal alterations involved in different clinical phases of CML, such as chronic phase, accelerated phase, and blast crisis, remains unclear. Here we review the methodological aspects of array-CGH analysis for studying CML and its related data analysis.

Key words: Chronic myeloid leukemia, Array comparative genomic hybridization, Copy number alteration, Tyrosine kinase inhibitor

1. Introduction

1.1. Molecular Characteristics of Chronic Myeloid Leukemia

Recent advances in our understanding of the biology of chronic myeloid leukemia (CML) have led to the development of remarkably effective therapies against this clonal stem cell disorder. Cytogenetically, CML is characterized by a specific chromosomal translocation, t(9;22)(q34;q11), resulting in a shortened chromosome 22, commonly referred as the Philadelphia (Ph) chromosome (1). For 90% of CML patients, the development of the Ph chromosome results in a new genetic sequence made up of

Diponkar Banerjee and Sohrab P. Shah (eds.), *Array Comparative Genomic Hybridization: Protocols and Applications*, Methods in Molecular Biology vol. 973, DOI 10.1007/978-1-62703-281-0_4, © Springer Science+Business Media, LLC 2013

breakpoint cluster region (*BCR*) from chromosome 22 and *c-ABL* (*Abelson murine leukemia viral oncogene homolog 1*) from chromosome 9. Hence, *BCR–ABL1* fusion gene is the hallmark of CML, and its product is a druggable target. The *BCR–ABL1* fusion gene encodes a constitutively active tyrosine kinase, *ABL1*, that is necessary and thought sufficient to drive malignant transformation (2). An attachment of the *BCR* sequences to *ABL1* results in two additional functional changes that contributes to a malignant potential other than constitutive activation of *ABL1*: (1) the DNA protein-binding activity of *ABL1* is attenuated and (2) the binding of *ABL1* to cytoskeletal actin microfilaments is enhanced (3, 4). Currently, it is accepted that conventional cytogenetic G-banding and fluorescence in situ hybridization (FISH) could detect CML patients effectively.

From the treatment perspective, the *BCR–ABL1* fusion product is a druggable target against which many tyrosine kinase inhibitors (TKIs) have been proven to be effective in the treatment of CML. The most commonly used TKI for all phases of the disease [chronic phase (CP), accelerated phase (AP), and blast crisis (BC)], is imatinib (5). The IRIS trial demonstrated that imatinib was significantly more effective and better tolerated than the combination of interferon-α plus cytarabine in the treatment of newly diagnosed patients with CML-CP (6). Starting from IRIS trial, imatinib has become a standard frontline therapy for CML. In this era, however, acquired resistance to *BCR–ABL1* TKIs has become a pressing challenge in the treatment of CML (7). This acquired resistance is frequently associated with point mutations within the *BCR–ABL* kinase domain, and some mutations are only partially resistant to imatinib. Those could be overcome with a dose escalation (8–12). In this sense, for patients who failed to respond to the first-line standard dose imatinib therapy, an escalated dose of imatinib is still a reasonable option. In fact, clinical trials have shown the efficacy of escalated doses of imatinib in this setting (13, 14). Also, this challenge led to the development of the second-generation TKIs targeting *BCR–ABL1*, including dasatinib and nilotinib. These second-generation TKIs are more potent inhibitors of *BCR–ABL1* than imatinib, in vitro. Hence, for patients with resistance to standard dose imatinib, either the imatinib dose escalation or the use of the second-generation TKIs could be options. However, the best practice for patients who showed poor response to imatinib is not well known (15, 16). It would be very valuable if we could effectively select groups of patients who fail to the standard dose imatinib treatment and who benefit from escalated dose imatinib therapy. This strategy may not only prolong the survival of such patients but also reduce their economic burden. And, although second-generation TKIs are potent, they do not resolve the fundamental problem regarding the resistance of CML cells to imatinib. Some CML cells resistant to imatinib (e. g. , T315I mutations) are also resistant to dasatinib or nilotinib. Further research using recent genetic technologies is mandatory for revealing underlying mechanisms related to the imatinib resistance.

1.2. Array Comparative Genomic Hybridization and Chronic Myeloid Leukemia

Conventional cytogenetic analysis including karyotyping and FISH provides useful information in many hematological malignancies including CML, and it still is the most commonly used diagnostic tool in clinical fields. However, there are several limitations of conventional cytogenetics analysis. For example, the requirement of metaphase chromosomes from the patient and the lower resolution of the available data for pinpointing copy number alteration breakpoints have been obstacles. In 1992, Kallioniemi et al. (17) reported comparative genomic hybridization (CGH) technique which enabled assessing the copy number changes across the whole genome with a single hybridization reaction. This technique was based on the principle that DNAs from normal tissue (reference) and from tumor tissue (test), labeled with different fluorescent dyes, are competitively hybridized onto normal metaphase chromosomes. The intensity ratio of the fluorescence signal is indicative of the relative copy number difference in test versus reference DNA. CGH technology allows us to overcome the limitations of conventional cytogenetic analysis. However, CGH had practical problems that limited its widespread use. For example, it required a good quality metaphase chromosome spreads, and the hybridization conditions varied by metaphase spread condition or by individual. Most importantly, the resolution of CGH (5–10 Mb) is not high enough to localize regional chromosome alterations (18, 19).

Subsequently, a microarray-based CGH (array-CGH) was established, where arrays of genomic sequences replaced the metaphase chromosomes as hybridization targets (20). By combining the advantages of CGH and microarray, array-CGH solved the technical difficulties of CGH, and it can support genome-wide screening for copy number alterations with an increased resolution (21). Especially, the resolution of array-CGH has been improved to less than 1 kb across the whole genome. Figure 1 illustrates the principles of array-CGH-based molecular karyotyping analysis in CML.

Following the initial bacterial artificial chromosome (BAC) arrays, the introduction of oligonucleotide arrays has enabled even more precise (less than 3 kb interval) scanning of whole genome for copy number alteration. Molecular karyotyping could also be performed using single nucleotide polymorphism (SNP) array, which was developed for SNP genotyping. Due to its high density across the genome (>1 M probes), the resolution of the copy number analysis of SNP array is comparable with the resolution of the high-density oligoarrays (>1 M probes). In addition, SNP array has the advantage of detecting the loss of heterozygosity (CN-LOH) in the copy number neutral region and can identify somatic uniparental

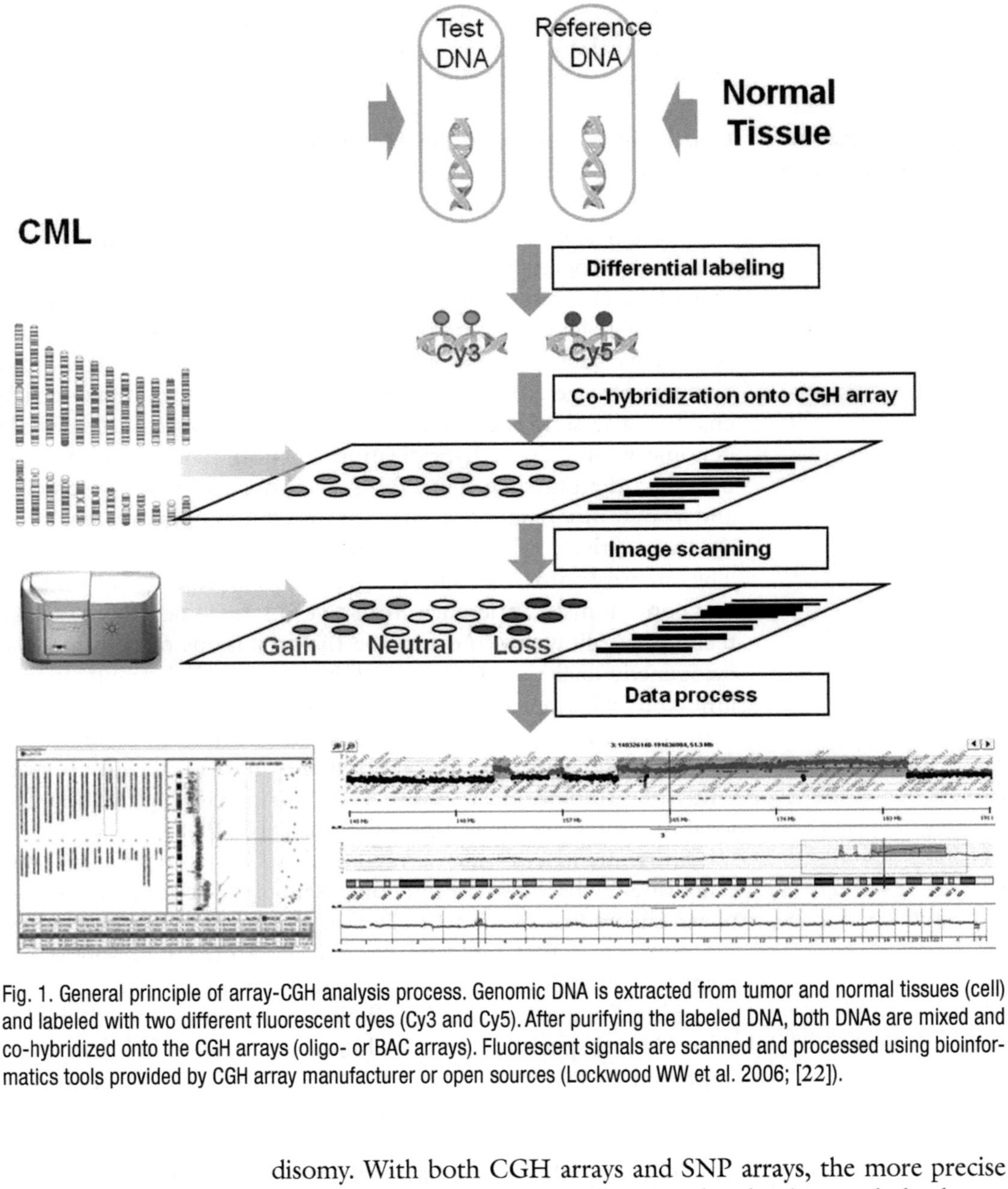

Fig. 1. General principle of array-CGH analysis process. Genomic DNA is extracted from tumor and normal tissues (cell) and labeled with two different fluorescent dyes (Cy3 and Cy5). After purifying the labeled DNA, both DNAs are mixed and co-hybridized onto the CGH arrays (oligo- or BAC arrays). Fluorescent signals are scanned and processed using bioinformatics tools provided by CGH array manufacturer or open sources (Lockwood WW et al. 2006; [22]).

disomy. With both CGH arrays and SNP arrays, the more precise detection of copy number alteration breakpoints and the better interpretation of copy number neutral cases are available.

Nowadays, the array-CGH is one of the major tools for genome-wide screening of DNA copy number profiles analyzing copy number variations (CNVs) associated with cancer risks as well as exploring somatic alterations in cancers for tumor classification or discovering new cancer-related genes (Fig. 2) (23–29).

In patients with CML, array-CGH studies have focused on two groups of patients: (1) Patients with CML-CP showing resistance to frontline therapy with *ABL1* TKI and (2) patients with CML-AP or

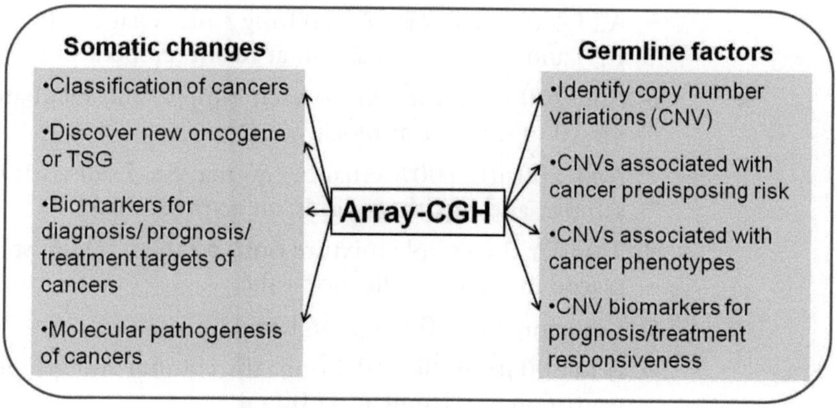

Fig. 2. Application of array-CGH technology for studying somatic changes in cancers and germline factors associated with cancer phenotypes.

CML-BC. However, the number of published studies is small, and more vigorous research in these two groups of patients is strongly warranted.

2. Materials

Genomic DNA of CML patients is necessary for array-CGH. Either a peripheral blood sample or a bone marrow sample can be used. However, for the proper data analysis, the unification of sample source (either peripheral blood or bone marrow) is preferred at least in a single study. To verify the existence and the proportion of CML cells in sample, a concomitant BCR–ABL FISH or cytogenetic test is recommended (see Note 1). At least 3 mL of sample in an EDTA tube is necessary. DNA can be isolated from frozen buffy coat specimens using conventional kits [e. g. , EZ1 tissue kit and robot (Qiagen, Inc. , Valencia, CA)].

3. Methods

3.1. DNA Extraction

1. Qiagen DNeasy Blood and Tissue Kit (Qiagen, Inc. , Valencia, CA).
2. Spin a maximum of 5×10^6 cells in a centrifuge for 5 min at $300 \times g$.
3. Resuspend the pellet in 200 μL PBS (Sigma, St. Louis, MO).
4. Add 20 μL proteinase K (Qiagen, Inc. , Valencia, CA).

5. Add 4 μL of RNase A (100 mg/mL, Qiagen, Inc., Valencia, CA) and incubate for 2 min at room temperature.
6. Add 200 μL buffer AL to each sample, and incubate at 70°C for 10 min in a heat block.
7. Add 200 μL 100% ethanol (Sigma, St. Louis, MO) to each sample, and mix thoroughly on a vortex mixer.
8. Transfer the sample mixture onto a DNeasy Mini spin column placed in a 2 mL collection tube.
9. Centrifuge at 6,000 × *g* for 1 min.
10. Add 500 μL buffer AW1 onto the column and spin in a microcentrifuge for 1 min at 6,000 × *g*.
11. Add 500 μL 80% ethanol onto the column, and centrifuge for 3 min at 20,000 × *g*.
12. Place the DNeasy Mini spin column in a clean 1. 5 mL microcentrifuge tube and pipet 100 μL of nuclease-free water directly onto the center of the DNeasy column membrane.
13. Incubate at room temperature for 1 min, and then spin in a microcentrifuge for 1 min at 6,000 × *g* to elute the DNA.

3.2. Oligoarray-CGH

3.2.1. DNA Fragmentation

1. For each reaction, combine 1 μg of genomic DNA (gDNA) and nuclease-free water to bring to the final volume 21 μL.
2. Incubate at 95°C in a PCR machine with a heated lid for 5 min to fragment the gDNA samples.
3. Transfer the sample tubes to ice and incubate on ice for 3 min.
4. Spin in a microcentrifuge for 30 s at 6,000 × *g* to drive the contents off the walls and lid.

3.2.2. DNA Labeling

1. Add 20 μL of Random Primers (Invitrogen, Carlsbad, CA) to each reaction tube containing 21 μL of digested gDNA to make a total volume of 41 μL. Mix well by pipetting up and down gently.
2. Incubate at 95°C in a PCR machine with a heated lid for 5 min and incubate on ice for 5 min.
3. Add 5 μL of 10× dCTP (Invitrogen, Carlsbad, CA).
4. Add 3 μL of Cy3-dCTP or Cy5-dCTP (Perkin Elmer, Waltham, MA).
5. Add 1 μL of exo-Klenow fragment (Invitrogen, Carlsbad, CA).
6. Incubate at 37°C in a PCR machine with a heated lid for 2 h.
7. Store at 4°C until the next step if necessary.

3.2.3. Clean-Up of Labeled Genomic DNA

1. Add 430 μL of 1× TE (pH 8. 0, Sigma, St. Louis, MO) to each reaction tube.

2. Place an Amicon 30 kDa filter (Millipore, Billerica, MA) into a 1. 5 mL microfuge tube and load each labeled gDNA into the filter.
3. Spin 10 min at 14,000 × *g* in a microcentrifuge at room temperature. Discard the flow through.
4. Add 480 μL of 1× to each filter.
5. Spin for 10 min at 14,000 × *g* in a microcentrifuge at room temperature. Discard the flow through.
6. Invert the filter into a fresh 1. 5 mL microfuge tube. Spin for 1 min at 1,000 × *g* in a microcentrifuge at room temperature to collect purified sample. The volume per sample will be approximately 21 μL.
7. Add 1× TE to bring to the sample volume 80. 5 μL.
8. Take 1. 5 μL of each sample to determine yield and specific activity.
9. The ideal values for dye incorporation are between 50 and 80; $(A_{260}/A_{550}) \times 23.15 = \text{nt/Cy3}$, $(A_{260}/A_{550}) \times 38.58 = \text{nt/Cy5}$.

3.2.4. Hybridization

1. Combine the test and reference sample using the appropriate cyanine 5-labeled sample and cyanine 3-labeled sample.
2. Add 50 μL of human Cot-1 DNA (1. 0 mg/mL, Connectagen, Seoul, Korea) to mixture tube.
3. Add 52 μL of blocking agent (Agilent Technologies, Santa Clara, CA).
4. Add 260 μL of 2× hybridization buffer (Agilent Technologies, Santa Clara, CA).
5. Transfer sample tubes to a heat block at 95°C. Incubate at 95°C for 3 min.
6. Immediately transfer sample tubes to a heat block at 37°C. Incubate at 37°C for 30 min.
7. Load a clean gasket slide (Agilent Technologies, Santa Clara, CA) into the Agilent SureHyb chamber base with the gasket label facing up and aligned with the rectangular section of the chamber base.
8. Slowly dispense 490 μL of hybridization sample mixture onto the gasket well.
9. Put a microarray slide onto the gasket slide.
10. Put the SureHyb chamber cover onto the sandwiched slides and slide the clamp assembly onto both pieces. Hand-tighten the clamp onto the chamber.
11. Put assembled slide chamber in the rotator rack in a hybridization oven (Agilent Technologies, Santa Clara, CA) set to 65°C.
12. Hybridize at 65°C for 40 h.

3.2.5. Microarray Washing

1. Completely fill slide-staining dish #1 with oligoarray-CGH wash buffer 1 (Agilent Technologies, Santa Clara, CA) at room temperature.
2. Fill slide-staining dish #2 with enough oligoarray-CGH wash buffer 1 at room temperature.
3. Fill slide-staining dish #3 with enough oligoarray-CGH wash buffer 2 (Agilent Technologies, Santa Clara, CA) at 37°C.
4. Disassemble the array chamber and separate the array slide and gasket slide while immersed in wash buffer 1.
5. Place slides in wash buffer 1 for 5 min at room temperature on a magnetic stir plate with 120 rpm.
6. Place slides in 37°C wash buffer 2 for 3 min on a magnetic stir plate with 120 rpm.
7. Remove the slide very slowly to dry.

3.2.6. Microarray Scanning

1. Assemble slides into appropriate slide holders and put assembled slide holders into scanner carousel.
2. Verify Scan Settings; Scan region is set to Scan Area (61 × 21. 6 mm).
3. Scan resolution is set to 5 μm.
4. Dye channel is set to Red and Green.
5. Green PMT and red PMT is set to 100%.
6. Clear the extended Dynamic Range check box.
7. Verify that the Scanner status in the main window says Scanner Ready.
8. Click Scan Slot.

3.2.7. Feature Extraction

1. Open the Agilent Feature Extraction program (Agilent Technologies, Santa Clara, CA).
2. Add the images (. tiff) to be extracted to the FE Project.
3. Add grid template file and protocol file.
4. Select start extraction.

3.2.8. Defining the Copy Number Changes

1. Open the Agilent Genomic workbench program (Agilent Technologies, Santa Clara, CA).
2. Import the FE files.
3. Apply "Intra Array Replicates" function.
4. Apply "ADM-2" copy number change detection algorithm.
5. ADM-2 threshold set 6. 0.
6. Apply "DefaultAberrationFilter_V2" filter; min number of probe: 3; min absolute average log ratio: 0. 25.

4. Data Analysis and Interpretation

In a study population of 45 patients who were enrolled in an imatinib dose escalation trial for their poor responses to frontline standard dose therapy, genome-wide screening for regions of genetic gains and losses was performed using array-CGH. Early molecular response (EMR), defined as >50% reduction in the ratio of *BCR–ABL1* to *ABL1* within 6 months after dose escalation, was a major endpoint for analysis. Twenty-nine recurrently altered regions were found: 23 with gains and 6 with losses (data not shown). Regional copy number changes between patients with EMR and patients without EMR were compared, and three recurrently altered regions were found to have *P*-values of <0. 1 (data not shown). Copy number gain of 16p11. 2 was observed dominantly in patients with EMR, compared with patients without EMR (*P*=0. 034). A tendency for increased copy number in 22q11. 23 for patients without EMR and decreased copy number in 17q12 for patients with EMR was observed, although the findings did not reach statistical significance (*P*=0. 072 and 0. 070, respectively).

After array-CGH analysis, a literature review using genome maps provided by the US National Center for Biotechnology Information (www. ncbi. nlm. nih. gov/sites/genome) was performed. The genes in the region of copy number alteration were identified, and those reported to be involved in cancer development or progression were selected among them. This procedure yielded four candidate genes: glutathione S-transferase theta 1 (*GSTT1*); sulfotransferase family, cytosolic, 1A, phenol-preferring, member 1 (*SULT1A1*); PYD and CARD domain containing (*PYCARD*); and transcriptional adaptor 2A (*TADA2A*).

For these four genes, validation studies were performed using real-time quantitative PCR assay in 52 patients whose baseline samples were available, of whom 44 had undergone array-CGH analysis. Copy number gain of *GSTT1*, on 22q11. 23, correlated well with copy number gain of 22q11. 23 (*P*<0. 001). *GSTT1* copy number gain was more frequently observed in patients without EMR (10/17) than in patients with EMR (10/35) (*P*=0. 035). *GSTT1* copy number gain also predicted short median time to treatment failure (TTFx), although this finding did not reach statistical significance (12. 0 months vs. not reached; *P*=0. 086) (Fig. 3). When analysis was performed for 24 patients without *BCR–ABL1* gene mutation, TTFx was significantly shorter in patients with *GSTT1* copy number gain than in patients without *GSTT1* copy number gain (10. 6 months vs. 18. 0 months; *P*=0. 007) (Fig. 3). When multivariate analysis was performed considering disease phase at study enrollment, baseline cytogenetic response to standard dose imatinib, and *GSTT1* copy number alteration, *GSTT1* copy number gain was an independent predictive factor for shorter TTFx (hazard ratio HR=1. 717, 95% CI=1. 088–2. 711; *P*=0. 020).

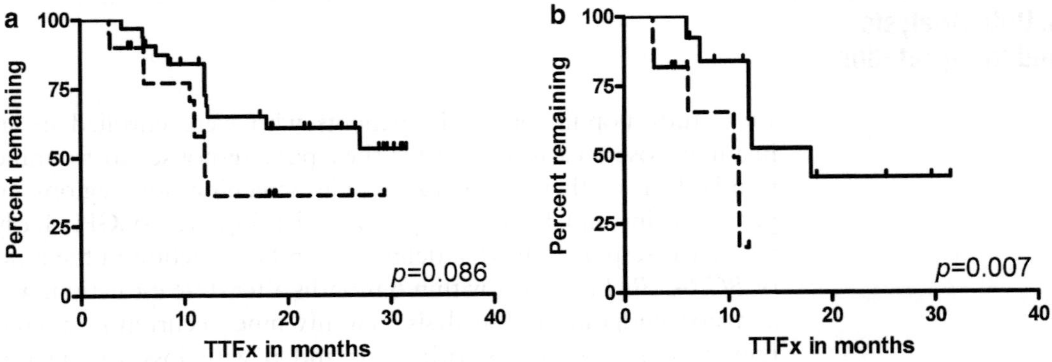

Fig. 3. Impact of *GSTT1* copy number gain on progression-free survival of CML patients receiving escalated dose imatinib. Progression-free survival of patients with *GSTT1* copy number gain (*dotted line*) seemed to be shorter ($P=0.086$) than patients without *GSTT1* copy number gain (*solid line*) (**a**). For patients lacking *BCR–ABL* mutations, *GSTT1* copy number gain predicted significantly shorter progression-free survival ($P=0.007$) (**b**).

5. Clinical Applications

In CML, the array-CGH analysis is useful in at least two patient groups. The first group of patients who are expected to have meaningful copy number alteration is the patients showing resistance to *ABL1* TKIs. Mechanisms underlying the resistance of CML cells to *ABL1* TKIs are not fully understood. It is proposed that approximately 50% of patients with acquired resistance to imatinib have point mutations in the kinase domain of *BCR–ABL1* (30, 31). However, for the rest of 50% patients showing acquired resistance to imatinib, the mechanism is not clearly elucidated. Defects in drug transport (32, 33) and the activation of other pathways such as Lyn pathway (34) have been suggested for these patients, but without definite evidences. Interestingly, the presence of mutations in *BCR–ABL1* kinase was associated with a clonal cytogenetic evolution, which may facilitate disease progression (31, 35). In concordance, the presence of mutations in *BCR–ABL1* kinase was associated with greater likelihood of progression to the blast phase, which suggests enhanced genomic instability in these cells (36–38). Considering the above, we can postulate that CML cells showing the resistance to TKIs might have profound genetic instabilities. These genetic instabilities would lead to the formation of mutations within the *BCR–ABL1* kinase and accumulations of additional genetic lesions. The additional genetic lesions may include point mutations, gene amplifications, genome loss, and chromosome translocations that are believed to be driving malignant process (39, 40). Hence, the identification of these alterations and the elucidation of genes implicated within these regions would provide a further understanding and improved outcomes for CML patients on TKIs. The array-CGH is one of the most powerful tool revealing genomic instabilities, and

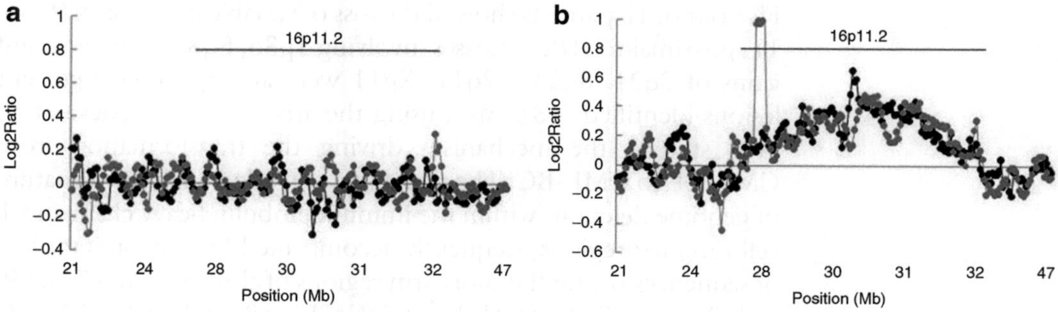

Fig. 4. Copy number profile of good responder to escalated dose imatinib and poor responder to escalated dose imatinib around 16p11. 2. Two non-EMR's (*black* and *gray dots*) showed no CNA on 16p11. 2 (**a**), and 2 EMR's showed copy number gain on this region (**b**). X-axis, linear position; Y-axis, test/reference intensity in $\log_2$ ratio.

it can be very helpful in the investigation of CML cells resistant to TKIs. In addition, it is highly probable that there are chances of having genetic changes other than point mutations, such as copy number alterations, that serve as a key factor related to TKI resistance in patients who do not harbor mutations in *BCR–ABL1* kinase. Hence, the array-CGH may play an important role in searching these potential genetic changes responsible for TKI resistance.

We performed the array-CGH analysis in CML patients who were to receive escalated dose imatinib for their resistance to standard dose (400 mg/day) imatinib (29). In this study, copy number gain in 16p11. 2 was more frequently observed in patients who showed favorable response to escalated dose imatinib (Fig. 4). Also, a tendency for frequent increased copy number in 22q11. 23 was observed in patients who showed poor response to the escalated dose imatinib. Tendency for decreased copy number in 17q12 was observed in patients who showed favorable response to escalated dose imatinib. We could also demonstrate that copy number gain of *GSTT1* in 22q11. 23 was frequently observed in patients who did not respond well to the escalated dose imatinib by real-time quantitative polymerase chain reaction assay for validation of the array-CGH results. *GSTT1* copy number gain was an independent predictive factor for short progression-free survival in these patients (Fig. 3).

The second group of patients is those with CML-AP or CML-BC. The recent high-resolution array-CGH analysis and SNP array analysis with extremely dense probes across the whole genome have detected multiple genetic aberrations in patients with CML-CP. The patients with CML-BC carried much more complex karyotypes (41, 42). Thus, genomic instability is an early event in patients with CML-CP, which accumulates in CML-BC. Still, however, not much is known about the genetics underlying CML-AP and CML-BC. Previous research using array-CGH in these patients revealed the following. In patients with CML-AP treated with imatinib, alterations flanking the *ABL1* and *BCR* genes on chromosomes 9 and 22, respectively, were the most commonly identified lesions.

Five out of 15 patients showed the loss of variable portions of 9q34. 11, proximal to *ABL1*. Losses involving 1p36, 5q31, 17q25, Y and gains of 3q21, 8q24, 22q11, Xp11 were among other recurrent lesions identified (43). Also, using the array-CGH, Nacheva et al. (44) studied the mechanism driving the transformation from CML-CP to CML-BC. The researchers revealed a unique signature of genome deletions within the immunoglobulin heavy chain and T cell receptor regions, frequently accompanied by concomitant loss of sequences within the short arm regions of chromosomes 7 and 9, including *IKZF1, HOXA7, CDKN2A/2B, MLLT3, IFNA/B, RNF38, PAX5, JMJD2C*, and *PDCD1LG2* genes.

6. Notes

1. Peripheral blood or bone marrow samples of patients with resistance to *ABL1* TKIs contain normal cells as well as CML cells. The percentage of CML cells among white blood cells can be estimated by *BCR–ABL* FISH or karyotyping. Because genetic aberrations generally are not observed in normal white blood cells, the coexistence of normal cells in peripheral blood or bone marrow samples does not generally influence laboratory results. However, when the proportion of CML cells is small, there are chances that genetic aberrations in CML cells would be missed. Hence, we recommend using samples that contain enough CML cells in array-CGH.

Acknowledgments

This study was supported by grants from the Korea Health 21 R&D Project, Ministry of Health & Welfare, Republic of Korea (0405-BC02-0604-0004)(A040002).

References

1. Nowell PC, Hungerford DA (1961) Chromosome studies in human leukemia. II. Chronic granulocytic leukemia. J Natl Cancer Inst 27:1013–1035
2. Faderl S, Talpaz M, Estrov Z et al (1999) The biology of chronic myeloid leukemia. N Engl J Med 341(3):164–172. doi:10. 1056/ NEJM199907153410306
3. Vigil CE, Griffiths EA, Wang ES et al (2011) Interpretation of cytogenetic and molecular results in patients treated for CML. Blood Rev 25(3):139–146. doi:S0268-960X(11)00019-1 [pii] 10. 1016/j. blre. 2011. 02. 001
4. Quintas-Cardama A, Cortes J (2009) Molecular biology of bcr-abl1-positive chronic myeloid leukemia. Blood 113(8):1619–1630. doi:blood-2008-03-144790 [pii] 10. 1182/ blood-2008-03-144790
5. Baccarani M, Saglio G, Goldman J et al (2006) Evolving concepts in the management of chronic myeloid leukemia: recommendations from an expert panel on behalf of the european

leukemianet. Blood 108(6):1809–1820. doi:blood-2006-02-005686 [pii] 10. 1182/blood-2006-02-005686

6. O'brien SG, Guilhot F, Larson RA et al (2003) Imatinib compared with interferon and low-dose cytarabine for newly diagnosed chronic-phase chronic myeloid leukemia. N Engl J Med 348(11):994–1004. doi:10. 1056/NEJMoa022457 348/11/994 [pii]
7. Diamond JM, Melo JV (2011) Mechanisms of resistance to BCR-ABL kinase inhibitors. Leuk Lymphoma 52(Suppl 1):12–22. doi:10. 3109/10428194. 2010. 546920
8. Chu S, Xu H, Shah NP et al (2005) Detection of BCR-ABL kinase mutations in CD34+ cells from chronic myelogenous leukemia patients in complete cytogenetic remission on imatinib mesylate treatment. Blood 105(5):2093–2098. doi:10. 1182/blood-2004-03-1114 2004-03-1114 [pii]
9. Corbin AS, La Rosee P, Stoffregen EP et al (2003) Several BCR-ABL kinase domain mutants associated with imatinib mesylate resistance remain sensitive to imatinib. Blood 101(11):4611–4614. doi:10. 1182/blood-2002-12-3659 2002-12-3659 [pii]
10. Morel F, Bris MJ, Herry A et al (2003) Double minutes containing amplified bcr-abl fusion gene in a case of chronic myeloid leukemia treated by imatinib. Eur J Haematol 70(4):235–239. doi:046 [pii]
11. Azam M, Latek RR, Daley GQ (2003) Mechanisms of autoinhibition and STI-571/imatinib resistance revealed by mutagenesis of BCR-ABL. Cell 112(6):831–843. doi:S0092867403001909 [pii]
12. Kantarjian HM, Talpaz M, O'brien S et al (2003) Dose escalation of imatinib mesylate can overcome resistance to standard-dose therapy in patients with chronic myelogenous leukemia. Blood 101(2):473–475. doi:10. 1182/blood-2002-05-1451 2002-05-1451 [pii]
13. Kantarjian HM, Larson RA, Guilhot F et al (2009) Efficacy of imatinib dose escalation in patients with chronic myeloid leukemia in chronic phase. Cancer 115(3):551–560. doi:10. 1002/cncr. 24066
14. Koh Y, Kim I, Yoon SS et al (2010) Phase IV study evaluating efficacy of escalated dose of imatinib in chronic myeloid leukemia patients showing suboptimal response to standard dose imatinib. Ann Hematol. doi:10. 1007/s00277-010-0910-8
15. Jabbour E, Cortes JE, Kantarjian HM (2009) Suboptimal response to or failure of imatinib treatment for chronic myeloid leukemia: what is the optimal strategy? Mayo Clin Proc 84(2):161–169. doi:84/2/161 [pii] 10. 4065/84. 2. 161
16. Lee TS, Potts SJ, Kantarjian H et al (2008) Molecular basis explanation for imatinib resistance of BCR-ABL due to T315I and P-loop mutations from molecular dynamics simulations. Cancer 112(8):1744–1753. doi:10. 1002/cncr. 23355
17. Kallioniemi A, Kallioniemi OP, Sudar D et al (1992) Comparative genomic hybridization for molecular cytogenetic analysis of solid tumors. Science 258(5083):818–821
18. Knuutila S, Aalto Y, Autio K et al (1999) DNA copy number losses in human neoplasms. Am J Pathol 155(3):683–694. doi:S0002-9440(10)65166-8 [pii] 10. 1016/S0002-9440(10)65166-8
19. Lee C, Rens W, Yang F (2001) Multicolor fluorescence in situ hybridization (fish) approaches for simultaneous analysis of the entire human genome. Curr Protoc Hum Genet. doi:10. 1002/0471142905. hg0409s24, Chapter 4:Unit4. 9
20. Pinkel D, Segraves R, Sudar D et al (1998) High resolution analysis of DNA copy number variation using comparative genomic hybridization to microarrays. Nat Genet 20(2):207–211. doi:10. 1038/2524
21. Albertson DG, Pinkel D (2003) Genomic microarrays in human genetic disease and cancer. Hum Mol Genet 12(Spec No 2):R145–R152. doi:10. 1093/hmg/ddg261 ddg261 [pii]
22. Lockwood WW, Chari R, Chi B, et al (2006) Recent advances in array comparative genomic hybridization technologies and their applications in human genetics. Eur J Hum Genet 14:139–48. doi:10.1038/sj.ejhg.5201531
23. Kallioniemi A (2008) CGH microarrays and cancer. Curr Opin Biotechnol 19(1):36–40. doi:S0958-1669(07)00148-6 [pii] 10. 1016/j. copbio. 2007. 11. 004
24. Shinawi M, Cheung SW (2008) The array CGH and its clinical applications. Drug Discov Today 13(17–18):760–770. doi:S1359-6446(08)00220-1 [pii] 10. 1016/j. drudis. 2008. 06. 007
25. Maciejewski JP, Tiu RV, O'keefe C (2009) Application of array-based whole genome scanning technologies as a cytogenetic tool in haematological malignancies. Br J Haematol 146(5):479–488. doi:BJH7757 [pii] 10. 1111/j. 1365-2141. 2009. 07757. x
26. Kim MY, Yim SH, Kwon MS et al (2006) Recurrent genomic alterations with impact on survival in colorectal cancer identified by genome-wide array comparative genomic hybridization. Gastroenterology 131(6):1913–1924. doi:S0016-5085(06)02255-4 [pii] 10. 1053/j. gastro. 2006. 10. 021
27. Kim TM, Yim SH, Lee JS et al (2005) Genome-wide screening of genomic alterations and their

clinicopathologic implications in non-small cell lung cancers. Clin Cancer Res 11(23):8235–8242. doi:11/23/8235 [pii] 10. 1158/1078-0432. CCR-05-1157

28. Kim TM, Yim SH, Shin SH et al (2008) Clinical implication of recurrent copy number alterations in hepatocellular carcinoma and putative oncogenes in recurrent gains on 1q. Int J Cancer 123(12):2808–2815. doi:10. 1002/ijc. 23901

29. Koh Y, Kim DY, Park SH et al (2010) GSTT1 copy number gain is a poor predictive marker for escalated-dose imatinib treatment in chronic myeloid leukemia: genetic predictive marker found using array comparative genomic hybridization. Cancer Genet Cytogenet 203(2):215–221. doi:S0165-4608(10)00485-1 [pii] 10. 1016/j. cancergencyto. 2010. 08. 022

30. La Rosee P, Hochhaus A (2010) Molecular pathogenesis of tyrosine kinase resistance in chronic myeloid leukemia. Curr Opin Hematol 17(2):91–96. doi:10. 1097/MOH. 0b013e32-83366be1

31. Willis SG, Lange T, Demehri S et al (2005) High-sensitivity detection of BCR-ABL kinase domain mutations in imatinib-naive patients: correlation with clonal cytogenetic evolution but not response to therapy. Blood 106(6):2128–2137. doi:2005-03-1036 [pii] 10. 1182/blood-2005-03-1036

32. Thomas J, Wang L, Clark RE et al (2004) Active transport of imatinib into and out of cells: implications for drug resistance. Blood 104(12):3739–3745. doi:10. 1182/blood-2003-12-4276 2003-12-4276 [pii]

33. Mahon FX, Belloc F, Lagarde V et al (2003) MDR1 gene overexpression confers resistance to imatinib mesylate in leukemia cell line models. Blood 101(6):2368–2373. doi:10. 1182/blood. V101. 6. 2368 101/6/2368 [pii]

34. Donato NJ, Wu JY, Stapley J et al (2003) BCR-ABL independence and LYN kinase overexpression in chronic myelogenous leukemia cells selected for resistance to STI571. Blood 101(2):690–698. doi:10. 1182/blood. V101. 2. 690 101/2/690 [pii]

35. Griswold IJ, Macpartlin M, Bumm T et al (2006) Kinase domain mutants of Bcr-Abl exhibit altered transformation potency, kinase activity, and substrate utilization, irrespective of sensitivity to imatinib. Mol Cell Biol 26(16):6082–6093. doi:26/16/6082 [pii] 10. 1128/MCB. 02202-05

36. Soverini S, Martinelli G, Rosti G et al (2005) Abl mutations in late chronic phase chronic myeloid leukemia patients with up-front cytogenetic resistance to imatinib are associated with a greater likelihood of progression to blast crisis and shorter survival: a study by the gimema working party on chronic myeloid leukemia. J Clin Oncol 23(18):4100–4109. doi:JCO. 2005. 05. 531 [pii] 10. 1200/JCO. 2005. 05. 531

37. Khorashad JS, De Lavallade H, Apperley JF et al (2008) Finding of kinase domain mutations in patients with chronic phase chronic myeloid leukemia responding to imatinib may identify those at high risk of disease progression. J Clin Oncol 26(29):4806–4813. doi:JCO. 2008. 16. 9953 [pii] 10. 1200/JCO. 2008. 16. 9953

38. Skorski T (2011) Chronic myeloid leukemia cells refractory/resistant to tyrosine kinase inhibitors are genetically unstable and may cause relapse and malignant progression to the terminal disease state. Leuk Lymphoma 52(Suppl 1):23–29. doi:10. 3109/10428194. 2010. 546912

39. Koptyra M, Cramer K, Slupianek A et al (2008) BCR/ABL promotes accumulation of chromosomal aberrations induced by oxidative and genotoxic stress. Leukemia 22(10):1969–1972. doi:leu200878 [pii] 10. 1038/leu. 2008. 78

40. Dierov J, Sanchez PV, Burke BA et al (2009) BCR/ABL induces chromosomal instability after genotoxic stress and alters the cell death threshold. Leukemia 23(2):279–286. doi:leu2008308 [pii] 10. 1038/leu. 2008. 308

41. Brazma D, Grace C, Howard J et al (2007) Genomic profile of chronic myelogenous leukemia: imbalances associated with disease progression. Genes Chromosomes Cancer 46(11):1039–1050. doi:10. 1002/gcc. 20487

42. Khorashad JS, De Melo VA, Fiegler H et al (2008) Multiple sub-microscopic genomic lesions are a universal feature of chronic myeloid leukaemia at diagnosis. Leukemia 22(9):1806–1807. doi:leu2008210 [pii] 10. 1038/leu. 2008. 210

43. Nadarajan VS, Phan CL, Ang CH et al (2011) Identification of copy number alterations by array comparative genomic hybridization in patients with late chronic or accelerated phase chronic myeloid leukemia treated with imatinib mesylate. Int J Hematol 93(4):465–473. doi:10. 1007/s12185-011-0796-9

44. Nacheva EP, Brazma D, Virgili A et al (2010) Deletions of immunoglobulin heavy chain and T cell receptor gene regions are uniquely associated with lymphoid blast transformation of chronic myeloid leukemia. BMC Genomics 11:41. doi:1471-2164-11-41 [pii] 10. 1186/1471-2164-11-41

Chapter 5

The Use of Cytogenetic Microarrays in Myelodysplastic Syndrome Characterization

Lisa G. Shaffer, Blake C. Ballif, and Roger A. Schultz

Abstract

Various microarray platforms, including BAC, oligonucleotide, and SNP arrays, have been shown to provide clinically useful diagnostic and prognostic information for patients with myelodysplastic syndromes (MDS). Clinically useful arrays are designed with specific purposes in mind and with attention to genomic content and probe density. All array types have been shown to detect genomic copy gains and losses, with SNP arrays having the added advantage of detecting copy neutral loss of heterozygosity (CNLOH). The finding of CNLOH has led to the identification of certain disease genes implicated in the initiation or progression of myeloid diseases. In addition, SNP karyotyping alone, or in conjunction with routine cytogenetics, can affect the outcome prediction and improve prognostic stratification of patients with MDS. Patients who were reclassified after array testing as having adverse-risk chromosomal findings correlated with poor survival. Results of over 25 published studies support the use of arrays in MDS testing. Because few balanced translocations are found in MDS, this disease is particularly amenable to microarray testing, and studies have shown better disease classification, identification of cryptic changes, and prognostication in this heterogeneous group of disorders. Novel genomic alterations identified by array testing may lead to better targeted therapies for treating patients with MDS.

Key words: Myelodysplastic syndrome, MDS, Microarray, aCGH, SNP, Copy number variant, CNV, Cytogenetics, Cancer

1. Introduction

The myelodysplastic syndromes (MDS) are a heterogeneous group of hematopoietic disorders that mostly affect patients ≥60 years of age (1). MDS patients have a propensity of leukemic transformation to acute myelogenous leukemia (AML). Recurrent, clonal cytogenetic abnormalities are identified in 30–50% of MDS cases using routine chromosome banding and fluorescence in situ

Diponkar Banerjee and Sohrab P. Shah (eds.), *Array Comparative Genomic Hybridization: Protocols and Applications*, Methods in Molecular Biology vol. 973, DOI 10.1007/978-1-62703-281-0_5, © Springer Science+Business Media, LLC 2013

hybridization (FISH). The identification of certain, clonal cytogenetic aberrations can confirm a diagnosis of MDS and predict patient prognosis and disease survival. The International Prognostic Scoring System (IPSS) (2) has three key prognostic factors, the number of peripheral blood cytopenias, the percentage of bone marrow blasts, and the chromosome abnormalities identified by cytogenetics. Approximately 90% of clonal aberrations are unbalanced, leading to gains or losses of whole chromosomes or chromosomal segments, making the detection of these genomic changes amenable to the new microarray technologies.

Chromosome abnormalities are classified as either numerical or structural. Cytogenetic analysis using banding can identify gains and losses of whole chromosomes and rearrangements visible under the light microscope that involve chromosomal segments of about 10 megabases (Mb) or greater. Such rearrangements include deletions, inversions, and translocations associated with certain types of cancers. Cytogenetic analysis requires the ability to culture cells and obtain metaphase chromosomes. Thus, a major limitation to the cytogenetic analysis of certain types of cancer is the ability to grow sufficient cells in culture and achieve analyzable metaphase chromosomes of satisfactory morphology necessary to visualize alterations. FISH can be performed on either cultured or uncultured cells, metaphase or interphase, so it has the added advantage of a rapid diagnosis even for cells that grew poorly in culture or resulted in metaphases of poor morphology. FISH "panels" are used for some diseases to screen for several different aberrations in a single specimen. These panels are labor intensive and require the ability to screen a large number of cells. It has been well documented that the identification of abnormalities by FISH can change a patient's risk factors, especially in those that initially had a normal karyotype by routine cytogenetic analysis (3, 4). However, FISH may yield complex hybridization patterns that are difficult to score and interpret. Finally, FISH detects only what is being interrogated—the rest of the genome cannot be assessed with this technique.

Although cytogenetic analysis and FISH have these limitations, identification of the precise chromosome abnormalities is important for predicting prognosis and prescribing specific, targeted therapies. As examples, deletion of 5q has a low to intermediate risk for MDS patients, and these patients show a high responsiveness to lenalidomide (5, 6), and MDS patients with monosomy 7 show high responsiveness to demethylating agents (7, 8). With the advent of new molecular cytogenetic technologies, such as microarrays, additional recurrent genomic changes may be identified that also provide prognostic value for certain therapies or become novel therapy targets. And indeed, recent studies using microarrays have uncovered recurrent aberrations that involve genes that may contribute to the pathogenesis of MDS as discussed below (9–13).

Thus, arrays can assess the whole genome and uncover complexity that cannot be identified with chromosome analysis or FISH.

Many recent studies have shown that both genome-wide and targeted microarray testing are powerful new technologies for detecting known cytogenetic alterations as well as cryptic, novel changes. This chapter will describe the various types of microarrays available for cytogenetic profiling of patients with MDS and will explain the advantages and disadvantages with each platform. Based on the medical literature published to date, microarrays are becoming an important tool to analyze patient samples and make decisions regarding the best treatments for better patient outcomes.

2. Materials and Methods

The cytogenetics laboratory has a number of options for implementing microarray testing for the detection of chromosome aberrations in MDS. There are two basic types of genomic microarrays: comparative genomic hybridization-based arrays (aCGH) and single-nucleotide polymorphism-based (SNP) arrays. There are benefits and challenges with each, and the laboratory must understand the differences between array types in order to interpret the results and to convey the diagnostic or prognostic implications to the healthcare provider.

Microarrays are constructed using genomic targets, immobilized on a solid support such as a piece of glass. They can be constructed using large genomic pieces of DNA such as that found in bacterial artificial chromosomes (BACs), which are 80–200 kilobases (kb) in size, or smaller targets such as synthetic oligonucleotides (oligos), which are typically 25–85 base pairs (bp) in length. The targets, representing various segments of the genome, can be thousands or a few million targets depending on the commercial vendor supplying the arrays and the needs of the laboratory. Both types of microarrays use DNA extracted directly from a specimen without culturing. Thus, either platform has the advantage over chromosome analysis and FISH of not needing viable, cultured cells. In addition, the resolution that microarrays can achieve is at least 100-fold better that cytogenetic analysis, depending on the genomic content and density of the probes used in the construction of the microarray.

Because DNA can be extracted directly from samples and does not require cell culture prior to extraction, many different types of samples can be used in microarray analysis including peripheral blood, bone marrow aspirate, lymph nodes, or solid tumors, depending on the suspected disorder or tissue involved. For MDS, bone marrow aspirates are the sample of choice. In addition, paraffin-embedded or formalin-fixed tissue is a rich source of archived specimens for research or a stabilized resource for retrospective clinical assessment.

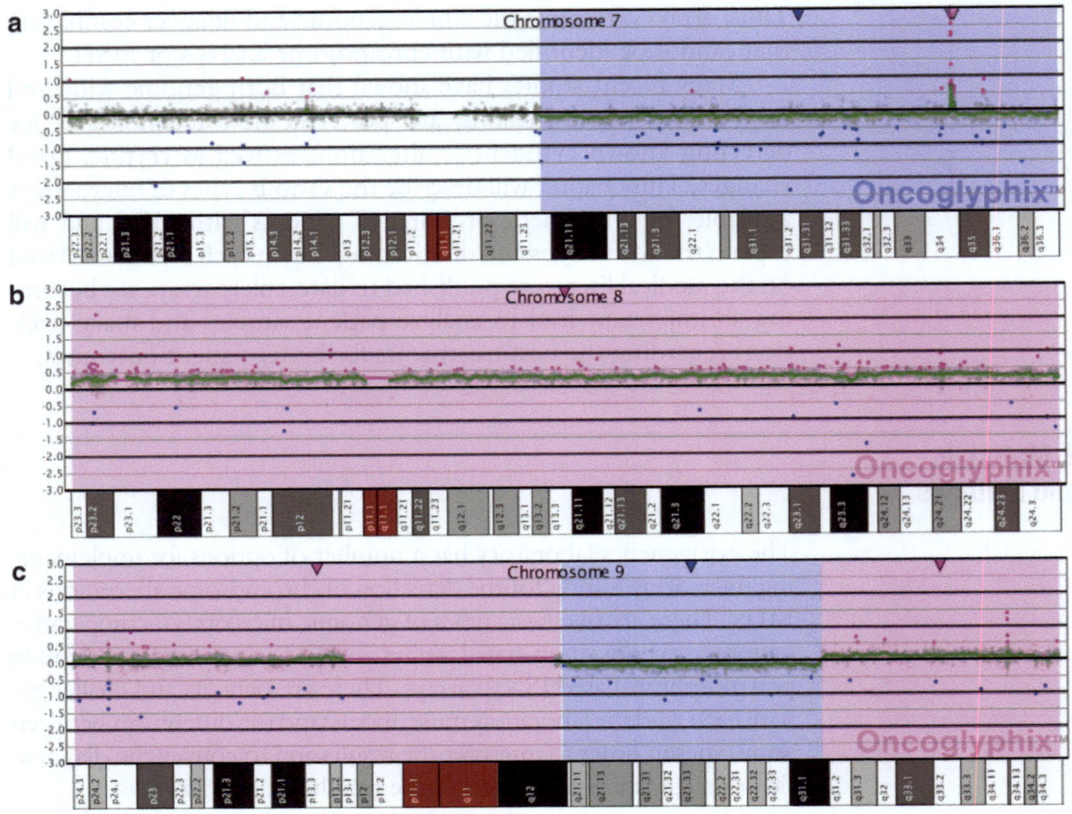

Fig. 1. Microarray plots from a patient with MDS and a complex karyotype. (**a**) Whole chromosome 7 plot showing a partial copy loss (*blue*-shaded region) of the q-arm and an interstitial copy gain (*pink* flag and dots) in the 7q34 region. (**b**) Whole chromosome 8 plot showing a copy gain (*pink*-shaded region) of the entire chromosome (trisomy 8 or +8). (**c**) Whole chromosome 9 plot showing a copy gain (*left pink*-shaded region) of the entire p-arm and a copy loss (*blue*-shaded region) and gain (*right pink*-shaded region) of the q-arm. This plot represents a trisomy 9 with a deletion in proximal 9q. The deletion on chromosome 9q21q22 is seen more often in AML but can be seen less commonly in MDS [56]. Probes are ordered on the *x*-axis according to physical mapping positions. Values along the *y*-axis represent $\log_2$ ratios of patient:control signal intensities. Results are visualized using Oncoglyphix (Signature Genomic Laboratories, Spokane, WA).

2.1. Comparative Genomic Hybridization-Based Arrays

Comparative genomic hybridization (CGH)-based arrays utilize the direct comparison of a patient sample with a known control sample. The patient DNA and a control DNA are each labeled with a distinct fluorescent dye, which are then hybridized in the same concentration to the same microarray. The fluorescent intensity of each dye is captured by computer imaging, and the ratios of the dyes are plotted on a graph in an order known to exist on the chromosome (14–19). Thus, when the patient has a gain or loss of DNA at a specific locus, as compared to the control, the differences between these genomes at this genomic location are visualized after graphing the ratios of the fluorescent intensities (Fig. 1). aCGH can be applied to both BAC-based arrays and oligo-based arrays. Depending on the genomic coverage contained within the targets on the array, in general, oligo arrays have the potential for higher-resolution analysis, with the ability to detect smaller alterations than can be identified by BAC arrays because of the large size of the genomic inserts in the BACs. Based on the current vendor offerings, oligo arrays for the purpose of CGH can be customized based on the needs of the laboratory. Clinical laboratories can design targeted arrays that cover certain genomic locations involved in cancer, or MDS, specifically. A higher density of probes within a gene(s) of interest will obviously increase the resolution of the aberrations that can be detected. Such a relationship could be extended to include all exons (exomic coverage). However, the value of such a strategy for clinical diagnostics would be limited by the expectation that small deletions are (1) sufficiently common causes of disease to warrant such high-density coverage and (2) not representative of common variation seen in normal noncancer controls. In addition to targeting specific loci, "backbone" coverage can be added to size alterations, for the identification of break points, and to identify novel aberrations that may not occur at the predetermined targets. For research applications seeking to identify novel alterations, the investigator may choose "whole-genome" coverage in which targets are placed on the array that represent some arbitrary coverage of the genome, such as at 1 kb intervals along the length of all chromosomes.

2.2. Single-Nucleotide Polymorphism-Based Arrays

Another type of synthetic oligo-based array is the single-nucleotide polymorphism (SNP) array. In contrast to aCGH, SNP arrays do not directly compare a patient and a control specimen on the same microarray. Instead, SNP arrays compare the dosage of a patient DNA specimen at any given locus to a database of control individuals at that same locus. As with aCGH, gains and losses of the genome are readily detectable using this method. SNP arrays have the added advantage of being able to detect specific genotypes for any given SNP. Sometimes called the B-allele frequency, the software can assign each probe or locus a genotype. For normal, two-copy dosage, the assignments would be AA, BB, or AB (Fig. 2). One-copy gains of chromosomal regions resulting from aneuploidy or duplications, for example, are assigned the combination of these genotypes, or AAA, BBB, ABB, or AAB. Single-copy losses will result in either an A or B genotype. The combination of multiple contiguous SNPs can show regions that have loss of heterozygosity (LOH), also referred to as copy neutral LOH (CNLOH), in the genome. In contrast to aCGH, SNP arrays readily detect ploidy changes, which are important in cancer diagnosis.

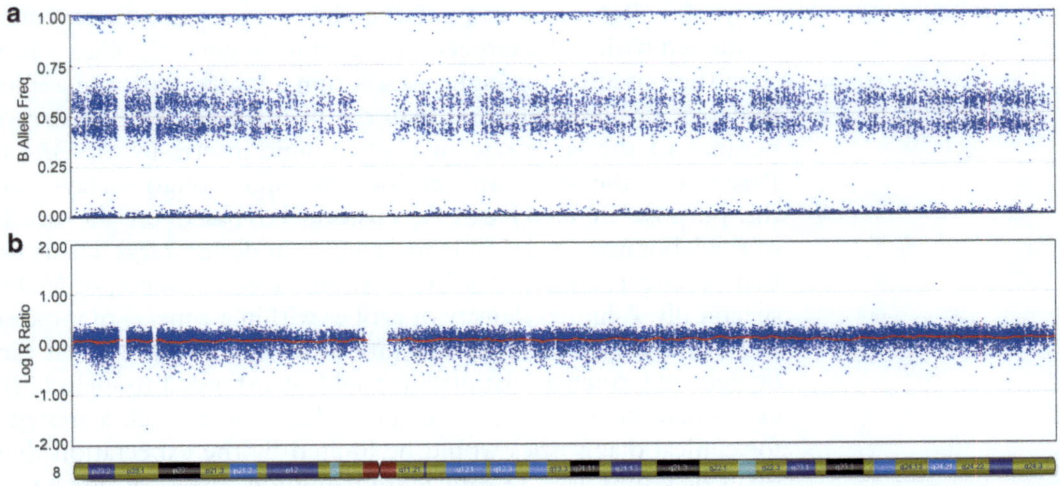

Fig. 2. SNP array plot of chromosome 8 from a patient with MDS. (**a**) Whole chromosome plot showing the B-allele frequencies. Four rows of data points indicate AAA, AAB, BBA, and BBB alleles. (**b**) Dosage for chromosome 8 showing a slightly elevated red line consistent with a gain of chromosome 8 (trisomy). Image was provided by Donna Wilmoth (The Children's Hospital of Philadelphia).

3. Data Analysis and Interpretation

The fluorescent intensities of the dyes used in microarray analysis are detected using laser scanners and are plotted on a graph using computer software programs. Many array vendors supply software developed specifically for cytogenetic applications. The capture and display of data have become the most streamlined aspect of array analysis. The most difficult features of the process are (1) the design of the array and (2) the interpretation of the genomic gains and losses found.

3.1. Array Design

For an array to be useful clinically to the healthcare provider and patient, it should be designed with a specific purpose. The specific genomic targets should be considered depending on the disease the array is designed to diagnose. Numeric abnormalities can be identified on most available microarrays, with reasonable density as long as the microarray represents each individual chromosome. Structural chromosomal changes are found in many cancers that involve specific genes or genomic regions that may require a more targeted approach for their detection.

In general, there are two classes of genes that are of concern in cancer diagnosis that may require targeting by the probe coverage on a microarray—oncogenes and tumor suppressor genes. During development, there are genes that are important for cell growth. Once the tissue type is mature, these genes are typically turned off or become inactivated. Structural changes in the genome may result in the reactivation of these embryologically important genes, leading to the inappropriate transcription of the gene, which leads to abnormal cell proliferation. These types of genes are called oncogenes. Many oncogenes become transcriptionally active after being translocated and becoming juxtaposed to an active gene. This mechanism has been observed for specific rearrangements associated with certain neoplastic disorders. The other class of genes important in cancer is that of the tumor suppressor genes. In normal cells, tumor suppressor genes control cell proliferation. The loss of a tumor suppressor gene, after deletion or formation of an unbalanced translocation, may lead to loss of regulation, and abnormal cell proliferation ensues resulting in neoplasia. Therefore, depending on the disease, it may be important to target both oncogenes and tumor suppressor genes. In addition, the design of the array should provide enough backbone coverage of the genome to identify numerical gains and losses, define break points, and size alterations.

3.2. Data Interpretation

Copy number alterations can be detected with either an aCGH-based approach or an SNP-based assay. With either platform, single-copy gains, amplifications, single-copy losses, or biallelic losses of individual chromosomes, chromosomal regions can be identified. However, because SNPs detect the genotype at a particular oligonucleotide, this type of array allows for the detection of LOH or copy number neutral LOH (CNLOH) in cancer that cannot be detected using aCGH. SNP arrays also provide more information regarding ploidy changes associated with certain types of cancers. The ability to detect SNPs will be important as our understanding of the contribution of LOH/CNLOH to disease and disease progression increases. In addition, because ploidy changes and clonal evolution can make the interpretation of arrays challenging, the ability to identify various alleles can aid in the identification of complex karyotypes that involve hyper- or hypodiploidy. The region(s) of LOH/CNLOH may contain tumor suppressor genes that may have homozygous mutations that can be detected by sequencing and other methods as have been demonstrated for *JAK2* mutations in polycythemia vera (20).

Copy number assessment has been performed in MDS by several investigators (9, 21–23). In all studies, the quantity of copy number alterations was considerably increased over that seen by routine cytogenetics, especially in those cases with normal karyotypes in which cryptic changes were identified by microarray analysis. The identification of numerical and structural changes can impact the prognosis and may identify new prognostic indicators (24). Microarray analysis can be used to further stratify patients to identify those patients that may have a more aggressive form of a disease or those that might benefit from alternative treatments because eventually microarray data will predict those to be nonresponders given the genotypes identified (25).

However, human genomes are subject to copy number variation (CNV) that may have no, or no known, clinical consequences. Although some copy number alterations (CNAs) can be clearly classified as pathogenic, others can be classified as benign, based on demonstration in normal individuals (26–28). In general, benign variants are identified in both patient and control populations in roughly equal frequencies. Benign variation may range from several base pairs of DNA to a few million base pairs, and the frequencies of each may depend on the individuals' ethnic background. Thus, in order to interpret microarray data, the diagnostician must have knowledge of normal variation in the genome and be able to recognize abnormal (pathogenic) alterations. The number of alterations found in normal individuals varies greatly, and their discovery is dependent on the tool used; higher-resolution arrays will uncover more variation than lower-resolution or targeted arrays. As much as 20% of the genome may be CNVs (29). There are some public databases of normal variation that can be used to help decipher normal variation from abnormal aberrations (e.g., (30– 35)). Databases specific to copy alterations that occur because of general instability of cancer genomes, but not associated with disease initiation or progression, would be helpful for the interpretation of array data from cancer patients. To overcome difficulties in interpretation, some laboratories compare a normal DNA specimen (such as cheek swabs or saliva) to the DNA from bone marrow or peripheral blood. In doing so, only differences between the cancer genome and the normal genome will be identified; thus, only acquired changes that are more likely to be involved in disease will be found (36). However, this approach will miss inherited copy number alterations that might have a predisposition to developing neoplasia because the presence of the same CNV in the two samples being compared will prevent identification.

4. Clinical and Research Applications of Arrays in MDS

The first microarrays to be applied to the study of MDS were BAC arrays. In 2006, Paulsson and coworkers (11) studied four subjects with MDS and trisomy 8 and found additional, cryptic changes by microarray. Likewise, O'Keefe and colleagues studied 38 MDS patients, 19 of which had normal karyotypes by routine cytogenetic analysis, failed to yield analyzable metaphase chromosomes, and verified known karyotypic changes and identified cryptic alterations in 32 subjects (21). Among subjects with known cytogenetic aberrations, 89.5% showed additional, novel genomic alterations.

The average number of abnormalities revealed by the array was 14.82, indicating that this method is superior to karyotyping for the comprehensive identification of cytogenetic abnormalities. The novel changes identified may lead to an understanding of their role in disease initiation and progression. Starczynowski and coworkers also studied 44 patients with a high-resolution BAC array and found that, although only 16 patients had anomalies detected by karyotyping, 36 of the patients were found to have alterations by microarray analysis (13). Among the changes identified, 21 were recurrent including gains of 11q24qter and 17q11.2q12 and losses at 2q33, 5q31, and 10q21.3. While these patients were initially characterized as relatively low risk, the additional aberrations identified correlated with poorer overall survival and were more frequently associated with transformation to AML. Low-risk subtypes of MDS represent about 50% of MDS patients and typically have normal karyotypes. The finding of cryptic changes by array analysis in patients with normal karyotypes by routine cytogenetics and their finding of 22 recurrent changes support the role of microarray testing as a better diagnostic and prognostic tool in the overall assessment of the patient with MDS. Continuing with the assessment of MDS patients using BAC arrays, others have also shown the exceptional utility of these arrays in further assessing pediatric patients (37) and in the identification of additional, cryptic anomalies in patients with known cytogenetic aberrations, placing them in a higher-risk category of MDS patients (38).

Oligo-based aCGH provides higher-resolution analysis than BAC arrays. One of the first publications utilizing oligo arrays in MDS showed that 5q break points could be refined based on microarray analysis as compared to the cytogenetically defined break points that banding affords (12). As an isolated aberration, deletion 5q confers a relatively good prognosis. However, patients that have deletion 5q and one or more additional chromosome abnormalities have a significantly shorter overall survival. In this publication, most of the aberrations found by array were larger than 20 Mb and should have been identified by routine cytogenetic testing. However, due to poor morphology or poor cell growth, these aberrations remained undetected by the conventional methods and required array testing to reveal these imbalances and to determine the poorer prognosis for these patients (12). Additional studies using oligo arrays confirmed these findings of higher quantities of copy number alterations and the finding of recurrent genomic alterations in MDS patients studied by arrays (39–41). Most importantly, the study by Thiel and coworkers showed that in 107 MDS patients with normal karyotypes, 39% had cryptic, novel, or recurrent genomic imbalances not detected by conventional cytogenetics (40). Four recurrent submicroscopic findings in that study involved regions previously implicated in MDS, including deletions of *TET2* (4q24), *RUNX1* (21q22.12),

5q31.2, and 7q22.1. Such results illustrate improved diagnostic yields achieved by the higher resolution of microarrays. In addition, this study showed that those patients with abnormal microarray findings had shorter survival as compared to the patients that were confirmed to have normal karyotypes by microarray testing. Furthermore, the study by Thiel et al. supports the use of microarray testing for patients with suspected or diagnosed MDS in order to provide more accurate assessment of the cancer genome and deliver more accurate prognostic information for patient management. Routine cytogenetic and microarray analyses provide complementing assessments of the genome because cytogenetics may miss cryptic and sometimes large genomic imbalances due to poor chromosome morphology and because arrays are not able to detect low-level (<20–30%) abnormal cell clonality. An integrated "cytogenomic" approach would improve the resolution and abnormality detection rate for MDS (41).

SNP-based arrays have also been used to detect gains and losses of the genome. In a study of 20q deletions in MDS patients, deletions were found in 23 patients, 3 of which were not detected by cytogenetic analysis (42); the arrays not only identified undetected deletion 20q but provided more precise break points and better assessment of deletion sizes. Deletion of 20q as a sole abnormality in MDS has a relatively favorable prognosis. The identification of additional genomic changes and karyotypic complexity by array analysis has important clinical implications as it predicts a poorer prognosis. In another study, SNP array analysis improved the detection of chromosome aberrations in 54% of MDS patients studied, with the finding of additional clonal abnormalities in 36% of patients with normal or noninformative cytogenetic studies (43). In addition, SNP arrays have the added advantage of detecting CNLOH. In one of the first uses of SNP arrays in cancer, Mohamedali and colleagues (10) studied 119 low-risk MDS patients using a whole-genome SNP array. Of these, patients had either normal karyotypes ($n=77$), isolated 5q deletions ($n=21$), or other ($n=21$). They found CNLOH in 46%, additional deletions in 10%, and amplifications in 8% of cases. In an interesting component of their study, 33 control DNAs were studied for CNLOH; 72% showed CNLOH with a range of 0.3–6.7 Mb in size. CNLOH of greater than 2 Mb was found in 12% of control individuals (10) and in a significant proportion of HapMap studies (44). Based on this information, Mohamedali et al. (10) excluded from their analysis all CNLOH of less than 2 Mb. They found that patient survival was not significantly impacted by regions of CNLOH found in the MDS patients. However, the regions of LOH in the controls as compared to patients were nonoverlapping, supporting that the CNLOH observed in MDS may be contributing to the pathogenicity of the disease.

Additional studies using SNPs and uncovering LOH has provided further insights into the clinical significance of these findings.

Gondek and colleagues (9) studied 94 MDS and 47 MDS/myeloproliferative disease (MPD) patients and found that 75% of patients showed aberrations by array analysis, whereas only 59% of MDS and 37% of MDS/MPD showed chromosome abnormalities by karyotyping. In addition, CNLOH was identified in 20% of MDS and 35% of MDS/MPD cases (9), although the frequency of such acquired CNLOH events is reported to be reduced for MDS seen in Fanconi anemia (45). The finding of regions with recurrent CNLOH may indicate the positions of novel genes associated with initiation or progression of MDS (46). Deletion of a region will also result in homozygosity, and these regions may represent genomic change associated with haploinsufficiency of a tumor suppressor gene. However, CNLOH may represent a region of homozygous mutation in an oncogene or tumor suppressor gene. SNP arrays can distinguish between deletion and CNLOH. Thus, the prognostic implications of SNP array findings are similar to visible cytogenetic deletions (9). Several other studies have also found CNLOH, as well as cryptic genomic alterations in patients with MDS studied with SNP arrays (22), including the study of 51 MDS patients, which identified clonal abnormalities in 41% of samples and CNLOH in ~8% (47). The finding of CNLOH may lead investigators to the regions in which to search for gene mutations that may help stratify patients and clinical outcomes to devise new treatment options. In one study (48), CNLOH led investigators to sequence *TET2* in 4q24 and identify mutations that were homozygous, heterozygous, compound heterozygous, or hemizygous (on the background of a deletion) in 58% of the MDS cases studied. The use of CNLOH to find additional genes associated with myeloid malignancies includes CNLOH13 and *FLT3* mutations, CNLOH1p and *MPL* mutations, CHLOH9p and *JAK2* mutations, and CNLOH11q and *c-CBL* mutations (reviewed in ref. (48)). In a recent study comparing MDS and progression of MDS to secondary AML with paired samples (49), CNLOH was observed in 11 out of 34 samples. Most CNLOH findings in this study appeared to be associated with a single recombination event and were seen in both MDS and secondary AML samples from the same patient, once again illustrating the potential for an inherited origin for some of these events. However, one case exhibited conversion of heterozygous *RUNX1* mutation to homozygosity in the AML specimen demonstrating the importance of this lesion in the progression of the disease (49). Thus, SNP array findings have a similar impact on clinical outcomes as other cytogenetic methods, yet can identify smaller defects and CNLOH that are not identifiable by metaphase chromosomes or FISH (50). Nonetheless, control populations (46) and paired normal samples from the patient (47) may be necessary for the interpretation of CNLOH found in cancer genomes.

Overwhelmingly, the studies reviewed here show that high-resolution genomic analysis improves prognostication in low-risk patients regardless of the array platform. In addition, intermediate-risk patients will also benefit from array testing because, as compared to chromosome analysis, arrays are faster and more comprehensive and results may place the patient into a different risk category. Many studies showed that even high-risk patients would benefit from array testing, even when the array results were no different than conventional cytogenetics and confirmed no additional complexity. Study after study confirmed that arrays outperform routine chromosome analysis and FISH for providing more information for determining the patient's prognosis (e.g., (9)). Thus, the identification of cryptic or novel aberrations by array analysis may have clinical significance by providing more information about disease progression and prognosis. In addition, these cryptic or newly identified changes may prove to be novel targets for future therapies.

5. Discussion

Microarrays can have very high-resolution detection of gains and losses of the genome based on the genomic content, array design, and density of the probes. Microarray analysis does not require the use of living cells and can be performed on DNA isolated from archived specimens. The results from array testing are objective and, with SNP arrays, can reveal CNLOH.

One criticism of array testing in cancer has been the potential identification of findings of unclear clinical significance or a high rate of false positives and misclassifying low-risk patients into higher-risk categories (9). However, Gondek et al. showed that patients who were reclassified as having adverse-risk chromosomal findings after array testing did in fact have a worse survival and, thus, were appropriately reclassified. In addition, Tiu and colleagues showed that SNP karyotyping alone, or in conjunction with routine cytogenetics, can affect the outcome prediction and improve prognostic stratification of patients with various types of MDS (24). The abnormalities found by SNP arrays were correlated with worse survival outcomes. In addition, arrays better demonstrated patients with truly normal karyotypes and provided better clinical prognoses in these cases. The high concordance rate between routine cytogenetics and SNP array detection abilities shows the reliability of SNP-based karyotyping and findings such as these have led Tiu and colleagues and others to propose a karyotyping algorithm that includes the use of SNP arrays in the diagnosis of MDS (24).

The use of microarrays in cancer is becoming an important tool for assessing the genome and identifying cryptic and novel changes. Microarrays are also becoming more accepted as another tool in the cytogeneticist's laboratory that can be used diagnostically to identify chromosome aberrations and provide a prognosis based on those abnormalities identified. More than 25 original research articles have been published just on the use of arrays in MDS. These papers have supported the use of microarrays to identify the abnormalities seen by routine karyotyping, those that are cryptic and require molecular cytogenetic methods to uncover, and aberrations that are novel. Microarrays are useful when cytogenetic methods fail to identify a chromosome abnormality, when cytogenetics are not possible due to culture failure, or when cytogenetic and FISH results are ambiguous, discrepant with each other, or discrepant with the pathology. Microarrays have shown results that reclassify a disease state, change a prognosis, and influence treatment decisions (50). In many instances, the identification of more complex karyotypes by arrays has resulted in improved prediction of prognosis and, in some cases, better therapeutic selections for the patient. Many studies have shown that arrays can stratify MDS patients into different risk categories and better predict prognosis. Microarrays are more comprehensive than routine cytogenetics and FISH and can have faster critical turnaround times. Because arrays are high resolution and comprehensive, they can be an adjunct or even a replacement for routine cytogenetics in certain cases, especially in those MDS cases with normal karyotypes. Because the majority of chromosome abnormalities in MDS are gains or losses of chromosomal segments or whole chromosomes rather than balanced translocations, cytogenetic assessment for MDS is well provided for by microarray analysis. As future treatment options become increasingly dependent on well-characterized cancer genomes, microarrays will play an essential role in the diagnosis and prognosis of individuals with suspected or diagnosed MDS. However, studies have shown that microarrays may not detect all abnormalities, especially those found in low-level clones. Thus, a combination of FISH, metaphase chromosome analysis, and microarrays may be necessary in some disorders (50). Nonetheless, array-based cytogenetics allows for objective disease classification and prognostic stratification in MDS, which represents a heterogeneous group of disorders. Ultimately, this clinically useful information will be important for targeted therapies.

6. Potential Pitfalls and Troubleshooting

Some specimens may have hyperdiploidy, hypodiploidy, clonal evolution, and other mechanisms leading to complex karyotypes that may be difficult to identify or interpret with microarray testing. The reason for this is that if a mixture of complex genomic changes or multiple clones exist in the specimen, all representative cells will be extracted together and may produce a complex (composite) pattern on the array that may not be interpretable with microarray testing. Therefore, the use of chromosome analysis and FISH may be more appropriate for highly complex cancer genomes.

Although a rare finding in MDS, neither aCGH nor SNP arrays will detect balanced translocations. If a translocation is truly balanced, there is not a net gain or loss of DNA. Thus, because microarray analysis relies on copy number changes between the patient and the control, in general these techniques cannot detect balanced translocations unless there is a deletion or duplication at the break point. A modification of aCGH, called translocation-CGH, provides a new approach to the detection of prognostically important translocations in neoplastic disorders (51, 52). Prior to labeling and microarray hybridization, a linear amplification is performed over potential translocation break points using primers in the vicinities of possible junctions. After labeling the reaction and hybridization to a specialized custom-designed array, a peak can be visualized at the site of the translocation break point when present in the specimen. In theory, by using the sequence of one of the known or potential partner chromosomes, the recipient chromosome involved in the translocation can be identified. Thus, a large pool of primers must be used in a multiplex assay to identify balanced translocations and stratify patients into specific diagnoses based on the translocations identified.

In addition to complex karyotypes and balanced translocations, microarrays are not sensitive enough to detect low-level clonal changes. Although one report suggests that aCGH can detect large cytogenetically visible abnormalities present in as few as 8% of cells (53), the smaller aberrations that should be identified by this technology must be in a higher minimal proportion of cells being studied; both aCGH and SNP arrays will reliably detect mosaicism in greater than about 20–30% abnormal cells (54, 55). Thus, chromosomal changes in a very minor, or low represented, clone may not be apparent in the DNA of a specimen after microarray analysis. Therefore, arrays should not be used to assess minimal residual disease (MRD). Monitoring for MRD for specific chromosomal changes can be accomplished using FISH or PCR during treatment or remission. However, microarray analysis could prove very important for the assessment of recurrent disease, in which a majority of cells will once again reveal the presence of an abnormal clone. The detailed characterization of abnormalities present in this new stage of disease, including those not observed at initial diagnosis may provide critical information regarding disease evolution, progression to AML, and appropriate choices for next-phase therapy. Such a second stage of microarray results may well benefit from that baseline data obtained at initial diagnosis, once again illustrating potential value for microarray analysis in the initial workup of all cases of MDS.

Acknowledgments

The authors thank Erin Dodge (Signature Genomic Laboratories) for her careful formatting of this manuscript. We thank Donna Wilmoth, (The Children's Hospital of Philadelphia) for the use of the SNP image.

References

1. Ma X, Does M, Raza A et al (2007) Myelodysplastic syndromes: incidence and survival in the United States. Cancer 109(8):1536–1542. doi:10.1002/cncr.22570
2. Greenberg P, Cox C, Lebeau MM et al (1997) International scoring system for evaluating prognosis in myelodysplastic syndromes. Blood 89(6):2079–2088
3. Bernasconi P, Cavigliano PM, Boni M et al (2003) Is fish a relevant prognostic tool in myelodysplastic syndromes with a normal chromosome pattern on conventional cytogenetics? A study on 57 patients. Leukemia 17(11):2107–2112. doi:10.1038/sj.leu.2403108 2403108 [pii]
4. Rigolin GM, Bigoni R, Milani R et al (2001) Clinical importance of interphase cytogenetics detecting occult chromosome lesions in myelodysplastic syndromes with normal karyotype. Leukemia 15(12):1841–1847
5. List A, Dewald G, Bennett J et al (2006) Lenalidomide in the myelodysplastic syndrome with chromosome 5q deletion. N Engl J Med 355(14):1456–1465. doi:355/14/1456 [pii] 10.1056/NEJMoa061292
6. Giagounidis AA, Germing U, Strupp C et al (2005) Prognosis of patients with del(5q) mds and complex karyotype and the possible role of lenalidomide in this patient subgroup. Ann Hematol 84(9):569–571. doi:10.1007/s00277-005-1054-0
7. Lubbert M, Wijermans P, Kunzmann R et al (2001) Cytogenetic responses in high-risk myelodysplastic syndrome following low-dose treatment with the DNA methylation inhibitor 5-aza-2′-deoxycytidine. Br J Haematol 114(2):349–357. doi:bjh2933 [pii]
8. Raj K, John A, Ho A et al (2007) Cdkn2b methylation status and isolated chromosome 7 abnormalities predict responses to treatment with 5-azacytidine. Leukemia 21(9):1937–1944. doi:2404796 [pii] 10.1038/sj.leu.2404796
9. Gondek LP, Tiu R, O'keefe CL et al (2008) Chromosomal lesions and uniparental disomy detected by snp arrays in mds, mds/mpd, and mds-derived aml. Blood 111(3):1534–1542. doi:blood-2007-05-092304 [pii] 10.1182/blood-2007-05-092304
10. Mohamedali A, Gaken J, Twine NA et al (2007) Prevalence and prognostic significance of allelic imbalance by single-nucleotide polymorphism analysis in low-risk myelodysplastic syndromes. Blood 110(9):3365–3373. doi:blood-2007-03-079673 [pii] 10.1182/blood-2007-03-079673
11. Paulsson K, Heidenblad M, Strombeck B et al (2006) High-resolution genome-wide array-based comparative genome hybridization reveals cryptic chromosome changes in aml and mds cases with trisomy 8 as the sole cytogenetic aberration. Leukemia 20(5):840–846. doi:2404145 [pii] 10.1038/sj.leu.2404145
12. Evers C, Beier M, Poelitz A et al (2007) Molecular definition of chromosome arm 5q deletion end points and detection of hidden aberrations in patients with myelodysplastic syndromes and isolated del(5q) using oligonucleotide array cgh. Genes Chromosomes Cancer 46(12):1119–1128. doi:10.1002/gcc.20498
13. Starczynowski DT, Vercauteren S, Telenius A et al (2008) High-resolution whole genome tiling path array cgh analysis of cd34+ cells from patients with low-risk myelodysplastic syndromes reveals cryptic copy number alterations and predicts overall and leukemia-free survival. Blood 112(8):3412–3424. doi:blood-2007-11-122028 [pii] 10.1182/blood-2007-11-122028
14. Kallioniemi A, Kallioniemi OP, Sudar D et al (1992) Comparative genomic hybridization for molecular cytogenetic analysis of solid tumors. Science 258(5083):818–821
15. Derisi J, Penland L, Brown PO et al (1996) Use of a cdna microarray to analyse gene expression patterns in human cancer. Nat Genet 14(4):457–460
16. Lucito R, Healy J, Alexander J et al (2003) Representational oligonucleotide microarray

analysis: a high-resolution method to detect genome copy number variation. Genome Res 13(10):2291–2305
17. Pinkel D, Segraves R, Sudar D et al (1998) High resolution analysis of DNA copy number variation using comparative genomic hybridization to microarrays. Nat Genet 20(2):207–211
18. Schena M, Shalon D, Davis RW et al (1995) Quantitative monitoring of gene expression patterns with a complementary DNA microarray. Science 270(5235):467–470
19. Solinas-Toldo S, Lampel S, Stilgenbauer S et al (1997) Matrix-based comparative genomic hybridization: biochips to screen for genomic imbalances. Genes Chromosomes Cancer 20(4):399–407
20. Kralovics R, Teo SS, Buser AS et al (2005) Altered gene expression in myeloproliferative disorders correlates with activation of signaling by the v617f mutation of jak2. Blood 106(10):3374–3376. doi:2005-05-1889 [pii] 10.1182/blood-2005-05-1889
21. O'keefe CL, Tiu R, Gondek LP et al (2007) High-resolution genomic arrays facilitate detection of novel cryptic chromosomal lesions in myelodysplastic syndromes. Exp Hematol 35(2):240–251. doi:S0301-472X(06)00637-0 [pii] 10.1016/j.exphem.2006.09.016
22. Akagi T, Ogawa S, Dugas M et al (2009) Frequent genomic abnormalities in acute myeloid leukemia/myelodysplastic syndrome with normal karyotype. Haematologica 94(2):213–223. doi:haematol.13024 [pii] 10.3324/haematol.13024
23. Kim JE, Woo KS, Kim KE et al (2010) Duplications of the long arm of both chromosome 1, dup(1)(q21q32), leading to tetrasomy 1q in myelodysplastic syndrome. Leuk Res 34(8):e210–212. doi:S0145-2126(10)00111-6 [pii] 10.1016/j.leukres.2010.02.028
24. Tiu RV, Gondek LP, O'keefe CL et al (2011) Prognostic impact of snp array karyotyping in myelodysplastic syndromes and related myeloid malignancies. Blood 117(17):4552–4560. doi:blood-2010-07-295857 [pii] 10.1182/blood-2010-07-295857
25. Koken MH, Daniel MT, Gianni M et al (1999) Retinoic acid, but not arsenic trioxide, degrades the plzf/raralpha fusion protein, without inducing terminal differentiation or apoptosis, in a ra-therapy resistant t(11;17)(q23;q21) apl patient. Oncogene 18(4):1113–1118. doi:10.1038/sj.onc.1202414
26. Iafrate AJ, Feuk L, Rivera MN et al (2004) Detection of large-scale variation in the human genome. Nat Genet 36(9):949–951
27. Sebat J, Lakshmi B, Troge J et al (2004) Large-scale copy number polymorphism in the human genome. Science 305(5683):525–528
28. Sharp AJ (2009) Emerging themes and new challenges in defining the role of structural variation in human disease. Hum Mutat 30(2):135–144. doi:10.1002/humu.20843
29. Cooper GM, Coe BP, Girirajan S et al (2011) A copy number variation morbidity map of developmental delay. Nat Genet. doi:10.1038/ng.909
30. (2007) Database of genomic variants—a curated catalogue of structural variation in the human genome. The Centre for Applied Genomics. http://projects.tcag.ca/variation/. Accessed 31 Aug 2011
31. International hapmap project (2011) International HapMap Project. http://hapmap.ncbi.nlm.nih.gov/. Accessed 31 Aug 2011
32. Wellcome trust case control consortium (2009) Wellcome Trust. http://www.wtccc.org.uk/index.shtml. Accessed 31 Aug 2011
33. Hgdp-ceph human genome diversity cell line panel (2011) Human Polymorphism Study Center. http://www.cephb.fr/en/hgdp/diversity.php. Accessed 31 Aug 2011
34. Wholley D, Manolio T, Brooks L et al (2011) Genetic association information network (gain). National human Genome Research Institute. http://www.genome.gov/19518664. Accessed 31 Aug 2011
35. Illumina icontroldb. (2011) Illumina, Inc. http://www.illumina.com/science/icontroldb.ilmn. Accessed 31 Aug 2011
36. Heinrichs S, Li C, Look AT (2010) Snp array analysis in hematologic malignancies: avoiding false discoveries. Blood 115(21):4157–4161. doi:blood-2009-11-203182 [pii] 10.1182/blood-2009-11-203182
37. Praulich I, Tauscher M, Gohring G et al (2010) Clonal heterogeneity in childhood myelodysplastic syndromes-challenge for the detection of chromosomal imbalances by array-cgh. Genes Chromosomes Cancer 49(10):885–900. doi:10.1002/gcc.20797
38. Slovak ML, Smith DD, Bedell V et al (2010) Assessing karyotype precision by microarray-based comparative genomic hybridization in the myelodysplastic/myeloproliferative syndromes. Mol Cytogenet 3:23. doi:1755-8166-3-23 [pii] 10.1186/1755-8166-3-23
39. Borze I, Juvonen E, Ninomiya S et al (2010) High-resolution oligonucleotide array comparative genomic hybridization study and methylation status of the rps14 gene in de novo myelodysplastic syndromes. Cancer Genet Cytogenet 197(2):166–173. doi:S0165-4608(09)00664-5 [pii] 10.1016/j.cancergencyto.2009.11.012

40. Thiel A, Beier M, Ingenhag D et al (2011) Comprehensive array cgh of normal karyotype myelodysplastic syndromes reveals hidden recurrent and individual genomic copy number alterations with prognostic relevance. Leukemia 25(3):387–399. doi:leu2010293 [pii] 10.1038/leu.2010.293
41. Bajaj R, Xu F, Xiang B et al (2011) Evidence-based genomic diagnosis characterized chromosomal and cryptic imbalances in 30 elderly patients with myelodysplastic syndrome and acute myeloid leukemia. Mol Cytogenet 4:3. doi:1755-8166-4-3 [pii] 10.1186/1755-8166-4-3
42. Huh J, Tiu RV, Gondek LP et al (2010) Characterization of chromosome arm 20q abnormalities in myeloid malignancies using genome-wide single nucleotide polymorphism array analysis. Genes Chromosomes Cancer 49(4):390–399. doi:10.1002/gcc.20748
43. Afable MG 2nd, Wlodarski M, Makishima H et al (2011) Snp array-based karyotyping: differences and similarities between aplastic anemia and hypocellular myelodysplastic syndromes. Blood 117(25):6876–6884. doi:blood-2010-11-314393 [pii] 10.1182/blood-2010-11-314393
44. Gibson J, Morton NE, Collins A (2006) Extended tracts of homozygosity in outbred human populations. Hum Mol Genet 15(5):789–795. doi:ddi493 [pii] 10.1093/hmg/ddi493
45. Quentin S, Cuccuini W, Ceccaldi R et al (2011) Myelodysplasia and leukemia of fanconi anemia are associated with a specific pattern of genomic abnormalities that includes cryptic runx1/aml1 lesions. Blood 117(15):e161–170. doi:blood-2010-09-308726 [pii] 10.1182/blood-2010-09-308726
46. Wang L, Fidler C, Nadig N et al (2008) Genome-wide analysis of copy number changes and loss of heterozygosity in myelodysplastic syndrome with del(5q) using high-density single nucleotide polymorphism arrays. Haematologica 93(7):994–1000. doi:haematol.12603 [pii] 10.3324/haematol.12603
47. Heinrichs S, Kulkarni RV, Bueso-Ramos CE et al (2009) Accurate detection of uniparental disomy and microdeletions by snp array analysis in myelodysplastic syndromes with normal cytogenetics. Leukemia 23(9):1605–1613. doi:leu200982 [pii] 10.1038/leu.2009.82
48. Jankowska AM, Szpurka H, Tiu RV et al (2009) Loss of heterozygosity 4q24 and tet2 mutations associated with myelodysplastic/myeloproliferative neoplasms. Blood 113(25):6403–6410. doi:blood-2009-02-205690 [pii] 10.1182/blood-2009-02-205690
49. Flach J, Dicker F, Schnittger S et al (2011) An accumulation of cytogenetic and molecular genetic events characterizes the progression from mds to secondary aml: an analysis of 38 paired samples analyzed by cytogenetics, molecular mutation analysis and snp microarray profiling. Leukemia 25(4):713–718. doi:leu2010304 [pii] 10.1038/leu.2010.304
50. Makishima H, Rataul M, Gondek LP et al (2010) Fish and snp-a karyotyping in myelodysplastic syndromes: improving cytogenetic detection of del(5q), monosomy 7, del(7q), trisomy 8 and del(20q). Leuk Res 34(4):447–453. doi:S0145-2126(09)00428-7 [pii] 10.1016/j.leukres.2009.08.023
51. Greisman HA, Yi HS, Hoffman NG (2007) Transcgh: rapid identification and high-resolution mapping of balanced igh translocations in archival DNA using custom oligonucleotide arrays. Paper presented at the American Society of Hematology Annual Meeting, Atlanta, GA
52. Greisman HA, Greiner TC, Yi HS et al (2008) High-throughput cloning of t(11;14) breakpoints outside the major translocation cluster in mantle cell lymphoma. Paper presented at the American Society of Hematology Annual Meeting, San Francisco, CA
53. Valli R, Marletta C, Pressato B et al (2011) Comparative genomic hybridization on microarray (a-cgh) in constitutional and acquired mosaicism may detect as low as 8% abnormal cells. Mol Cytogenet 4:13. doi:1755-8166-4-13 [pii] 10.1186/1755-8166-4-13
54. Cross J, Peters G, Wu Z et al (2007) Resolution of trisomic mosaicism in prenatal diagnosis: estimated performance of a 50k snp microarray. Prenat Diagn 27(13):1197–1204. doi:10.1002/pd.1884
55. Neill NJ, Torchia BS, Bejjani BA et al (2010) Comparative analysis of copy number detection by whole-genome bac and oligonucleotide array cgh. Mol Cytogenet 3:11. doi:1755-8166-3-11 [pii] 10.1186/1755-8166-3-11
56. Sweetser DA, Peniket AJ, Haaland C et al (2005) Delineation of the minimal commonly deleted segment and identification of candidate tumor-suppressor genes in del(9q) acute myeloid leukemia. Genes Chromosomes Cancer 44(3):279–291. doi:10.1002/gcc.20236

Chapter 6

CGH Protocols: Chronic Lymphocytic Leukemia

Esteban Braggio, Rafael Fonseca, and Neil E. Kay

Abstract

Array-based comparative genomic hybridization (aCGH) is a powerful assay to identify copy number abnormalities underlying the pathogenesis of cancer. aCGH has become the gold standard for whole genome copy number analysis in medium and large cohorts in clinical and research laboratories. Identifying the best workflow is critical to achieving the optimal performance for this assay. Here we describe the aCGH protocol used by our group in the study of B-chronic lymphocytic leukemia (CLL). We also describe some initial applications of aCGH in association with clinical outcome for CLL.

Key words: CLL, aCGH, DNA digestion, DNA labeling

1. Introduction

For the past few years, a remarkable effort has been made in identifying DNA copy number changes associated with normal genetic variation (1, 2), genetic diseases (3, 4), and tumorigenesis (5–11). The success of this search has depended, to a great extent, on the technological advance of the exploration tools, going from the initial discoveries of whole chromosome changes (12, 13) and microdeletion/microduplication abnormalities (14) by conventional cytogenetic and chromosome banding methodologies to the examination of the whole genome at subgene level by more modern bacterial artificial chromosome (BAC) and array-based tools (15, 16).

Currently, there are two major, array-based, available assays for examining the genome for copy number abnormalities (CNAs), the single nucleotide polymorphism (SNP) and the array comparative genomic hybridization (aCGH) platforms. Next-generation sequencing (NGS) is a very powerful and promising technology to

Diponkar Banerjee and Sohrab P. Shah (eds.), *Array Comparative Genomic Hybridization: Protocols and Applications*, Methods in Molecular Biology vol. 973, DOI 10.1007/978-1-62703-281-0_6, © Springer Science+Business Media, LLC 2013

be used not only in mutational screenings but also in copy number analysis (17). However, large-scale NGS projects are still only available in very few centers, and the data analysis and storage is still challenging. For that reason, aCGH and SNP arrays are still the gold standard for whole genome copy number analysis in medium and large cohorts in clinical and research laboratories.

Besides the selection of the ideal microarray platform for each analysis, the achievement of the optimal performance greatly depends on other steps of the technique, such as DNA fragmentation, DNA labeling, and array hybridization. The establishment of the best workflow is especially crucial in the analysis of the cancer genome, where the simultaneous presence of (a) more than one genetic subclone and (b) contaminating normal cells in the tumor sample are common events and might significantly affect the data analysis. Thus, increasing the data quality is imperative to maximizing the detection of CNAs.

Because of the advantages afforded by high-resolution genome-wide assessment, aCGH is gaining acceptance as a clinical diagnostic tool to detect CNAs of prognostic significance in chronic lymphocytic leukemia (CLL).

1.1. Clinical Applications for aCGH in CLL

CLL is a very common hematologic malignancy being the most common leukemia in the Northern hemispheres including the United States. This common but largely indolent malignancy is characterized by a marked degree of clinical heterogeneity from highly stable disease and a nearly normal life expectancy, to patients with a rapidly progressive disease. It is currently estimated that for CLL patients, 70 of 100 with initial early Rai stage (0–1) disease will progress to the point of needing therapy. Counseling the exact clinical course for these patients is highly problematic even using the current traditional or novel prognostic tests available. Using genetic testing has been one of the most helpful tools in predicting a clinical course for CLL patients. Fluorescent in situ hybridization (FISH) has been widely used in the identification of chromosomal abnormalities, being deletions of 13q14, 17p13 (including *TP53*), and 11q22–q23 (including *ATM*), and trisomy of chromosome 12 the most frequent alterations found in CLL (18). With the exception of "high-risk" abnormalities by FISH (i.e., 17p13 and 11q23 deletions), which are associated with more aggressive clinical courses decreased patient survival (19), there are no obvious definitive predictive algorithms.

The major limitation of the FISH approach is that can only be focused on the analysis of few chromosomal abnormalities simultaneously. Thus, FISH is not a discovery tool and does not account for all the global genetic events that may occur in CLL patients whether stable or progressive. More recently, we and others (20, 21) have used aCGH and SNP arrays to generate more global genetic information and to better understand the implications of genomic instability and outcome on the CLL disease course and our initial approach is discussed here. We sought to address genomic complexity as a predictor of sensitivity to the powerful approach of chemoimmunotherapy (CIT) and ultimately clinical outcome in this disease. We used aCGH with a one million probe array and identified gains and losses of genetic material in 48 patients treated on a CIT clinical trial. We found a median of 4 CNAs per patient (range 0–32), with an overall ratio of 3:1 of copy number deletions to gains. Interestingly only one patient had no CNAs detected. Higher genomic complexity, as a mechanism favoring clonal selection, was associated with shorter progression-free survival and predicted a poor response to treatment (20). Genomic complexity was evident in patients with deletion of 17p13.1 and in the more favorable genetic subtype characterized by the presence of 13q14.1 deletion. The aCGH study was consistent with what was found recently by SNP-based array (21) where genomic complexity is strongly associated with poor trial response even with CIT. The use of aCGH in sequential CLL samples by us and others, also demonstrated increasing genetic complexity for previously untreated CLL as they progress and raised the question of whether those patients are now at higher risk for poor responses to CIT (22, 23). Thus, we believe that there is a place for genomic analysis of CLL patients using aCGH for both generating more information of the global changes in CLL B cells and generating unique prognostic information.

2. Materials

Prepare all the solutions with ultrapure, DNase- and RNase-free, distilled water and analytical grade reagents. Prepare and store all reagents at room temperature (unless indicated otherwise).

2.1. DNA Fragmentation

1. TAE buffer 50×: Add 242 g of Tris base, 57.1 mL of glacial acetic acid, and 100 mL of 500 mM EDTA (pH 8.0). Make up to 1 L with water. The stock solution is diluted 50:1 with water to make a 1× working solution.
2. 10× Buffer: Add 500 μL of 1 M Tris HCL (pH 8.0) (Invitrogen, Carlsbad, CA, USA), 500 μL of 1 M $MgCl_2$ (Ambion, Austin, TX, USA) and 25 μL of 2 M $CaCl_2$. Make up to 5 mL with DNase- and RNase-free water. Aliquot in 0.5 mL fractions and store at −20°C. The 10× buffer can be frozen and thawed several times.
3. Enzyme Buffer: Add 50 μL of 1 M Tris HCL (pH 8.0), 50 μL of 1 M $MgCl_2$, 2.5 μL of 2 M $CaCl_2$, and 25 μL of 10 mg/mL

bovine serum albumin (Promega, Madison, WI, USA). Make up to 5 mL with DNase- and RNase-free water. Aliquot in 50 μL fractions and store at −20°C. Do not reuse the enzyme buffer aliquots once they were thawed.

4. DNase 1: Recombinant bovine DNase I (RNase-free) enzyme (Ambion). Store at −20°C.
5. Agarose gel: Weigh 1.8 g of molecular grade agarose (Ambion). Make up to 300 mL with TAE buffer 1×. Boil in microwave until a homogeneous solution is obtained. Store at 56°C.
6. Minihorizontal units for electrophoretic systems (Fisher Scientific, Pittsburgh, PA, USA).
7. Ethidium bromide 1% (Fisher Scientific).
8. Power supply EC250-90 (Thermo Electron, Waltham, MA, USA).

2.2. DNA Labeling

1. BioPrime Total for Agilent aCGH labeling kit (Invitrogen, Carlsbad, CA, USA) (see Note 1).

2.3. DNA Clean Up

1. Use the columns and solutions provided by the BioPrime Total for Agilent aCGH labeling kit (Invitrogen).

   Binding buffer B2: add 6 mL of isopropanol before use. Store at RT.

   Wash buffer: add 40 mL of ethanol 96% before use. Store at RT.
2. NanoDrop 1000 Spectrophotometer (Fisher Scientific).

2.4. Microarray Hybridization

1. Human Cot-1 DNA 1 μg/μL (Invitrogen). Store at −20°C.
2. 10× blocking agent (Agilent Technologies, Palo Alto, CA, USA). Add 1.375 mL of ultra pure, DNase- and RNase-free, distillated water. Incubate at 37°C for 1 h. Store at −20°C.
3. 2× Hi-RPM buffer (Agilent). Store at RT.
4. Microarrays (Agilent). Store at RT.
5. Microarray slide gaskets (Agilent). Store at RT (see Note 2).
6. Microarray chambers (Agilent).
7. Heat block with holder for 1.7 mL Eppendorf tubes.

2.5. Microarray Washes

1. CGH & ChIP-on-Chip wash buffers #1 and #2 (Agilent) (see Note 3).
2. Glassware (see Note 4).
3. Tweezers.
4. Water bath.
5. Nitrogen tank with spray gun.

2.6. Microarray Scanning

1. Microarray scanner (Agilent).
2. Scan control software (Agilent).
3. Slide holders (Agilent).

2.7. Data Extraction and Initial QC

1. Feature Extraction Software (Agilent).

3. Methods

3.1. DNA Fragmentation

1. Thaw the 10× and enzyme buffers at room temperature (see Note 5). Fifteen minutes before using, add 0.5 μL of bovine DNase I enzyme in one aliquot of enzyme buffer. Mix by flicking.
2. Add 1.2 μg of each tumor DNA in separate 0.5 mL Eppendorf tubes (see Notes 6, 7, and 8). The total amount of reference DNA needed for the entire experiment should be pooled and fragmented in a unique tube (see Note 9).
3. Make up to 20 μL with TE buffer in each tube with tumor sample (pH 8.0). See Note 8 for the amount of TE buffer to be added to the reference DNA reaction.
4. Add 2 μL of 10× buffer in each tube with tumor sample. See Note 9 for the amount of 10× buffer to be added to the reference DNA reaction.
5. Mix DNA, water, and 10× buffer by vortexing for 5 s and then spin down to collect.
6. Add 2 μL of enzyme buffer (with 0.5 μL of enzyme previously added) and flick the tube for 5 s to mix. See Note 9 for the amount of enzyme buffer to be added to the reference DNA reaction.
7. Incubate at RT for 12 min (see Note 10).
8. Inactivate the reaction at 70°C for 30 min. Store on ice.
9. Mix 40 mL of 0.6% agarose gel and 4 μL of Ethidium bromide in a 50 mL conical flask. Mix by inversion.
10. Load the agarose on a minihorizontal unit for electrophoresis. Insert a 10-well gel comb immediately without introducing air bubbles. Wait until the agarose solidifies.
11. Add digested DNA samples in every lane of the agarose gel. Add 1 kb Plus Ladder (Invitrogen) in one well. Electrophorese at 20 mAmps for 40 min.
12. The target is to visualize fragmented DNA with sizes ranging from 300 bp to 2 kb (Fig. 1).
13. Depool the fragmented reference DNA into individual tubes for the labeling step.

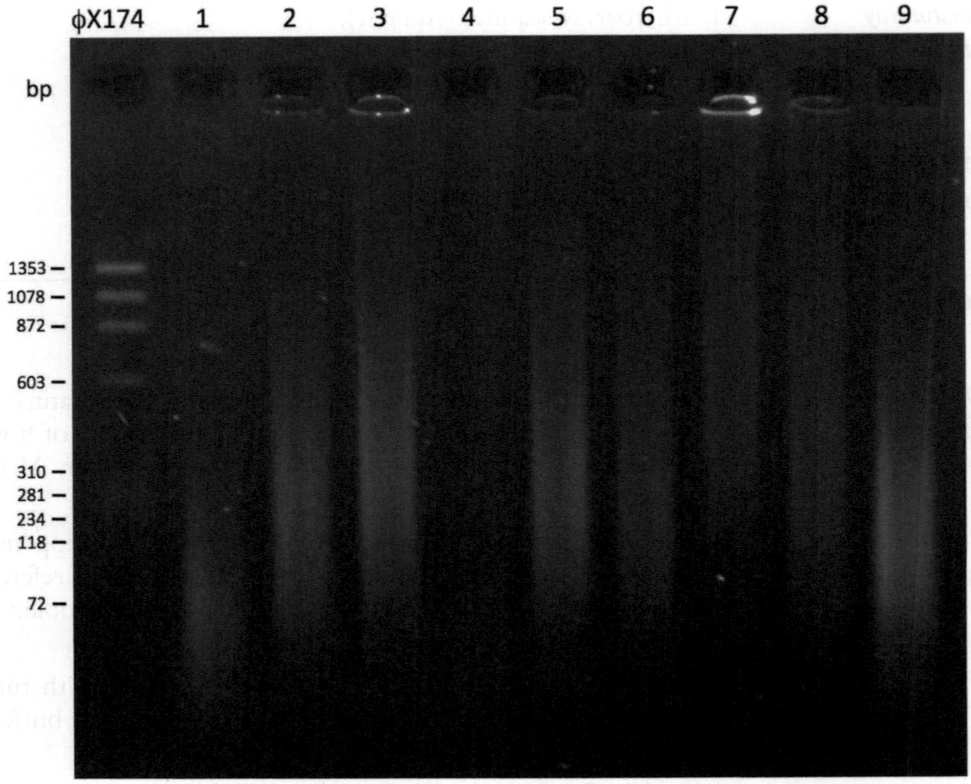

Fig. 1. Agarose gel showing examples of DNA fragmentation using Recombinant bovine DNase I enzyme. Samples in lanes 2, 3, 5, 6, 8, and 9 are in the desired fragment size range. Samples 1 and 4 are overdigested; samples 3 and 7 are underdigested.

3.2. DNA Labeling

1. Briefly vortex the 2× Reaction Mix (included in the BioPrime Total for Agilent aCGH labeling kit) and centrifuge to collect the contents of the tubes. Place the tubes on ice.
2. Add 25 μL of Alexa 5 Fluor 2× Reaction Mix to each tumor sample and 25 μL of Alexa 3 Fluor 2× Reaction Mix to each reference sample.
3. Gently pipet up and down to mix and incubate at 95°C, protected from light, for 5 min. Immediately cool on ice for 5 min.
4. On ice, add 3 μL of exo-Klenow fragment (included in the BioPrime Total for Agilent aCGH labeling kit) to each tube, for a final reaction volume of 50 μL.
5. Briefly vortex and spin down the tubes.
6. Incubate in a heat block or water bath (protected from light) at 37°C for 2 h.
7. If the reaction are storing for any length of time prior to purification, add 5 μL of stop buffer (included in the BioPrime

Total for Agilent aCGH labeling kit) to each tube to quench the reaction. If proceeding directly to the purification step, there is no need of adding stop buffer.

3.3. Cleanup

1. Add 200 μL of binding buffer B2 to each tube.
2. Briefly vortex and spin down the tubes.
3. Load each sample onto a PureLink spin column (included in BioPrime Total for Agilent aCGH labeling kit), pre-inserted in a collection tube.
4. Centrifuge at 10,000 × *g* for 1 min. Discard the flow through and place the column back in the collection tube.
5. Add 650 μL of wash buffer (included in BioPrime Total for Agilent aCGH labeling kit) to the column.
6. Centrifuge at 10,000 × *g* for 1 min. Discard the flow through and place the column back in the collection tube.
7. Spin at maximum speed for 2 min to remove any residual wash buffer. Discard the flow through. Place the spin column in a new tube.
8. Add elution buffer E1 to the center of the column and incubate at RT for 1 min. The volume to be added depends on the array format used (see Table 1).
9. Centrifuge at maximum speed for 2 min. The flow through contains the purified labeled DNA.
10. Take 1.5 μL of each sample to determine the yield and specific activity in NanoDrop spectrophotometer. The following results are expected: (a) yield DNA >5 μg, (b) dye incorporation >200 pmol (see Note 11).
11. Pool and depool the labeled reference DNA in order to reduce the labeling bias among reactions.

3.4. Hybridization

1. Add to a 1.7 mL Eppendorf tube: labeled tumor DNA, labeled reference DNA, Cot-1 DNA (1 μg/μL), 10× blocking agent, and 2× Hi-RPM buffer (Agilent Technologies). The volumes to be added depend on the array format used (see Table 1).
2. Gently mix the sample by pipetting up and down, avoiding the production of bubbles and then quickly spin.
3. Transfer samples tubes to a heat block at 95°C. Incubate for 3 min.
4. Incubate in a water bath at 37°C for 30 min.
5. Spin at 12,000 × *g* for 3 min.
6. During the centrifugation, put the gasket slides in bottom of on the hybridization chamber with the barcode label facing up.
7. Dispense the mix onto the center of the gasket slide.

Table 1
Volume reaction and hybridization time for different Agilent microarray formats

	Clean up	Hybridization					
Array format	Elution buffer (μL)	Tumor and reference samples (μL of each)	Cot-1 DNA (μL)	10× Blocking agent (μL)	2× Hi-RPM buffer (μL)	Total volume (μL)	Hyb time (hours)
1 array per slide	81	79	50	52	260	520	40
2 arrays per slide	41	39.5	25	26	130	260	24
4 arrays per slide	21	19.5	5	11	55	110	24
8 arrays per slide	11	8	2	4.5	22.5	45	24

8. Place microarray slide with the active side (the one with Agilent barcode on) facing down (see Note 12).
9. Tighten the top portion of the chamber. Make sure all bubbles move within the chamber.
10. Place in hybridization oven set at 65°C, rotating at 20 rpm. The hybridization time will depend on the array format used (Table 1).

3.5. Washing

1. Remove hybridization chambers from oven and disassemble them.
2. Transfer slides to wash buffer #1. Gently separate the microarray slide from the gasket slide with tweezers. The tweezers should be introduced in the barcode end of the slides in order to avoid damaging the active region.
3. Agitate slide for few seconds to remove bubbles.
4. Quickly transfer to fresh glassware also containing wash buffer #1. Wash on a magnetic stir plate at RT for 5 min.
5. Transfer slides to wash buffer #2 in a water bath at 37°C for 3 min. Hand agitate for 2–3 s every 30 s (see Note 13).
6. Pull slides out very slowly avoiding any kind of bubble on the surface of the array. If bubbles remain in the slide, put in and pull out the buffer again. If bubbles were not totally removed, use the nitrogen gun to clean the slides. Touch only the barcode region of the microarray slide.
7. Repeat step 6 for additional slides (see Note 14).
8. Put the microarray slides in the slide holders with the active side (Agilent barcode) facing up.
9. Scan slides immediately to minimize impact of environmental oxidants on signal intensities.

3.6. Scanning

Turn on scanner and scan control software several minutes before use to warm up laser (see Note 15).

1. On the scan control software select the start and end slot, corresponding with the position of the first and last microarray location in the scanner carousel, respectively.
2. Select: (a) Slide ID = Auto Detect; (b) Channels = R + G; (c) Scan region = Agilent HD (61 × 21.6 mm); (d) R PMT and G PMT = 100%; (e) XRD = No XRD; (f) Tiff = 16 bit; and (g) Output Path = drive location desired.
3. Select the scan resolution based on the array format that you are using: 10 μm for 1 × 44K arrays; 5 μm for 1 × 244K, 2 × 105K, 4 × 44K, or 8 × 15K arrays; and 3 μm for 1 × 1 M, 2 × 400K, 4 × 180K, or 8 × 60K microarrays.

4. Scan arrays (takes 8 min per slide at 5 μm and 15 min at 3 μm resolution).

3.7. Feature Extraction

1. Use the Agilent feature extraction (FE) software.
2. Select Add New Extraction Set(s) > Add Extraction.
3. Browse to the location of the tiff files and select them.
4. The FE program automatically assigns a default grid template.
5. Select the corresponding protocol (CGH).
6. Select the Project Properties tab.
7. In the Output and Data Transfer section, change the outputs MAGE and JPEG to "None" and the Text-Output Package to "Compact."
8. Select Project > Start Extracting.
9. The generated text files are ready to be imported into analytic software.

4. Notes

1. Cyanine 5 is sensitive to ozone degradation. Ozone levels >5 ppb (approximately 10 μg/m^3) can affect dye signal and compromise microarray results. One of three options needs to be considered: (a) use ozone-barrier slide cover during the scanning, (b) use ozone filters to keep the ozone level of the room bellow 5 ppb, or (c) use Alexa dyes instead of Cyanine dyes. We incorporate the option (c) in our practice.
2. The type of slide gaskets to be used will depend on the array format (1, 2, 4, or 8 arrays per slide).
3. Rinse the spigots that come with the wash buffers with ethanol 70% before installation.
4. Rinse slide-staining dishes, slide racks, and stir bars thoroughly with high-quality milli-Q water before use and in between washing groups. *Never* use tap water.
5. The 10× buffer can be frozen and thawed several times. The enzyme should be added to the buffer 15 min before use. Use new enzyme dilution every time.
6. DNA samples with more than 30% of contaminant non-tumor DNA may have the results compromised. It is strongly recommendable to enrich tumor B cells by using magnetic bead columns anti-CD19+ or double selection of anti-CD19+/CD5+

(Miltenyi Biotec, Auburn, CA, USA) or flow sorting of the CD19+/CD5+ fraction.

7. For a successful performance, use high-quality genomic DNA free of RNA and protein contamination. High-quality genomic DNA samples should have an A260/A280 ratio of 1.75 to 1.9, indicating the absence of contaminating proteins and RNA. An A260/A230 ratio of >2.0 is indicative of pure genomic DNA without organic contaminants such as alcohol, phenol, lipids, and carbohydrates.

   Also it is recommended that all stock DNAs should be in concentrations ranging from 250 to 400 ng/μL to ensure accuracy.

8. Since obtaining normal reference DNA for each tumor analyzed is very challenging, we recommend using a common DNA reference for all experiments. To create our reference DNA, we pool DNAs obtained from nine lymphoblastoid cell lines from the Coriell repository.

9. To avoid the DNA fragmentation bias of reference samples, we recommend pooling the reference DNA in a single for the DNA fragmentation step. The total amount of reference DNA required per experiment will be $= n \times 1.2$ μg, where n is the number of tumors to be analyzed. Make up to $n \times 20$ μL with TE buffer. Next, add $n \times 2$ μL of 10× buffer. Finally, add $n \times 2$ μL of enzyme buffer.

10. The time of DNA fragmentation to reach the desired fragment sizes may vary per lot of DNase I enzyme.

11. *Yield DNA* (*μg*): DNA concentration (ng/μL) × 0.08. *Dye Incorporation Alexa* 5 (*pmol*): (A650–A750)/19.2. *Dye Incorporation Alexa* 3 (*pmol*): (A555–A650)/12. Where A is absorbance in the different wavelengths.

12. Agilent barcodes of microarray and gasket slide have to be face-to-face.

13. Prewarming the wash buffer #2 in water bath before using for at least 2 h (overnight is recommended).

14. A maximum of five disassembly procedures yielding five microarray slides is advised at one time in order to facilitate uniform washing.

15. An Agilent B scanner is required for 1 × 244K, 2 × 105K, 4 × 44K, and 8 × 15K density microarrays. An Agilent C scanner and Agilent scanner Control software v8.3 or higher is required for 1 × 1 M, 2 × 400K, 4 × 180K, and 8 × 60K density microarrays.

References

1. Redon R, Ishikawa S, Fitch KR et al (2006) Global variation in copy number in the human genome. Nature 444(7118):444–454
2. Mccarroll SA, Kuruvilla FG, Korn JM et al (2008) Integrated detection and population-genetic analysis of snps and copy number variation. Nat Genet 40(10):1166–1174
3. Lee JA, Lupski JR (2006) Genomic rearrangements and gene copy-number alterations as a cause of nervous system disorders. Neuron 52(1):103–121
4. Veltman JA, Jonkers Y, Nuijten I et al (2003) Definition of a critical region on chromosome 18 for congenital aural atresia by arraycgh. Am J Hum Genet 72(6):1578–1584
5. Braggio E, Keats JJ, Leleu X et al (2009) Identification of copy number abnormalities and inactivating mutations in two negative regulators of nuclear factor-kappab signaling pathways in waldenstrom's macroglobulinemia. Cancer Res 69(8):3579–3588
6. Keats JJ, Fonseca R, Chesi M et al (2007) Promiscuous mutations activate the noncanonical nf-kappab pathway in multiple myeloma. Cancer Cell 12(2):131–144
7. Mao X, Orchard G, Lillington DM et al (2003) Amplification and overexpression of junb is associated with primary cutaneous t-cell lymphomas. Blood 101(4):1513–1519
8. Mullighan CG, Goorha S, Radtke I et al (2007) Genome-wide analysis of genetic alterations in acute lymphoblastic leukaemia. Nature 446(7137):758–764
9. Clark J, Edwards S, Feber A et al (2003) Genome-wide screening for complete genetic loss in prostate cancer by comparative hybridization onto cdna microarrays. Oncogene 22(8):1247–1252
10. Martinez-Climent JA, Alizadeh AA, Segraves R et al (2003) Transformation of follicular lymphoma to diffuse large cell lymphoma is associated with a heterogeneous set of DNA copy number and gene expression alterations. Blood 101(8):3109–3117
11. Pollack JR, Sorlie T, Perou CM et al (2002) Microarray analysis reveals a major direct role of DNA copy number alteration in the transcriptional program of human breast tumors. Proc Natl Acad Sci USA 99(20):12963–12968
12. Edwards JH, Harnden DG, Cameron AH et al (1960) A new trisomic syndrome. Lancet 1(7128):787–790
13. Lejeune J, Turpin R, Gautier M (1959) Mongolism; a chromosomal disease (trisomy). Bull Acad Natl Med 143(11–12):256–265
14. Weiss LA, Shen Y, Korn JM et al (2008) Association between microdeletion and microduplication at 16p11.2 and autism. N Engl J Med 358(7):667–675
15. Pollack JR, Perou CM, Alizadeh AA et al (1999) Genome-wide analysis of DNA copy-number changes using cdna microarrays. Nat Genet 23(1):41–46
16. Albertson DG, Pinkel D (2003) Genomic microarrays in human genetic disease and cancer. Hum Mol Genet 12 Spec No 2:R145–R152
17. Metzker ML (2010) Sequencing technologies—the next generation. Nat Rev Genet 11:31–46
18. Stilgenbauer S, Bullinger L, Lichter P et al (2002) Genetics of chronic lymphocytic leukemia: genomic aberrations and v(h) gene mutation status in pathogenesis and clinical course. Leukemia 16(6):993–1007. doi:10.1038/sj.leu.2402537
19. Dohner H, Stilgenbauer S, Benner A et al (2000) Genomic aberrations and survival in chronic lymphocytic leukemia. N Engl J Med 343(26):1910–1916. doi:MJBA-432602 [pii] 10.1056/NEJM200012283432602
20. Kay NE, Eckel-Passow JE, Braggio E et al (2011) Progressive but previously untreated CLL patients with greater array CGH complexity exhibit a less durable response to chemoimmunotherapy. Cancer Genet Cytogenet 203(2):161–168. doi:S0165-4608(10)00492-9 [pii] 10.1016/j.cancergencyto.2010.09.003
21. Ouillette P, Collins R, Shakhan S et al (2011) Acquired genomic copy number aberrations and survival in chronic lymphocytic leukemia. Blood 118(11):3051–3061. doi:blood-2010-12-327858 [pii] 10.1182/blood-2010-12-327858
22. Braggio E, Kay NE, Vanwier S, Tschumper RC, Smoley S, Eckel-Passow JE, et al (2012) Longitudinal genome-wide analysis of patients with chronic lymphocytic leukemia reveals complex evolution of clonal architecture at disease progression and at the time of relapse. Leukemia 26(7):1698–1701
23. Knight SJ, Yau C, Clifford R, Timbs AT, Sadighi Akha E, Dréau HM, Burns A, Ciria C, Oscier DG, Pettitt AR, Dutton S, Holmes CC, Taylor J, Cazier JB, Schuh A (2012) Quantification of subclonal distributions of recurrent genomic aberrations in paired pre-treatment and relapse samples from patients with B-cell chronic lymphocytic leukemia. Leukemia 26(7):1564–1575

Chapter 7

Analysis of Acquired Genomic Copy Number Aberrations and Regions of Loss of Heterozygosity in Acute Myelogenous Leukemia Genomes Using Affymetrix SNP 6.0 Arrays and Supporting Software Tools

Peter D. Ouillette, Kerby A. Shedden, Cheng Li, and Sami N. Malek

Abstract

The application of SNP array technology to the analysis of cancer genomes has greatly advanced our knowledge of the incidence and functional consequences of acquired genomic copy number aberrations (aCNA) and LOH in various malignancies. The major challenges of using SNP arrays are accurately identifying acquired genomic DNA aberrations in the raw array data with very high sensitivity and specificity and meaningfully assessing the associations between these aberrations and biological characteristics or patient outcomes. Critical to the success and valid interpretation of data derived from SNP array profiling are (1) the purity of cells used as a source of template DNA; (2) the analysis of paired DNA samples (tumor and normal); (3) use of validated software tools for data analysis; (4) access to an acceptable gold standard for aCNA and LOH, including FISH data, cytogenetic results, and Q-PCR data; and (5) statistical support to employ or develop algorithmic approaches to SNP array data analysis. Overcalling of lesions including lack of validation and undercalling of lesions that display low fractional allelic representations are common problems. This guide should help the reader establish this powerful technology in the laboratory and aims to stimulate transition of SNP array profiling into clinical applications.

Key words: Affymetrix SNP 6.0 arrays, Genomic copy number analysis, Loss of heterozygosity, Copy-neutral loss of heterozygosity, Paired DNA analysis, Polymorphic copy number variants, Genomic complexity

1. Introduction

1.1. AML Blast Cells and Flow Cytometry Sorting

The application of SNP array technology to cancer cells in recent years has yielded much novel information about the nature and scope of genome-wide DNA copy number changes and loss of heterozygosity (LOH) (1–8). The impact of normal cell contamination in tumor DNA preparations upon the ability to detect such genetic lesions, in particular LOH, is a concern. In AML, the

Diponkar Banerjee and Sohrab P. Shah (eds.), *Array Comparative Genomic Hybridization: Protocols and Applications*, Methods in Molecular Biology vol. 973, DOI 10.1007/978-1-62703-281-0_7, © Springer Science+Business Media, LLC 2013

percentage of blasts in the peripheral blood and bone marrow varies widely, from 20% to over 90%, and in our laboratory's experience, AML mononuclear cell isolation via Ficoll-Paque gradient centrifugation preparatory to cryopreservation occasionally yields samples with more than 10% erythroid precursors and mature erythrocytes. AML blast cells have a characteristic expression pattern of CD45 relative to other leukocytes (9), and this, when combined with analysis of other known myeloid antigens, can be exploited to achieve a blast-derived DNA sample of very high purity from virtually any AML sample by using a flow cytometry sorter.

1.2. SNP Array Data Quality Control and Generation of SNP Calls Using Affymetrix Genotyping Console Software

We employ the Affymetrix Genotyping Console (GTC) software suite for performing quality control checks on SNP chip data files and for generating genotyping calls. Downstream analysis of genomic copy number and LOH is carried out through other programs, described below. The CEL file is the data file which is used for microarray analysis and for generating SNP calls through GTC. CEL files and ancillary ARR files are added to GTC, and the software performs an initial quality control computation based on the Dynamic Model algorithm. For SNP 6.0 arrays in GTC 2.0, 3022 SNPs are used for QC. The result is a "bounds call" in which sample files are classified as either "In Bounds" or "Out of Bounds." Out-of-bounds files are rejected from further analysis. In-bounds files may proceed to the next step, which is the generation of genotyping calls through Affymetrix's "Birdseed" algorithm. The Birdseed genotyping call rates which result are the criterion by which array files are considered fit or unfit for downstream copy number and LOH analysis. GTC version 2.0 with Birdseed version 1 has no minimum required number of CEL files in order to perform genotyping. However, the software manual recommends a batch size of 44 samples (A). For large data sets composed of hundreds of CEL files, the Birdseed algorithm process may take almost 2 days to complete in a 32-bit operating system environment, even on a computer with 4 GB of RAM. Upon completion of the Birdseed algorithm, a SNP summary table will display that contains the computed call rate for all SNPs. Our experience indicates that a call rate of at least 95% (most call rate results are >98%) is required for an acceptable signal-to-noise ratio in downstream analysis. Affymetrix cautions that an occasional sample may pass the initial QC step but have a suboptimal genotyping performance (B), which we have also observed. Export of the completed genotyping files must be done through GTC. CEL files whose final SNP call rates are acceptable may then be imported into dChip for copy number analysis.

1.3. Analysis of Genomic Copy Number Changes Using dChip Software

The program dChip (C-F) (10, 11) has been freely available since 2002 (see Note 1) and has undergone continuous improvements and refinements in the time since. In the 64-bit computer operating system environment, dChip can process over 600 SNP arrays at a time.

Affymetrix chip description (CDF) and genome information files (see Note 2) appropriate for the chip platform type must be downloaded, and the end user must create text files listing names and describing the characteristics of the samples to be analyzed. The computer platform required to perform these functions will depend upon the number of SNP 6.0 arrays involved. Our lab currently uses a Dell Precision T7400 workstation with the Windows XP Professional x64 edition operating system and 32 GB RAM. A less powerful standard 32-bit operating system with 4 GB RAM should support data sets of ~100 SNP arrays (a conservative estimate).

1.4. Analysis of Loss of Heterozygosity Using Two-Step Software Analysis with the Software Programs Pre-LOH Unification Tool and the LOH Tool 2.0

Once genotyping calls (AA, BB, AB, or NoCall) for each SNP have been generated for the data set, it is possible to perform analysis for LOH and copy-neutral LOH (also called acquired uniparental disomy [aUPD]). aUPD occurs in approximately 20% of AML and is undetectable with current cytogenetic or copy number analysis through SNP arrays. Our laboratory has utilized the most basic possible method of detecting LOH, which is to simply pair the resulting SNP calls from one patient's buccal DNA with those from the leukemic blast DNA, and enumerate all instances where a heterozygous SNP call in the buccal becomes AA or BB in the blast DNA. By using Java to develop a flexible graphical user interface (GUI) capable of displaying a simple hash mark on a line representative of the chromosome for every detected occurrence of LOH, the large amount of information contained in the SNP array can be presented in such a manner that regions of continuous LOH become immediately apparent. The program was then given the capability to display data from over 100 patients at once. This is the program we call the LOH tool (12). With the advent of the SNP 6.0 array, the amount of available information has increased tenfold, and the LOH tool GUI would have been overwhelmed with random platform "noise" calls of LOH. It was necessary to develop a filter feature for the LOH tool which would limit the number of LOH hash marks it would display using proximity criteria defined in the software code but adjustable by the user. The upgraded version is termed LOH tool 2.0.

The continuing evolution of software for SNP array processing also presented a challenge with the actual content of the SNP call output files. CHP files resulting from Birdseed algorithm processing contain only a SNP-A identification number, followed by the SNP call itself. There is no accompanying information on the chromosomal location, the physical position within the chromosome, or the associated rsID number, which allows cross-referencing with the NCBI database dbSNP. This information is provided separately on the Affymetrix SNP 6.0 annotation file. Another program needed to be written which could rapidly align the information in the Birdseed output CHP file with the appropriate genomic information contained within the Affymetrix annotation file. The result was the PLUT, also Java based. Parsing of all CHP files through PLUT must first take place before data are presented to the LOH tool 2.0. And, as with dChip, with increasing data set size, it became necessary to give both programs the capability to access additional RAM within the 64-bit operating system environment. A copy of both software programs is available upon request. Please note that it is important to use the same Affymetrix annotation file "GenomeWideSNP_6.na24.annot.csv" for both dChip and PLUT/LOH tool. A form of this file appropriately formatted for dChip is directly downloadable through the Web site, discussed later in this protocol.

1.5. Algorithmic aCNA Analysis

The raw data for an oligonucleotide-based Affymetrix SNP 6.0 array hybridization consists of quantitative fluorescence intensity measures corresponding to oligonucleotide sequences matching each of the two alleles at a SNP. Several algorithms have been developed to convert this oligonucleotide-level data to allele-level data, which can be used to quantitatively assess aCNA, LOH, and acquired DNA damage, as well as to produce genotype calls for use in genome-wide association studies and other genetic mapping efforts. DChip-SNP is a widely used software implementation of one such algorithm that we have used in all our work. Several other algorithms have been reviewed (13). The output of DChip-SNP used in our work is a quantitative value reflecting the copy number estimates for each SNP position.

When the focus is on copy number aberrations, the allele-level summaries can be summed to produce a locus-level summary that reflects the total amount of genomic DNA at a given chromosomal position. This locus-level summary will have a value close to 2 when the diploid DNA is intact. In our work, we have always obtained two SNP array measurements per subject, one for DNA extracted from cancer cells and one for normal DNA from buccal cells or normal lymphocytes. Thus, we are able to differentiate inherited polymorphic copy number variations (inherited CNVs) from copy number variations (acquired CNAs) acquired by the cancer cells. Details of how this is done are given below. A major challenge in this type of work is summarizing the degree of genomic complexity (sum of all aCNAs) at very high sensitivity and specificity. In fact, as we work with pools of approximately 10^5 cells in our assays, we are looking at a mixture of clones with varying levels of genomic complexity. Thus, at best we obtain the average level of genomic complexity/instability in the cell population under study.

aCNAs span from a kilobase to multiple megabases; small amplifications or deletions are invisible in the current generation of SNP arrays due to the absence of a sufficient number of probes in the deleted/amplified region. The Affymetrix SNP 6.0 array that is widely used in molecular cytogenetic research contains over 900,000 probes and is marketed as a "high-density" platform. For example, the median spacing between consecutive probes on chromosome 1 on this platform is 1,390 bases. In our AML work, we have chosen to focus on aberrations spanning 25 or more probes. Calling smaller lesions with high specificity requires individual validation using a genomic gold standard like copy number Q-PCR.

It has been noted that estimation of copy numbers from SNP arrays is more challenging and subject to a greater variety of artifacts than genotype calling using the same platform (13). In particular, researchers have observed various presumably artifactual "waves" and "bands" (14, 15), possibly related to GC content, but unlikely to reflect true DNA aberrations. Like other researchers, we have generally found homozygous deletions to be the easiest aberration to detect, with heterozygous deletions and amplifications being more challenging to resolve with high accuracy. Subclonal aCNAs are only detectable at high sensitivity when present in >25% of the clone (16).

Software and algorithms for processing SNP array data can produce either locus-level copy number estimates or a list of predicted lesion instances (amplifications or deletions) defined by their starting and ending chromosomal position. The latter type of algorithm is sometimes referred to as a "segmentation" algorithm. There have been a number of approaches developed for segmentation, including algorithms based on Hidden Markov Models, sparse regression, and running moments and quartiles.

Comparative evaluations of algorithms for CNV detection have been published (13). Such assessments may focus on various performance parameters, including bias and accuracy of the quantitative level of copy number change, accuracy in estimating the endpoints of the lesion, and sensitivity/specificity for detecting a lesion in the vicinity of a true lesion. Assessing the true number of DNA breaks that resulted in distinct aCNAs puts a high priority on correctly characterizing a region of changed copy numbers as constituting one larger lesion versus multiple smaller lesions. An algorithm that over-fragments such regions into multiple small lesions may still provide accurate quantitative copy number estimates at most loci (if the spacing between the called lesions is small) and may do a good job of detecting at least one lesion in close proximity to each true lesion. However, algorithms that over-fragment may be strongly upwardly biased when used to assess the number of aCNA and therefore would not be sufficient to assess prognostic importance of aCNA as various lesion thresholds.

Much of our work has used visual lesion calling to identify lesions which are then summed to produce a summary score of genomic complexity. The potential for bias is minimized by using software that blinds the person doing the visual assessment to the identities or characteristics of the samples and by using two or more independent assessors whose calls are then reconciled or averaged.

More recently, we have developed a "sliding window" strategy (described further below) that is relatively simple but highly accurate for our data. Using this algorithmic lesion calling provides faster and arguably more objective assessments of genomic complexity compared to manual lesion calling. There are four key ideas to this sliding window algorithm (2), described as follows.

First, we consider the mean/variance relationship in the raw data. In general, a mean/variance relationship is a situation in which the variability of a quantity depends on its mean value (usually with a positive dependency so that the variance is greater when the values are larger). Most quantitative molecular biology assays have a mean/variance relationship, and nearly all effective statistical methods for analyzing such data employ a "variance stabilization" to remove the mean/variance relationship. A log transform is commonly applied, but we found that the fourth-root transform removes the mean/variance relationship more completely. This finding may be specific to the Affymetrix SNP 6.0 platform or to our experimental protocols, so in practice the mean/variance relationship should be reassessed in each setting. One benefit of variance stabilization is that it tends to symmetrize the distribution of data around the modal value of 2, making tuning parameters for gain detection and loss detection more similar (although for optimum performance we still find that gains and losses should be treated somewhat differently, as discussed further below).

Second, we combine the paired tumor and normal data by subtracting the quantitative smoothened locus-level copy number estimates of the normal cells from those of the tumor cells. In our assessments, combining the quantitative normal and tumor data by subtraction produces superior results to alternative approaches, such as separately segmenting the tumor and normal data and then excluding tumor lesions that overlap with (presumably inherited) lesions in the normal cells.

Third, we consider both the magnitude of the apparent change in copy number and a measure of its statistical significance relative to the background variation. We have found that incorporating these two distinct factors is important. If only the Z-scores are considered, then regions of the genome with low variability can produce lesion calls even when the magnitude of the change is small (e.g., a region with mean copy number 1.8 may be deemed a loss). Similarly, in regions of the genome with high variability, even a copy number shift of one unit may not be statistically distinguishable from the background variation.

The magnitude of the copy number change is quantified as the mean value within a block of consecutive SNPs defined to span a fixed number of SNPs (blocks spanning 20–30 SNPs give the most meaningful values in our experience). Note that since all our analysis is done using tumor/normal differences, this magnitude is interpreted as the mean difference on the variance-stabilized scale of the tumor compared to the normal copy numbers within a particular block. The statistical significance of the copy number change is quantified using a Z-score, defined as the ratio of the mean difference to the standard deviation of the subtracted, variance-stabilized data. Since variances are intrinsically more difficult to estimate than means, we found that more meaningful results are obtained when calculating the variance using a window that is four times wider than the window used to calculate the mean (e.g., if the mean is calculated using 20 consecutive SNPs, the variance is calculated using 80 consecutive SNPs).

Preliminary lesion calls (subject to further refinement as described below) are based on thresholds for the mean difference and the Z-score. For losses, we require a mean difference of −0.12 and a Z-score of 12. Both these thresholds were determined to work well empirically in the sense of largely recapitulating visual findings and also gave largely correct results on positive and negative control data (determined from FISH assays). Further statistical justification of the Z-score threshold is given below. Due to the fourth root transform, the mean difference of −0.12 is equivalent to an average tumor copy number estimate of 1.3, indicating that a lesion that on average presents as a heterozygous (single copy) deletion, with a mean copy number of 1.2 (empirically determined), should be detectable in our analysis.

The Z-score of 12 may seem excessively conservative, but this number, initially identified by trial and error, turns out to have a strong statistical basis. Due to multiple comparisons of over 900,000 probes, a conventional Z-score threshold of 2 would yield many false positives. A Bonferroni correction for 900,000 tests would require a threshold of 5.4 to achieve a family-wise error rate of 0.05. A second important factor is that the copy numbers within a window are positively correlated. To assess the effect of this correlation, we calculated autocorrelations up to lag 30 (the window size) and used it to define a correlation matrix for the data in a window. Based on this correlation matrix, we can determine an "effective number of SNPs" for mean estimation as 30^2/sum(C), where sum(C) denotes the sum of all elements of the 30×30 correlation matrix. Our effective number of SNPs, averaged over 254 AML samples, was 5.9. For comparison, if the data were perfectly uncorrelated, we would have an effective number of SNPs of 30. To account for this dependency, we should multiply the Z-score threshold by sqrt(30)/sqrt(5.9) ~ 2.2. Thus, the threshold of 5.4 derived from Bonferroni adjustment becomes $5.4 \times 2.2 \sim 12$, which is exactly the value determined empirically to work well in that it gave good agreement with visual calling.

The fourth and final step of our sliding window approach to algorithmic lesion calling is particularly adapted to the challenge of accurate summarization of genomic complexity in a single score, as described above. Following the three steps outlined above, we obtain lesion calls that are internally strong both in terms of magnitude and statistical significance. However, in some instances, what appears visually as a single large lesion can for a small part of its span drop below the stringent thresholds that must be used to minimize false positives in the initial detection step. This may indicate that two or more lesions are truly present, but in many cases, we observe that even within this "gap" the copy number deviation and significance level remain elevated. Thus, our last step is to bridge gaps between consecutive lesions of up to 500 bases, as long as the mean Z-score within the gap exceeds two in magnitude. We note that this strategy of identifying key components of a pattern based on a stringent condition and then filling in the rest of the pattern using a weaker condition is widely used in pattern analysis, such as in the "sketch" algorithms for image analysis.

2. Materials

2.1. Flow Cytometry Sorting of AML Blast Populations, Including Optional Presort Microbead-Based Depletion of Non-blast Cells

Note: Items 1 through 5 are needed only if doing presort bead-based depletion and are therefore optional.

1. BSA/EDTA buffer: 1× phosphate-buffered saline with 0.5% bovine serum albumin (BSA) and 1 mM EDTA (must be degassed prior to use).
2. LS separation columns, QuadroMACS magnet, and magnetic microbeads for human CD3, CD19, CD14, and CD235a (Miltenyi Biotec, Auburn, CA).
3. HBSS buffer with 1% fetal calf serum (FCS).
4. RPMI cell culture medium with 10% fcs and penicillin–streptomycin antibiotics.
5. Prior information regarding degree of leukemic blast expression (if any) of CD3, CD19, or CD14.
6. This protocol assumes that leukemia samples to be analyzed by this method have been cryopreserved following mononuclear cell separation using Ficoll-Paque PLUS (GE Healthcare, Piscataway, NJ).
7. Cryopreserved normal human Ficoll-isolated peripheral blood mononuclear cells (pbmc).
8. Large-bore pipet tips, 1,000 μl volume (Axygen, Union City, CA).
9. Normal mouse IgG (Southern Biotech, Birmingham, AL).
10. 100 U/μl DNase I (Sigma-Aldrich, St. Louis, MO).
11. Antihuman CD33 FITC, CD13 PE, and CD45 APC (eBioscience, San Diego, CA).

12. 1 mg/ml propidium iodide (PI) (Sigma-Aldrich, St. Louis, MO). Prepare in ddH_2O and store at 4°C.
13. 70 μm cell strainers (Fisher Scientific, Waltham, MA).
14. FACSAria flow cytometry sorter (BD Biosciences, San Jose, CA).

2.2. Affymetrix SNP 6.0 Array Data Analysis

1. CEL files generated through the complete Affymetrix SNP 6.0 DNA processing, chip hybridization, and chip scanning steps.
2. GTC software version 2.
3. dChip software.
4. PLUT and LOH tool 2.0 software (see Note 3).
5. Computer workstation with sufficient RAM and hard disk space to accommodate all SNP array files expected to be generated from experiments (see above).

3. Methods

3.1. Optional Presort Microbead-Based Depletion of Non-blast Cells

While flow cytometry sorting of primary human AML samples is sufficient in most cases to yield a sorted population of over 95% blast cells, certain samples, such as those with a large number of lingering erythrocytes (even following mononuclear cell separation with Ficoll) or peripheral blood samples with low blast percentages, may benefit from pre-enrichment using magnetic microbeads targeted to cell antigens not expressed by blasts. If pre-enrichment is desired, perform steps 1 through 19. Otherwise, proceed to Subheading 3.2.

The end user will be directed to the standard Miltenyi Biotec human microbead and LS column product package inserts at relevant points in this protocol. Information relevant to the handling of cryopreserved human AML samples is included here.

1. Warm RPMI medium with 10% calf serum and pen-strep antibiotics to 37°C. 25 ml is needed for each sample being thawed.
2. Thaw cells and add dropwise to 25 ml warmed RPMI/10% fcs.
3. Centrifuge 10 min at 290 × *g* at room temperature.
4. Cool centrifuge to 4°C following completion of step 4.
5. Aspirate supernatant but leave approximately 0.5 ml atop pellet.
6. Resuspend pellet with large-bore pipet tip.
7. If pellet fails to resuspend completely, add 2 μl of 100 U/μl DNase I.

8. Incubate 10 min at room temperature with occasional mixing. Clumps should dissolve.
9. Transfer to 15 ml conical centrifuge tube and add 14 ml cold 1× HBSS with 1% fcs.
10. Centrifuge 10 min at 290×*g* and 4°C.
11. Determine whether AML blasts in any sample express CD3, CD19, or CD14. If blasts express any of these markers, do not use that microbead for that particular sample.
12. Use microbeads in the following volumes per ten million cells: CD3—20 μl; CD19—10 μl; CD14—10 μl; CD235a—10 μl.
13. Place cells on ice and discard all supernatant.
14. Use wide-bore 1,000 μl pipet tips for resuspension of all cell pellets.
15. At this point, please refer to the Miltenyi Biotec human microbead and LS column product package inserts for the standard protocols of (1) combining and incubating microbeads and cells and (2) purifying cells over the LS columns with the QuadroMACS magnet.
16. Following collection of leukemic cell fractions, centrifuge 10 min at 290×*g* and 4°C.
17. Remove all of the supernatant by vacuum aspiration.
18. Resuspend in ice-cold HBSS/1% FCS.
19. Count cells by an established method.

3.2. Flow Cytometry Sorting

1. Thaw and wash cells as described in Subheading 3.1 if presort microbead-based depletion is not performed.
2. Place cells on ice.
3. Count cells by an established method and adjust concentration to ten million cells per ml in HBSS with 1% FCS.
4. Block by adding normal mouse IgG to a final concentration of 200 μg/ml and incubate on ice for 10 min.
5. While incubating, label five separate tubes for flow cytometry compensation controls: (1) unstained, (2) propidium iodide (PI) only, (3) CD33 FITC only, (4) CD13 PE only, and (5) CD45 APC only.
6. Distribute 100 μl normal human pbmc (at ten million/ml) to the five control tubes.
7. Add eBioscience brand fluorochrome-conjugated antibodies in the ratio of 1 μl antibody solution per 200,000 cells (if the vial states 20 μl/test) or 1 μl antibody per 800,000 cells (if the vial states 5 μl/test). For the control tubes, use 5 μl antibody and 1.2 μl antibody, respectively.
8. Vortex briefly and gently, then incubate on ice 15 min in the dark.

9. Add 12 ml HBSS + 1% fcs to all tubes.
10. Centrifuge 10 min at 290 × *g* and 4°C.
11. Aspirate supernatant and repeat steps 9 and 10, resuspending cells with large-bore pipet tips.
12. Calculate final volumes to use for each sample:
    (a) Final volume should be 1 ml per 1.0×10^7 cells present, but this depends upon the capabilities and setup of the sorter (see Note 4).
    (b) For fewer than 5.0×10^6 cells, use 500 μl.
13. While centrifuging, dilute 1 mg/ml PI stock to 20 μg/ml in HBSS + 1% FCS.
14. Aspirate all supernatant.
15. Resuspend cells in appropriate amount of buffer. For controls, use 500 μl.
16. Filter all AML samples (not necessary to filter compensation control tubes) using 70 μm cell strainers.
17. Add 20 μg/ml PI to filtered sample cell suspensions such that the final concentration of PI is 1 μg/ml.
18. Each of the single-color compensation controls is run first.
19. When the cytometer has been properly calibrated and compensated, 10–30,000 cells from each primary AML are analyzed to determine gate positions, following the method of Borowitz et al., and CD33 and CD13 are used to discriminate any erythroid cells (9) (see Note 5).
20. At this point the sort must be managed in order to get through all cell suspensions in a timely manner with the greatest possible efficiency. Several factors affect this process (see Note 4).
21. At the conclusion of the sort appointment, a small number of cells may be examined using cytospins and Wright–Giemsa stain to visualize the type of cells actually sorted.

3.3. SNP Array Data Quality Control and Generation of SNP Calls Using GTC Software (Please Note: The Following Procedure Is for GTC Version 2.0; Some Steps May Be Different in Later Versions of the Software)

1. When CEL and ancillary data files are received, immediate backup is recommended.
2. Install GTC and set up a workspace. Double-click the workspace icon.
3. At "workspace" prompt click OK.
4. If starting a new project, right click "data set," then click "create new data set." Name the new project and change the "array" option to genotyping_SNP 6.0.
5. If adding more data to an existing data set, select that data set name, right click, and then select "add data."
6. When "Add data to [data set name]" prompt is up:

(a) Under "select data to add to data set," check "sample files (ARR, XML)," "Intensity and QC files (CEL, GQC)," and the button "Select files."

(b) Under "automation," select "auto-add sample files," "auto-add intensity and QC files," and "auto-QC intensity files." Click OK.

7. Navigate to the folder containing the CEL and ancillary files you want to add and click "Open." Select files individually if necessary.

8. QC analysis will run automatically on newly added files and will display the results. On the tab "Intensity Data," you can see the QC score of select groups of files. You may view all files ("ALL"), only those files which passed QC ("In Bounds"), or only those which failed QC ("Out of Bounds").

9. Genotyping and call rate determination through the Birdseed algorithm.

10. Right click on the "In Bounds" group and select "Perform genotyping."

11. Select analysis configuration as "Default."

12. Click OK to start. If any errors appear regarding lack of annotation or library files, they can be directly downloaded by selecting "File," then "download annotation or library files." Be certain that the annotation and library files are the same ones used for dChip analysis. Our lab has only used the file "GenomeWideSNP_6.na24.annot.csv." We recognize that several annotation updates have since taken place. However, for the purposes of this protocol, we recommend only the use of the na24 version.

13. The length of time required for Birdseed analysis will vary greatly, depending upon the number of CEL files involved. The computer must be protected from automatic shutdowns or restarts during this time.

14. When complete, right click on the file under "Genotyping results" and select "Export genotyping results."

15. Note that you do not need to select any filtering options (make sure all are unchecked).

16. To view the genotyping call rates, double-click on the batch name under "Genotyping results."

3.4. Analysis of Genomic Copy Number Changes Using dChip Software

1. Download dChip from the dChip Web site: http://www.dchip.org. Our laboratory uses the build date of February 25, 2009.

2. Create the sample file list. This tells dChip which CEL files are to be "read in." It is simply a list of the CEL filenames and the appropriate file directory path.

Table 1
dChip sample information file: information columns and their purposes

Column heading	Purpose
Array	The name of the CEL file, *without* accompanying directory path information or the .CEL extension
Sample	Alternate name, if desired, for the CEL file. The name entered into this field will be the name displayed in the dChip heatmap. This field is useful for presenting uniform filename syntax and including additional information about the sample, e.g., date or disease subtype
Type	Specify either "tumor" or "normal" for each sample. When dChip computes copy number for each SNP, the signal values of all of the normal samples are averaged to obtain the mean signal of two copies, and the observed copy number is defined as (observed signal/mean signal of two copy)×2 (17)
Gender	Enter M or F. This information is important, as dChip will multiply male ("M") X chromosome signal values by 2 as part of its copy number calculation (D). If this field is left blank, dChip will assume the sample is female
Ploidy(numeric)	Enter "2" for each "normal" sample; for "tumor," leave blank
RefBatch	Leave blank to consider all samples as one batch. Additional information about using this field and the array list file to help control "batch effect" is available on the dChip Web site (E-F)

(a) In each row, beginning with row 1, list the file path of each CEL file, ending with the actual CEL filename and the .CEL extension. Example: C:\Documents and Settings\Administrator\Desktop\\AMLproject\AML-1-T_(GenomeWideSNP_6).CEL.

3. Create the sample information file (see Table 1). Note that the column headers must appear *exactly* as shown in Table 1, or dChip will not recognize the column of information.
4. From Excel, save as text file format (.txt).
5. Download the Affymetrix CDF file, if not already done (see Note 2).
6. Before beginning file addition to dChip, restart the computer and allow only dChip to be open, as a precaution to avoid any out-of-memory errors.
7. Double-click the dChip icon.
8. From the top drop-down menu, choose "Analysis," then "Open Group."

9. Type a name for your group in the "Group Name" field (our lab uses one group name for each cancer type).
10. Click the "Data file list (.txt)" button and select the sample file list created earlier.
11. "File type": select CEL.
12. "Suffix of TXT/CHP call"—set to CHP—but this is only if dChip will be used to analyze genotyping call files. Our lab uses a different software for this purpose. Warnings will pop up during read-in that the CHP files cannot be found. Sample copy number analysis through heatmaps can still be performed, so these warnings should be ignored.
13. At the top of the pop-up window, click the "Other Information" tab.
14. Next to "CDF File," click "None" and navigate to the CDF file downloaded earlier.
15. Set "array type" to "SNP 6.0."
16. Within the "Information Files" box, next to "Sample info," click "None" and then navigate to the sample information file created earlier.
17. Return to the tab titled "Data files" and click the "Options" button at the bottom.
18. In the "Analysis" tab, in the "Open Group" box:
    (a) Set MBEI memory to 500 MB. This amount can be increased as necessary to handle larger numbers of CEL files. For approximately 600 arrays in our 64-bit OS computer, we use a setting of 2,500 MB.
    (b) Click the "Working Directory" button and select the directory where the appropriate CEL files are stored.
    (c) Check the box next to "Check CEL file array type."
    (d) Make sure that "Load probe data in memory" is *not* checked and click OK.
    (e) Once back in the "Data files" tab click OK.
19. dChip will now perform the "read in" step. Check carefully for any red-colored alert messages. Warnings that state CHP files are not found can be ignored.
20. When read-in has satisfactorily completed, from the top drop-down menu "Analysis," choose "Normalize and Model…."
21. Uncheck the box labeled "view normalization plot."
22. Check the box titled "Ignore existing normalization data," or else dChip will not renormalize previous data to any new data that are being added.

23. Check the box titled "Compute model-based expression/signal values" to proceed immediately from normalization to MBEI, and click OK.
24. Once normalization and MBEI have satisfactorily completed, the remaining step is to construct an array list file prior to viewing heatmap data.
25. From the main top drop-down menu, click "Tools," then "Array list file."
26. In the Array List File control panel, a list of all arrays read in will be shown on the left. The array names that appear are from the "Sample" column in the sample information file (see Table 1). Arrays to be displayed are placed in the right-hand window. Any number of the available arrays can be displayed, and they can be in any order. A "standardize" separator is included at the end of the list to assure all samples are treated as one batch.
27. When the array list file is complete, name the file in the bottom field and click "Save and Use" button.
28. From the main top drop-down menu, click "Analysis," then "Chromosome."
29. At "Genome Information," click "None" and navigate to the SNP 6.0 genome information file downloaded earlier (see Note 2).
30. For "Analysis Method," choose "Copy number and LOH."
31. To analyze only one chromosome, type that chromosome number into the "Chromosome (0 for all)" box. For the X chromosome, type "23." For the Y chromosome, type "24."
32. Click OK.
33. Press the "D" key to change between displays. dChip's default display is copy number, with various shades of red. Press the "D" key, and the display will change to $\log_2$ ratio, with blue indicating copy loss, white 2n copy, and red copy gain. Degree of copy change is indicated by the intensity of the colors.
34. Zoom in on the chromosome display by pressing the down arrow button. Zoom out by pressing the up arrow button. Widen the display by pressing the right arrow button. Narrow by pressing the left arrow key.
35. Change chromosomes by pressing either "Home" or "End." Scroll through a chromosome using the mouse wheel or jump sections by using "Page Up" or "Page Down."

3.5. Analysis of Loss of Heterozygosity Using Two-Step Software Analysis with PLUT and the LOH Tool 2.0

1. PLUT and the LOH tool were designed to analyze large numbers of samples in the context of *paired* (normal and tumor) primary human samples.
2. Before beginning it is recommended to restart the computer and have only PLUT open for the duration of data processing.
3. PLUT requires very specific filename syntax in order to recognize and pair individual patient normal and tumor samples. Alphanumeric filenames are acceptable; however, the end user is cautioned that LOH tool 2.0 is currently only capable of displaying numeric values for samples.
4. When PLUT completes its process, the result is one text document which will then be imported into the LOH tool. If this file is opened, the user will see that each filename, no matter what the alpha prefix originally was, is now "CLL." This is an uncorrected design requirement of the LOH tool and does not affect the data points or numeric values associated with the original alpha prefix.
5. Goal filename syntax (example): AML001-T.txt.
6. An indefinite number of alpha characters are allowed for each filename. However, all alpha prefixes must be exactly the same in each batch of samples analyzed by PLUT.
7. The number associated with each sample must be composed of three digits. For example, AML001-T.txt is acceptable. AML1-T.txt is not. There cannot be a space between the alpha prefix and the file number.
8. A hyphen must follow the number, with no space before or after it.
9. A "N" or "T" must follow the hyphen to indicate either normal or tumor.
10. Each patient sample must have two files—one N and one T.
11. The three-letter file extension must be ".txt."
12. SNP call files exported from GTC carry the original filename as part of the file content, so despite necessary renaming for PLUT, the original filename is always accessible.
13. If not previously done, download the Affymetrix SNP 6.0 annotation file "GenomeWideSNP_6.na24.annot.csv." Please note that PLUT was written in the context of this annotation file. It is unknown whether more recent versions of this file will interface properly with PLUT.
14. Launch the PLUT program by double-clicking the Windows Batch file "PLUT_091908.bat."
15. Please note that while functions or capabilities other than those described in this protocol may appear to be possible in PLUT

due to menu choices, these may only represent functions whose development was never fully completed. Therefore, the end user should consider the only function of PLUT to be exactly that which is described in this protocol.

16. Click "New SNP ID File Load" button.
17. Navigate to the Affymetrix annotation file "GenomeWideSNP_6.na24.annot.csv" and click the "CSV file" button.
18. A window will then appear, titled "Add Patient Files." Navigate to the folder where the patient SNP call files are stored. Highlight all files to process. Only paired normal and tumor samples can be processed, and each individual tumor must have a normal. When all files are highlighted, click the "Add Patient Files" button.
19. A window will then appear titled "Select Headers." You must select the following three column headings: "Probe Set ID," "Chromosome," and "Physical Position." Do not select any other column headings.
20. Click "Submit."
21. PLUT will now process all files, and it will ultimately produce a text file titled "SNP_File.txt," which will be located within the PLUT folder. This file can be renamed in any fashion and stored in any convenient location.

3.6. LOH Tool Version 2

1. Launch the LOH tool by double-clicking on the Windows Batch file titled "findlohamp_64bit.bat," and from the "File" menu, select "Import."
2. Navigate to the "SNP_File.txt" produced by PLUT earlier and select.
3. LOH tool 2.0 will rapidly go through a step titled "Combining All Data." There will be an approximate 2-min pause, and then the "Import Progress" window will appear, with a progress bar for "Reloading Data…." Loading time will be proportional to the number of SNP array files in the batch. Please note that although the progress bar may pause for 3–5-min periods at 7% intervals of completion, the software is not frozen and has always resumed progress successfully.
4. When the "Reloading Data" progress bar reaches 100% and the window disappears, close LOH tool 2.0 and restart it by double-clicking the .bat file.
5. Do not yet attempt to change the graphic display. Now, a separate file listing all SNP rsID values must be incorporated. From the "File" menu, select "Load Rs."
6. Navigate to the file located within the LOH tool folder called "rsiddef." This is a 57 MB text file which lists approximately 930,000 SNPs on the SNP 6.0 array by SNP ID, rsID, and physical position values. If the file is opened in a program

like Wordpad, SNP call information will be seen for four samples titled "CLL." This is a vestigial remnant of earlier programming iterations. These SNP calls serve no purpose in LOH tool, nor do they incorporate any false data into the display.

7. The LOH tool will display "Loading Rs" with a progress bar pop-up window. Currently, the progress bar will reach 100% long before the software actually finishes doing the necessary data "crunching." Progress may be checked by clicking the chromosome selector in the LOH tool and moving the pointer over hash marks of LOH and checking to see if there is a display of rsID value in addition to the physical position.
8. Do not close the LOH tool until the loading of rsID values is complete.
9. Once the rsID file has been loaded for a given set of data, it does not need to be reloaded with each successive launch of the software. It will be necessary when importing new data.
10. To manage multiple data sets, a new copy of the existing LOH tool folder can simply be placed elsewhere in the hard drive. This new copy of LOH tool is launched, the previously existing data deleted (through selecting "File," then "Clear Data," followed by deleting the existing "alldata.txt" file within the folder), and a new output file from PLUT imported.
11. The remaining steps describe the functions of the LOH tool 2.0 GUI.
12. *Caution*: Note that while functions or capabilities other than those described in this protocol may appear to be possible in LOH tool 2.0 due to menu choices, these may only represent functions whose development was never fully completed. Therefore, the end user should only attempt to perform data analysis in LOH tool 2.0 using the GUI controls described in this subsection.
13. When data has first been imported into LOH tool 2, the GUI will by default display unfiltered data for both LOH and NoCalls.
14. From the drop-down menu next to the "NoCall" button, choose "Leave Out." All red marks in the GUI should disappear.
15. The "transparency" slider should be moved all the way to the left. All beige color should disappear.
16. From the drop-down menu next to the "LOH" button, select "Filter."
17. In the "Filter" field, enter the value "3,000" and click "Go" (see Note 6).

18. To change chromosomes, use the drop-down menu titled "Chromosome." Chromosome 13 is shown first by default.
19. To fit more samples into the LOH viewing area of the GUI, move the slider bar at the extreme right of the display up or down.
20. To lengthen or shorten the displayed region of the chromosome, move the slider bar between the LOH viewing window left or right.
21. To view physical position and SNP rsID number, click a hash mark and scroll, if necessary, to the top of the window.

4. Notes

1. dChip is downloadable from http://sites.google.com/site/dchipsoft/ (accessed September 13, 2011).
2. Affymetrix CDF (library) files can be downloaded from http://www.affymetrix.com/support/technical/libraryfilesmain.affx (accessed September 13, 2011); the specific genome information file formatted for dChip can be accessed through http://sites.google.com/site/dchipsoft/snp-array-analysis/snp-array-data-processing/genome-information-files-for-snp-arrays and following the SNP 6.0 link adjacent to "2/12/08" (accessed September 13, 2011).
3. Copies of PLUT and LOH tool 2.0 are available by contacting the corresponding author.
4. Numerous factors affect an investigator's final throughput and cell yield from a single flow cytometry sort. The FACSAria sorter can process thousands of events per second. However, the maximum evt/sec is determined by both the instrument's flow cell tip diameter and sheath fluid pressure and the concentration of the sample. It is important to note that no cytometer is 100% efficient at sorting cells and up to 40% of processed events may not enter the collection tube. Generally, the rarer the cell population, the less efficient the sort process will be. Slowing down the flow rate of cells will improve efficiency; however, it will take longer to process the same volume of cell suspension. Higher pressures and narrower flow cell tips, when such adjustments are possible with a given instrument, will allow higher event rates and will actually provide greater sorting efficiencies.
5. The utility of CD13 and CD33 staining varies with a number of factors. Blasts may express either of these markers brightly, dimly, variably, or not at all. The use of CD235a microbeads to subtract erythroid cells prior to sorts, while adding to cost,

virtually obviates the need for a secondary plot besides CD45 versus side scatter. Lymphocytes nearly always cleanly separate from the blast population on CD45 versus side scatter. In cases of AML with monocytic differentiation, the predominant leukemic cell population may have a degree of CD45 expression equal to that of lymphocytes. The higher side scatter and frequent CD13 or CD33 expression, however, is sufficient to discriminate these cells from lymphocytes. On rare occasions, abnormal granulocytes may fail to be eliminated at the Ficoll gradient step and may fall into the blast gate. These are easily seen if a Wright–Giemsa-stained cytospin is performed following the sort.

6. The filter feature of LOH tool 2.0 is very important for proper interpretation of the data display. Our lab has standardized on a setting of 3,000 for scanning for regions of LOH, with the chromosome view "stretched" out to near maximum using the bottom slider bar. Filtering at this setting restricts the LOH tool 2.0 from displaying hash marks unless they are within 3,000 base pairs of another call.

Acknowledgments

We acknowledge the important contributions to this protocol and to the LOH tool/PLUT development by Mr. Sajid Shakhan and Ms. Whitney Wright.

References

1. Mullighan CG, Goorha S, Radtke I et al (2007) Genome-wide analysis of genetic alterations in acute lymphoblastic leukaemia. Nature 446(7137):758–764
2. Parkin B, Erba H, Ouillette P et al (2010) Acquired genomic copy number aberrations and survival in adult acute myelogenous leukemia. Blood 116(23):4958–4967. doi:blood-2010-01-266999 [pii] 10.1182/blood-2010-01-266999
3. Parkin B, Ouillette P, Wang Y et al (2010) Nf1 inactivation in adult acute myelogenous leukemia. Clin Cancer Res 16(16):4135–4147. doi:1078-0432.CCR-09-2639 [pii] 10.1158/1078-0432.CCR-09-2639
4. Walter MJ, Payton JE, Ries RE et al (2009) Acquired copy number alterations in adult acute myeloid leukemia genomes. Proc Natl Acad Sci U S A 106(31):12950–12955
5. Tiu RV, Gondek LP, O'keefe CL et al (2011) Prognostic impact of SNP array karyotyping in myelodysplastic syndromes and related myeloid malignancies. Blood 117(17):4552–4560. doi:blood-2010-07-295857 [pii] 10.1182/blood-2010-07-295857
6. Fitzgibbon J, Smith LL, Raghavan M et al (2005) Association between acquired uniparental disomy and homozygous gene mutation in acute myeloid leukemias. Cancer Res 65(20):9152–9154
7. Raghavan M, Lillington DM, Skoulakis S et al (2005) Genome-wide single nucleotide polymorphism analysis reveals frequent partial uniparental disomy due to somatic recombination in acute myeloid leukemias. Cancer Res 65(2):375–378
8. Dunbar AJ, Gondek LP, O'keefe CL et al (2008) 250k single nucleotide polymorphism array karyotyping identifies acquired uniparental disomy and homozygous mutations, including novel missense substitutions of c-Cbl, in myeloid malignancies. Cancer Res 68(24):10349–10357

9. Borowitz MJ, Guenther KL, Shults KE et al (1993) Immunophenotyping of acute leukemia by flow cytometric analysis. Use of cd45 and right-angle light scatter to gate on leukemic blasts in three-color analysis. Am J Clin Pathol 100(5):534–540
10. Li C, Wong WH (2001) Model-based analysis of oligonucleotide arrays: expression index computation and outlier detection. Proc Natl Acad Sci U S A 98(1):31–36. doi:10.1073/pnas.011404098011404098 [pii]
11. Li C, Hung Wong W (2001) Model-based analysis of oligonucleotide arrays: model validation, design issues and standard error application. Genome Biol 2(8):RESEARCH0032
12. Ross CW, Ouillette PD, Saddler CM et al (2007) Comprehensive analysis of copy number and allele status identifies multiple chromosome defects underlying follicular lymphoma pathogenesis. Clin Cancer Res 13(16):4777–4785
13. Eckel-Passow JE, Atkinson EJ, Maharjan S et al (2011) Software comparison for evaluating genomic copy number variation for affymetrix 6.0 snp array platform. BMC Bioinformatics 12:220. doi:1471-2105-12-220 [pii] 10.1186/1471-2105-12-220
14. Diskin SJ, Li M, Hou C et al (2008) Adjustment of genomic waves in signal intensities from whole-genome snp genotyping platforms. Nucleic Acids Res 36(19):e126. doi:gkn556 [pii] 10.1093/nar/gkn556
15. Marioni JC, Thorne NP, Valsesia A et al (2007) Breaking the waves: improved detection of copy number variation from microarray-based comparative genomic hybridization. Genome Biol 8(10):R228. doi:gb-2007-8-10-r228 [pii] 10.1186/gb-2007-8-10-r228
16. Kujawski L, Ouillette P, Erba H et al (2008) Genomic complexity identifies patients with aggressive chronic lymphocytic leukemia. Blood 112(5):1993–2003
17. Zhao X et al (2004) An integrated view of copy number and allelic alterations in the cancer genome using single nucleotide polymorphism arrays. Cancer Res 64:3060–3071

Further Readings

Affymetrix Genotyping Console 2.0 User Manual pg 70, available by request through Affymetrix (1-888-DNA-CHIP)

Affymetrix Genotyping Console 2.0 User Manual pg 55, available by request through Affymetrix (1-888-DNA-CHIP)

"dChip Software," http://sites.google.com/site/dchipsoft/home. Accessed 7 Sept 2011

"Copy number analysis," http://sites.google.com/site/dchipsoft/snp-array-analysis/loh-and-copy-number-analysis/copy-number-analysis. Accessed 13 Sept 2011

"Adjust batch and individual effect," http://sites.google.com/site/dchipsoft/snp-array-analysis/loh-and-copy-number-analysis/adjust-batch-and-individual-effect. Accessed 7 Sept 2011

Chapter 8

Clinical Applications of BAC Array-CGH to the Study of Diffuse Large B-Cell Lymphomas

Cristina Robledo, Juan Luis García, and Jesús M. Hernández

Abstract

BAC array-CGH is a powerful method to identify DNA copy number changes (gains, amplifications and deletions) on a genome-wide scale, and to map these changes to genomic sequence. It is based on the analysis of genomic DNA isolated from test and reference cell populations, the differential labelling with fluorescent dyes and the co-hybridization with a genomic array. BAC array-CGH has proven to be a specific, sensitive, and reliable technique, with considerable advantages compared to other methods used for the analysis of DNA copy number changes. The application of genome scanning technologies and the recent advances in bioinformatics tools that enable us to perform a robust and highly sensitive analysis of array-CGH data, useful not only for genome scanning of tumor cells but also in the identification of novel cancer related genes, oncogenes and suppressor genes.

Cytogenetic analysis provides essential information for diagnosis and prognosis in patients with hematologic malignancies such as lymphomas. However, the chromosomal interpretation in non-Hodgkin lymphoma (NHL) is sometimes inconclusive. Copy number aberrations identified by BAC array-CGH analyses could be a complementary methodology to chromosomal analysis. In NHL the genomic imbalances might have a prognostic rather than a diagnostic value. In fact, the diagnosis of NHL is based on pathological and molecular cytogenetics data. Furthermore genetic variations and their association with specific types of lymphoma development, and elucidation of the variable genetic pathways leading to lymphoma development, are important directions for future cancer research. Array-CGH, along with FISH and PCR, will be used for routine diagnostic purposes in near future.

Key words: BAC array-CGH, Non-Hodgkin lymphoma, BAC (bacterial artificial chromosome) clone, Diffuse large B-cell lymphoma, MegaCHOP

1. Introduction

Diffuse large B-cell lymphoma (DLBCL) is the most common type of malignant lymphoma, accounting for 25–30% of adult non-Hodgkin lymphomas (NHL). DLBCL represents a heterogeneous

Diponkar Banerjee and Sohrab P. Shah (eds.), *Array Comparative Genomic Hybridization: Protocols and Applications*, Methods in Molecular Biology vol. 973, DOI 10.1007/978-1-62703-281-0_8, © Springer Science+Business Media, LLC 2013

group of tumors with different genetic abnormalities, morphological, phenotypic, molecular and clinical characteristics (1–4). Furthermore, this group of lymphomas is heterogeneous in terms of their response to therapy with half of all patients achieving a complete response with standard anthracycline based chemotherapy (3, 5). The International Prognostic Index (IPI) is the most important tool for predicting the response to treatment and for classifying patients into subgroups with different prognoses (6). However, even within these IPI risk groups, variability in outcome of DLBCL has been observed (7).

In DLBCL multiple breakpoints and translocations with a multiple of numerical abnormalities (gains and losses) have been identified (8). Thus, DLBCL are characterized by clonal chromosomal rearrangements often involving the immunoglobulin (IG) gene loci and somatic hypermutation at the IG heavy-chain locus, which are indicative of a germinal center (CG)-derived mature B cell origin (9). Up to 50% of DLBCL harbor translocations involving the immunoglobulin genes (10, 11). Most common translocations in DLBCL bring genes including *BCL6* (35–50%), *C-MYC* (15%), and *BCL2* (13%) under the inappropriate control of an immunoglobulin regulatory element. Rearrangements of *MYC* can be found in 5–20% of DLBCL. In more than half of these cases additional abnormalities, most commonly the t(14;18), can be identified. These "double-hit" or "dual-hit" lymphomas often have complex cytogenetic alterations, including occasional breakpoint at 3q27/*BCL6* ("triple-hit" lymphomas). These complexities can be best detected by conventional cytogenetics, as it represents a genome-wide method by which many structural and numerical abnormalities can be detected simultaneously, albeit at low resolution (8, 12–16). Conventional cytogenetics methods need fresh tumor tissue, which must be processed within a few hours after removal from patient. In addition, the need for a relatively large volume of tissue and the high level of ability required to culture NHL cells, to prepare high-quality metaphase cells and to analyze complex karyograms make the use of conventional cytogenetics difficult in a routine setting. The fluorescence in situ hybridization (FISH) method could be complementary to conventional cytogenetics. FISH is relatively easily applicable technique to formalin-fixed and paraffin-embedded tissues and does not need vital, growing cells or specific culture systems (8, 17, 18). The major limitation of FISH is that this method is targeted to the detection of specific abnormalities and the genome-wide perspective offered by conventional cytogenetics is lost. This may be clinically important, as additional aberrations (or their absence) could be diagnostic or predictive for the course of disease.

Genomic complexity in DLBCL is usually high. Thus in a recent study of aggressive B cell lymphomas with MYC-IG breakpoint three DLBCL showed more than 21 cytogenetic abnormalities (19–22).

Based on these data in DLBCL, it is now justifiable to regard the presence of a complex karyotype as a diagnostic criterion for distinguishing DLBCL from other lymphomas. The importance of this distinction argues for the retention of techniques that provide a genome-wide survey of the genetic abnormalities of lymphoma. Genome-wide methods that do not need dividing tumors cells are conventional comparative genome hybridization (Conventional CGH) and array comparative genomic hybridization (array-CGH). A major disadvantage of both techniques is that balanced chromosomal alterations such as translocations cannot be detected. Conventional CGH is based on competitive hybridization between normal DNA and tumor cell-derived DNA, where chromosomes obtained from the individuals are used as template for hybridization. The resolution of this technique is not much better than conventional cytogenetics. Therefore conventional CGH has been replaced by array-CGH, in which the DNA fragments used for competitive hybridization are spotted onto glass slide. Depending on the number of probes used to create the platform, array-CGH can achieve a very high resolution. Of interest, DNA isolated from paraffin tissue blocks can be used successfully for hybridization (8). The application of genome scanning technologies such as conventional CGH and array-CGH, has been useful not only for genome scanning of tumor cells but also for the identification of novel oncogenes and tumor suppressor genes (23–27).

Until now only a few reports on the application of array-CGH in mature B cell and T cell lymphomas have been published. Overall the results showed distinct patterns of gains and losses characteristic of particular subtypes of lymphomas (28–34). However information regarding DLBCL is scarce.

In DLBCL, array-CGH confirmed the genetic instability detected by conventional CGH. Array-CGH studies have shown the numerous regions gained or lost in these lymphomas and, moreover, these changes have been associated with germinal center (GCB) or activated B cells (ABC) groups determined by gene expression arrays. Thus gains on 3q23-q28 appear to be associated to ABC group. In contrast, gains on 7q22-q36 were detected in GCB group and in a few cases of ABC group. These differences showed that both groups are genetically different (35).

However, the role of genomic gains or losses in the outcome of DLBCL remains unclear. Thus the breakpoints at 1q21 and deletions of 6q could be associated with poor outcome, while chromosomal-CGH suggests an association between 17p loss and short survival (9, 36). In addition, array-CGH studies have focused on target gene identification, showing that the loss of chromosomes 2 and 16 could be prognostic indicators of poor survival. Other array-CGH studies in DLBCL detected gains on 6p21 (CCND3 gene), 18q21, and on chromosomes 2 and 20. In some of these regions *BCL2*, *MDM2*, *JAK2*, and *REL* genes are located.

The losses in DLBCL were showed on chromosome 1, 3p14 (*FHIT* gene) and on 10q21. However, these studies have been hampered by considerable bias as they have not been performed on tissue from uniformly treated patients (Table 1) (10, 35, 37–39).

In order to identify regions of DNA copy number changes associated with response to the therapy and outcome, we analyzed a series of 49 newly diagnosed NHL patients treated with dose-escalated CHOP and intensification prior to high dose therapy/autologous stem cell transplantation, by means of a comprehensive array-CGH analysis.

2. Materials

A total of 49 NHL were studied with BAC arrays. Most of them (45 patients) had a DLBCL, three were NHL-T and the remaining case was an aggressive Follicular Lymphoma. All samples were classified according to the WHO classification. All the patients received dose escalated CHOP and were evaluated as previously described (10, 35, 37–39). All patients had aggressive lymphomas with an age adjusted IPI > 1 or IPI = 1 plus high β2 microglobulin (β2M). Transformed lymphomas were excluded. The study protocol was approved by local ethics committees and written informed consent was obtained. The main characteristics of these patients are shown in Table 2. The median age was 53 years (range 24–68). Most patients were in advanced clinical stages, showing B-symptoms and high LDH levels. The median follow-up was 39 months.

3. Methods

3.1. DNA Isolation

In this study it was possible to obtain optimal quality DNA for hybridization on array-CGH of all patients included in the study. Genomic DNA was extracted from fresh-frozen samples (24 cases) and formalin-fixed/paraffin-embedded (FFPE) samples (25 patients). For extraction of DNA from FFPE samples tissue sections of 4 μm thickness were used. These sections were deparaffinized three times in 1 ml xylene and three times in 1 ml 100% ethanol for 10 min each. After air-drying, samples were suspended in 200 μl of DNA extraction buffer, composed of 180 μl of ATL buffer and 20 μl of proteinase K (both QIAamp DNA Mini kit by Qiagen), and were incubated at 55 °C overnight. After incubation, the DNA was extracted according to the manufacturer's recommendations. DNA from fresh-frozen samples was extracted using the standard phenol–chloroform method (40).

Table 1
Summary of the most relevant genomic imbalances in DLBCL

Authors (year) (Ref)	Number of samples	Type of CGH-array	Genomic changes: Regions gains		Regions losses
Wessendorf (2003) (27)	13	496 BAC/PAC	18q21		
			12q12-q24		
			9p13-q24		
			2p13		
Tagawa (2004) (26)	70	2.088 BAC/PAC	1q21.2-q32.2	8q24.13-q24.21	1p36.31-p36.32
			3p25.2-q29	11q23.2-q24.3	3p14.2
			5p13.1-p13.2	12q13.2-q21.2	6p14.1-q27
			6p21.1-p25.3	16p13.3	9p21
			7p22.2-q31.1	18	17p13.3-p11.2
Kameoka (2004) (10)	66	Array specific of 3p14.2 *(FHIT)*			3p14.2
Kasugai (2005) (39)	17	Array specific of 6p21 (*CCND3*)	6p21		
Tagawa (2005) (35)	99	2.304 BAC/PAC		*ABC*	
			3p23-q28		6p22.31-q24.1

(continued)

Table 1 (continued)

Authors (year) (Ref)	Number of samples	Type of CGH-array	Genomic changes Regions gains	Regions losses
			18q11.2-q23	9p21.3
			19q13.41-q13.43	
			CG	
			1q21.1-q23.3 7q22.1-q36.2	
			1q31.1-q42.13 12q13.1-q14	
			2p15-p16.1	
Chen (2006) (37)	64	1.552 BAC/PAC	2	1
		2.621 BAC	20	10q21.1
Diertamm (2008) (38)	116	2.799 BAC/PAC	18q21	

Table 2
Clinical characteristics of patients with high risk NHL at diagnosis

Variable	No of patients (n=49)
Age	53 years (24–68)[a]
Sex (F)	25
Ann Arbor stage III–IV	86%
B-symptoms	55%
2 extranodal sites	35%
BM involvement	25%
Bulky mass	31%
ECOG PS >1	47%
High LDH	76%
High β2M	49%

F female, *BM* bone marrow, *ECOG PS* eastern cooperative oncology group performance status, *LDH* lactate dehydrogenase; *β2M* beta-2-microglobulin
[a]Median (range)

Normal DNA was prepared from the placentas of healthy donors (41). All DNA, from fresh-frozen and FFPE tissue, was quantified using the Nanodrop spectrophotometer (ND-1000, NanoDrop Technologies, Wilmington, DE, USA). DNA quality was assessed by ratio of the absorbance at 260 and 280 nm (*A*260/280) in an ultraviolet spectrophotometer and its integrity by agarose gel ethidium bromide visualization (42).

3.2. BAC Array-CGH Fabrication

BAC array-CGH allows the analysis of all possible genomic changes detected in the total DNA samples from patients enrolled in the study (41, 43). Our slides containing 3,528 BACs spanning the genome were produced at the "Centro de Investigación del Cáncer" (Salamanca, Spain) (41, 43). The particular bacterial artificial chromosome (BAC) and P-1 derived artificial chromosome (PAC) set used to produce this array was distributed to academic institutions by the Wellcome Trust Sanger Institute (Cambridge, UK) and contains targets spaced at 1 Mb density over a full genome, a set of subtelomeric sequences for each chromosome arm, and a few hundred probes selected for their involvement in oncogenesis. The clone content is available in the "Cytoview" windows of the Sanger Institute mapping database site (http://www.ensembl.org/) and in the UCSC Genome Browser (http://genome.ucsc.edu/). According to these databases, clones were ordered along the chromosomes.

These clones were isolated from their bacterial cultures and the DNA was extracted using the standard protocol of the Wellcome Trust Sanger Institute (Cambridge, UK). Ten nanograms of DNA (BAC/PAC) were used as a template for three degenerate oligonucleotide primers PCR (DOP-PCR) (44). These products were ethanol precipitated and dissolved in an aqueous DMSO buffer as a spotting solution (45). For the spotting of BAC clones on microarrays, glass has become the most widely used substrate in view of the combination of desirable properties it possesses: rigidity, non-porous, low background, low fluorescence, suitability for chemical modification to allow different surface chemistries, resistance to heat and impact. Table 3 indicates the types of surface chemistries possible and what type of nucleic acid (PCR product/larger DNA fragments or oligonucleotides) is most effectively retained by the surface (46). In our BAC array-CGH a minimum of three replicates per clone were printed by *Microgrid II* (Biorobotics) on each slide (*Ultragaps Coated Slides*, Corning). This strategy results in the loss of the least number of BAC clones analyzed in the BAC array-CGH. During spotting the failure of pins can produce an irregular morphology and position of the spot. Inadequate humidity can cause spots in the form of donuts while a lower efficiency DOP-PCR can result a dark spots (47, 48). After the slides for BAC array-CGH have been spotted, they must be stored in a dark and dry environment.

For the microarray hybridization the first step was the labelling of the samples, test and control DNA. Briefly, to test for labelling reactions, 2 μg of nonamplified genomic DNA, test (DNA of tumor) and 2 μg of reference material (DNA of control) were digested separately with DpnII restriction enzyme (New England Biolabs, Beverly, MA). For samples embedded in paraffin DNA digestion with the DpnII restriction enzyme is not recommended. The DNA extracted from paraffin-embedded samples was degraded and had the appropriate size for label with fluorescent nucleotides (Cy3 and Cy5) (49). The digested DNA was separately labelled using random primers (Bioprimer labelling kit, Invitrogen) and Cy3-dCTP and Cy5-dCTP fluorescent dye for paired hybridization samples (Amersham Biosciences). The labelled nucleotide incorporation was quantified using a NanoDrop spectrophotometer (ND-1000). Equal amounts of labelled test (Cy5) and reference DNA (Cy3) were combined with 100 μg Cot-1 Human DNA (Roche, Indianapolis, IN) followed by an ethanol precipitation. Optimization of the amount of Human Cot-1 DNA (1 μg/μl) is important as both excess and insufficient amounts of Cot-1 Human DNA increased the standard deviation (SD) in the analysis of array-CGH (50). Resuspension of the pellet was done in hybridization solution (50% Formamide, 10% Dextran sulfate, 2× standard saline citrate, 10 mM Tris–HCL pH 7.6, 2.7% sodium dodecyl sulfate and 10 μg/μl of yeast tRNA). DNA mixtures were co-hybridized

Table 3
Characteristics of the slides used in BAC CGH-arrays

Company	Product	Surface chemistry	Optimized use
Amersham Biosciences	CodeLink activated slides	Proprietary	PCR products/ oligonucleotides
Apogent Discoveries	EZ-rays aminosilane	Aminosilane	PCR product
	EZ-rays Pure	Proprietary	Acrydite modified oligonucleotides
Asper Biotech Ltd	Genorama	Aminosilane	Unmodified DNA
		Aminosilane with linker	Short amino modified DNA
Bio Express	EasySpot universal	Proprietary	PCR product/ oligonucleotides
	EasySpot oligo	Epoxy mixture	Oligonucleotides
Bioslide Technologies	Precision CT-clean	None	–
	Precision CT-amine	Amine covalently bound to glass	PCT product/ oligonucleotides
	Precision CT-aldehyde	Aldehyde group covalently bound to glass	Amino modified PCR product/ oligonucleotides
	Precision CT-epoxy	Epoxy coated	Amino modified DNA
Coming	Gaps II	Aminosilane	PCR product
	UltraGaps	Aminosilane	Oligonucleotides
Eppendorf	Creative chip PCR	Proprietary	PCR product
	Creative chip oligo	Proprietary	Oligonucleotides
Erie Scientific	UltraClean	None	–
	SuperChip	Aminopropylsilane	PCR product/ oligonucleotides
Full Moon Biosystems	cDNA slides	Proprietary	PCR product
	Oligo slides	Proprietary	Oligonucleotides
GeneScan	ArrayLink	Epoxy silane	Amino modified DNA
Genetix	Amine coated slides	Amine coated	PCR product/ oligonucleotide
	Aldehyde coated slides	Aldehyde coated	Amino modified PCR product and oligonucleotides
PerkinElmer Life Sciences	Micromax SuperChip 1	Aminosilane	PCR product

(continued)

Table 3 (continued)

Company	Product	Surface chemistry	Optimized use
	UltraClean I	None	–
Quantifoil	QMT-epoxy	Epoxy coated	Amino modified DNA
	QMT-aldehyde	Aldehyde coated	Amino modified DNA
	QMT-amino slides	Amino coated	PCR product
Schleicher and Schuell	Fast slides	Nitrocellulose on glass	PCR product/ oligonucleotides
Telechem	Superamine	Amine covalently bound to glass	PCR product/ oligonucleotides
			PCR product/ oligonucleotides
	Superaldehyde	Aldehyde group covalently bound to glass	Amino modified PCR product and oligonucleotides
	Superclean	None	–

to the arrays in the GENETAC hybridization station (Genomic Solutions) for 48 h at 42 °C according to the manufacturer's recommended protocol.

3.3. Data Analysis and Interpretation

After hybridization and washing of BAC array-CGH co-hybridization with test DNA and reference DNA, images and signal intensities were acquired using the GenePix 4000B (Axon Instruments, Burlingame, CA, USA) dual laser scanner in combination with the GenePix Pro 4.0 (Axon Instruments) imaging software (37, 41, 43). For analyzed data obtained from GenePix Pro 4.0, the Cy5/Cy3 intensity ratios of every spot were converted into $\log_2$ ratios. The $\log_2$ ratio of each clone was normalized to the median $\log_2$ ratio of the 17 control hybridizations, after which the median of triplicate spots was calculated. Data from two-color hybridizations for both sets of DNA was normalized using the Diagnosis and Normalization of MicroArray Data (DNMAD) module of the Gene Expression Profile Analysis Suite (GEPAS) software (http://www.gepas.org) (51–53).

There were several types of normalization to the data obtained of the hybridizations BAC array-CGH (Table 4). Some methods of normalization were within-slide locations such us global normalization, intensity dependent normalization or within-print-tip-group normalization. Global methods assumed that the red and green intensities were related by a constant factor. A common choice for the location parameter $c = \log_2 k$ was the median or mean of the log-

Table 4
Normalization methods used in the analysis of BAC CGH-arrays

Name of normalization	Description
None	No normalization
Median	Global or median normalization
Lowess	Global normalization dependent on A
Within-print-tip-group normalization	Normalization dependent on A, within print-tip groups with lowess
Two D	Spatial normalization lowess
Scale print tip MAD	Normalization dependent on A, within print-tip groups with lowess followed of scale print-tip within-slide with the MAD function

intensity ratios for a particular gene set. Global normalization methods were mentioned as preprocessing steps in a number of papers on the identification of differentially expressed genes in a single-slide array-CGH experiment. In one of the first treatments of the problem, Chen et al. (54) assumed that $R = k \times G$ and proposed an iterative method for estimating the constant normalization factor k and cut-offs for the red and green intensity ratio R/G. In some software packages (GenePix) a constant normalization factor was estimated such that the arithmetic mean of the intensity ratios of all genes on a given microarray was one. This method is still the most widely used methods in spite of the evidence of spatial or intensity dependent dye biases in numerous experiments (55–57). Other type of normalization was intensity dependent normalization. In this method, the dye bias appears to be dependent on spot intensity, as revealed by plots of the log-ratio M versus overall spot intensity A. An intensity or A-dependent dye normalization method may thus be preferable to global methods. This method was employed in this study with GEPAS software. And finally, the within-print-tip-group normalization was a method which every grid in an array was printed using the same print-tip. Different experiments may be done using different printing set-up depending on the layout of the tips in the print-head of the arrayer. Some systematic differences existed between the print tips, such as slight differences in the length or in the opening of the tips, and deformation after many hours of printing. Alternatively, print-tip groups were proxies for spatial effects on the slide. Within-print-tip-group normalization was simply a *print-tip + A*-dependent normalization. Furthermore other methods of normalization as within-slides scale have been described. In this normalization, after within-print-tip-group normalization, all the normalized log-ratios from the different print-tip groups were centered around zero. However, it was possible that the log-ratios from the various print-tip groups had different spreads and some scale adjustment was required (47, 48, 55–59).

In order to interpret the data, regions of copy number gain and loss for the BAC array-CGH were identified by creating sample specific thresholds (60). The clones with $\log_2$ ratios above or below a control sample's threshold value were considered as gains or losses, respectively. At least two contiguous BAC clones with a $\log_2$ ratio of −0.4 or less was defined as a lost region and $\log_2$ ratio of +0.4 or more was defined as a gained region. For the paraffin samples, tissue from normal controls with normal DNA was used. Following the strategy used with the fresh-frozen samples, at least two contiguous BAC clones with a $\log_2$ ratio of −0.7 or less was defined as a lost region and $\log_2$ ratio of +0.7 or more was defined as a gained region. Furthermore, spots with weak Cy3 or Cy5 intensity (below $R^2 < 0.2$) were excluded from analysis. In addition, clones with a standard deviation of more than 0.3 in the triplicate spots were also excluded. In total, approximately 10% of clones were excluded. All data sets were carefully reviewed for frequently affected chromosomal sites of normal copy number variations (CNV) or polymorphisms. CNVs can be identified on arrays as those clones for which the intensity ratios of independent experiments repeatedly fall outside the above-defined cut-off level. Therefore the identification of these polymorphic clones is an essential first step toward proper interpretation of array-CGH data. When abnormal values are obtained in a number of separate experiments either using DNA from individuals without an obvious clinical phenotypic or in different experiments with DNA from individuals with very different clinical phenotypes, it can be assumed that these clones are polymorphic (50). Therefore every clone on the array was compared with the "Database of Genomic Variants" (January 2011) (www.project.tcag.ca/cariation/) and that of chromosomal imbalances and phenotype in humans using Ensembl Resources (DECIPHER: http://www.sanger.ac.uk/PostGenomics/decipher/) (60–62).

3.4. Statistical Analysis

The relationship between clinical or biological data and genomic characteristics was analyzed using an independent sample *t*-test. The overall survival (OS) was calculated as the time from the start of treatment to death or the last follow-up. Survival and response duration were analyzed using the Kaplan–Meier method (63) and compared with the log-rank test (64). Multivariate analysis with the variables that proved to be significant in the univariate analysis was performed according to the Cox proportional hazard regression model (65). All *p*-values reported were two-sided and statistical significance was defined as p-values <0.05. Statistical evaluation was carried out using the SPSS statistical software.

4. Results

4.1. Analysis of 49 NHL Specimens

After making the BAC arrays and obtaining the threshold for the genomics gains and losses for control cases, we proceeded to hybridization of NHL samples on BAC arrays. In 43 out of the 49 patients genomic changes were observed (88%). The most frequent such changes were chromosomal gains detected on 3p21, 11q12-q13.1, 11q23.3 (43%), 6p21 (41%), 1q21-q22 (39%), 1q32, 12q13 (31%), 1q23.3 (29%), 16q23-ter, 7p22.1 (27%), 7q22.1 (25%), 1p36, 9q34, 12q24, 20qcent-q13 (22%), 1p35-34 (20%), 2p16 (16%), while the genomic losses were located on 5q21.1 (29%), 8q23.1, 11q14.3 (20%), 3q13.31-32, 17p13 (18%), 1p31.1 (16%). Figure 1 shows all the changes detected by BAC array-CGH.

All patients included in this study were uniformly treated with dose escalating CHOP. Therefore the correlation of the clinical and biological characteristics with genomics changes could be ascertained. It should be noted that patients displaying loss on 1p35-p34 and 17p13.1 had high levels of LDH ($p<0.05$) while patients with advanced Ann-Arbor stage (III–IV) had gains on 1p36, 1q21-q22, 1q32, 3p25.3, 3p21.1, 6p21.1, 7p22.1, 16p11.2-pcent, 16q23-qter, 20qcent-q13, 22q, as well as losses on 1p35-p34, 5q21.1, 10q23.1 (*PTEN*) ($p<0.05$). In addition, the presence of bulky disease correlated with gains on 3p21, 11q23.3, 12q14.1, 12q24 ($p<0.05$) while the patients with less than 60 years had gains on 6p21, 11q12.2-q13.1, 11q23, 12q13, 12q14.1, 16q23-qter and losses on 11q14.3 ($p<0.005$). Table 5 illustrates the correlation between gains and losses assessed by BAC array-CGH and the most relevant clinical and biological characteristics of DLBCL.

In addition, DNA copy number changes were associated with the response to treatment with dose escalating CHOP. After three

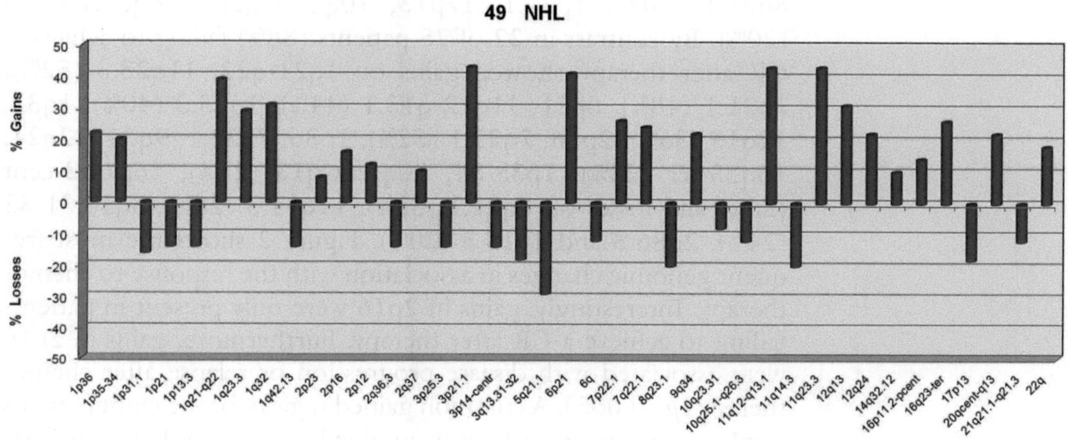

Fig. 1. Common regions of genomic gains or losses in 49 patients analyzed by BAC array-CGHs. The tree shows the genomic regions that exhibited gains (*up*) or losses (*down*). Y axis shows the frequency of change for each region.

Table 5
Correlations between genomics imbalances assessed by CGH-array and the more salient clinical and biological characteristics in high risk NHL

Clinical or biological characteristics		Genomics regions				
Ann-Arbor stage III–IV	Gains	1q21-q22	1q32	3p25.3	3p21.1	6p21.1
		7p22.1	16p11.2	16q23	20q	22
	Losses	1p35-p34	5q21.1	10q23.3		
Age (<60 year)	Gains	6p21	11q12.2-13.1	16q23	11q23	12q13
		12q14.1	12q24			
	Losses	11q14.3				
B-symptom	Gains	12q13				
	Losses	2p12	2q36.3	11q14.3		
Bulky mass	Gains	3p21	11q23.3	12q14.1	12q24	
↑LDH	Losses	1p35-p34	17p13.3			

p-value < 0.05 for all the correlations, *LDH* lactate dehydrogenase

MegaCHOP courses 41% of the patients (20 out of 49) were in CR, 31% in PR, while the remaining ten patients did not respond. The changes more frequent observed in 17 of 20 (85%) patients reached CR after MegaCHOP were: gains in chromosomes 11q23.3, 3p21.1, 6p21, 11q12-q13.1 (40%), 1q32, 7p22.1 (30%), 12q13, 16q23-ter (25%), 1q21-q22, 1q23.3, 7q22.1, 1p36, 12q24, 1p35-34, 20qcent-q13, 22q (20%) and losses on 5q21.1, 8q23.1 (30%), 1p31.1, 17p13, 10q25.1-q26.3, 6q, 1q42.13 (20%). By contrast in 22 of 25 patients (88%) failing to achieve a CR after therapy showed gains on 1q21-q22, 11q23.3 (52%), 3p21.1 (48%), 6p21, 11q12-q13.1 (44%), 1q23.3 (40%), 1q32, 12q13 (36%), 2p16, 7q22.1 (32%), 1p36, 7p22.1, 9q34, 12q24, 16q23-ter (28%), 1p35-34, 20qcent-q13 (24%), 16p11.2-cent (20%) and losses on 5q21.1 (32%), 11q14.3 (28%), 3q13.31-32 (24%), 2q36.3 and 1p13.3 (20%). Figure 2 shows the most frequent genomic changes in association with the response to chemotherapy. Interestingly, gains in 2p16 were only present in patients failing to achieve a CR after therapy. Furthermore, gains in 2p16 were associated with disease progression or relapse after chemotherapy ($p = 0.005$). A common gained region for the eight patients displaying gains on 2p16 was confined to 634 kb, between BAC clones bA440P5 and bA373L24. This region contains *REL* and

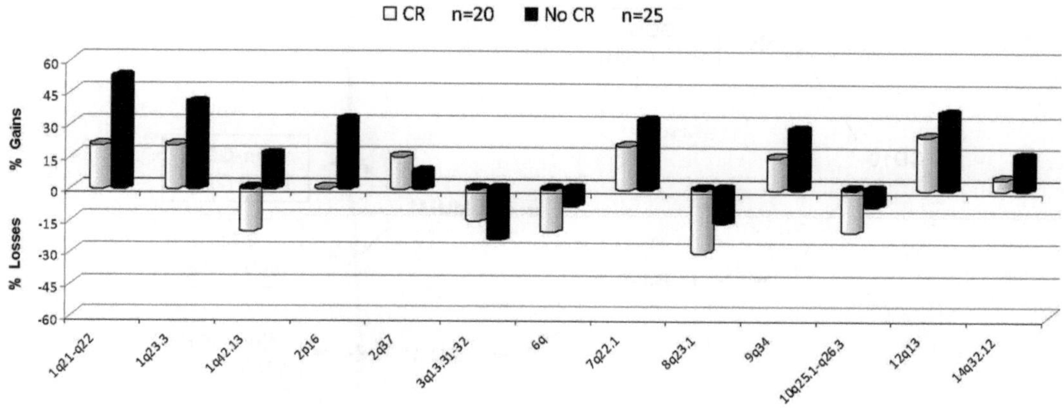

Fig. 2. Most common genomic imbalances in DLBCL in relation to the response to the therapy with dose escalating CHOP. The frequency of gains (*up*) and losses (*down*) of patients achieving a complete remission (CR) after MegaCHOP are shown in white bars while non-responding patients (No CR) are shown in black bars.

BCL11A as candidate genes. In addition, six of eight patients showing a loss in 17p13.1 (containing *TP53* gene) succumbed to disease after chemotherapy ($p=0.042$). In the univariate analysis only gains on 2p16 and losses on 17p13.1 were related to the failure of response to high-dose CHOP. Both variables (gains on 2p16 and losses on 17p13.1) were selected in the regression model for response to therapy.

4.2. Correlation Between Genomic Changes and Survival in DLBCL

Regarding clinical variables, only age (more than 60 years) correlated with a short survival whereas the IPI had no impact on survival. Of note, patients showing losses in either 10q23.31 or in 17p13.1 ($p=0.04$ and $p=0.03$, respectively) also had a short survival (median survival of 10 months for loss of 10q23.31 and not reached for those with normal 10q23.31; and a median survival of 11 months in patients showing loss in 17p13.1 and median survival not reached for those patients with normal 17p13.1). No other data correlated with survival. The Cox regression model showed that only age ($p=0.01$) and loss in 10q23.31 ($p=0.03$) correlated with a short survival.

4.3. Genomics Changes in Germinal Center B-Cell Like and Non-germinal Center B-Cell Like Detected by Array-CGH

All the DLBCL included in the present study were classified by the Hans algorithm into CGB (Germinal Center B-cell like) and non-CGB (non-Germinal Center B-cell like) types depending on the expression of CD10, Bcl-6 and MUM1 inmunophenotype markers (66). Forty of the forty-five patients with DLBCL included in the study were classified as CGB ($n=18$) and the remaining 22 patients were classified as non-CGB (Fig. 3). Seventeen out of eighteen patients (94%) classify in CGB showed genomics alterations by BAC array-CGH. The gains affected 3p21.1, 6p21.1, 11q12.2-13.1 (56%), 11q23.3 (50%), 1q21-q22, 1q32, 7p22.1,

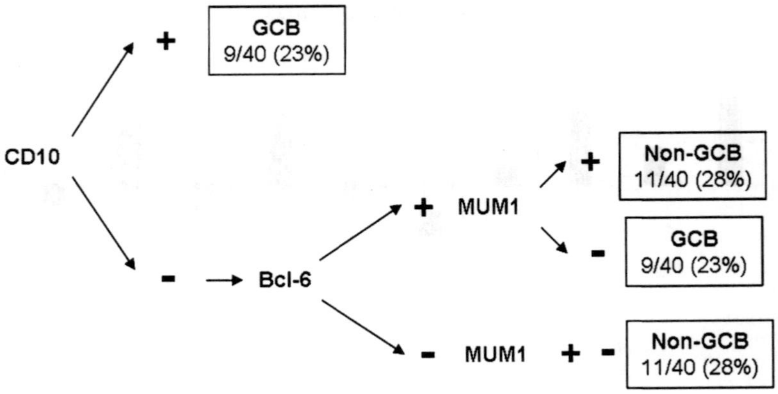

Fig. 3. Summary of the DLBCL including in the study, classified according to Hans algorithm. Most of the patients showed BCL6, but lack in the expression of CD10.

12q13 (44%), 1p36, 1q23.3, 7q22.1, 20qcent-q13 (39%), 2p16.1, 12q24, 16q23-qter, 17p13.2 (33%), while losses involved 5q21.1 (33%). In addition, 20 out of 22 patients (91%) classified in non-CGB showed genomics changes detected using array-CGH. The more frequent gains were located on 11q23.3 (41%), 1q32, 3p21.1 (32%), 1q21-q22, 1q23.3, 6p21.1, 11q12.2-13.1 (27%) and the losses on 5q21.1 and 17p13.1 (27%). The losses on 10q23.3 (*PTEN* gene) were less frequent in GCB (6%) subgroups than in non-GCB subgroup (14%), although this differences did not reach statistical significance. Between both subgroups of DLBCL (GCB and non-GCB) only the gains on 2p16 (*BCL11A* and *REL* genes) were associated with GCB DLBCL ($p = 0.004$) (Table 6).

5. Discussion

Chemotherapy cures more than a half of the patients with aggressive B-cell non-Hodgkin's lymphoma (NHL). However patients showing adverse prognostic factors have a dismal prognosis. Therefore high-dose chemotherapy with autologous stem cell support (HDT/ASCT) has been widely used in these patients (67). Recently, The Spanish GELTAMO Group has reported the results concerning 86 patients with poor prognosis DLBCL treated with escalated CHOP and tailored intensification with IFE prior BEAM/ASCT. This strategy resulted in an overall PFS and OS at 5 years of 56 and 64%, respectively (68). Therefore, new predictors such as biological markers are needed in order to better identify this poor prognostic group of patients. In the present study we report the results of a BAC array-CGH analysis in a series of 49 patients. Our results show a relationship between genomic imbalances and the outcome of the patients.

Table 6
Genomic imbalances detected in DLBCL in relation to gene expression profile group

% Gains						% Losses					
Regions	GC (%)	$n=18$	Non-GC (%)	$n=22$	p	Regions	GC (%)	$n=18$	Non-GC (%)	$n=22$	p
1p36	39	(7/18)	14	(3/22)	0.1	1p35-p34	6	(1/18)	23	(5/22)	0.2
1p35-p34	28	(5/18)	14	(3/22)	0.2	1p31.1	22	(4/18)	14	(3/22)	0.7
1q21-q22	44	(8/18)	27	(6/22)	0.5	1p21	22	(4/18)	14	(3/22)	0.7
1q23.3	39	(7/18)	27	(6/22)	0.7	1p13.3	17	(3/18)	18	(4/22)	0.9
1q32	44	(7/18)	3	(7/22)	0.6	1q42.13	11	(2/18)	18	(4/22)	0.8
1q42.13	17	(3/18)	14	(3/22)	0.8	2q36.3	28	(5/18)	5	(1/22)	0.1
2p23	0	(0/18)	14	(3/22)	0.1	3p14-cent	22	(4/18)	14	(3/22)	0.7
2p16	33	(6/18)	0	(0/22)	**0.004***	3q13.31-q32	27	(5/18)	9	(2/22)	0.2
2q37	17	(3/18)	9	(2/22)	0.7	5q21.1	33	(6/18)	27	(6/22)	0.6
3p25.3	22	(4/18)	9	(2/22)	0.4	6q	11	(2/18)	18	(4/22)	0.6
3p21.1	56	(10/18)	32	(7/22)	0.2	8q23.1	17	(3/18)	23	(5/22)	0.4
6p21.1	56	(10/18)	27	(5/22)	0.1	10q23.1	6	(1/18)	14	(3/22)	0.6
7p22.1	44	(8/18)	18	(4/22)	0.1	10q25.1-q26.3	6	(1/18)	23	(5/22)	0.1
7q22.1	39	(7/18)	23	(5/22)	0.2	11q14.3	28	(5/18)	18	(4/22)	0.3
9p21.3	0	(0/18)	9	(2/22)	0.2	17p13.2	11	(2/18)	27	(6/22)	0.1
9q34-ter	28	(5/18)	18	(4/22)	0.7	17p13.1	17	(3/18)	27	(6/22)	0.5
11q12.2-q13.1	56	(10/18)	27	(6/22)	0.1	20qcent-q13	6	(1/18)	14	(3/22)	0.07

(continued)

Table 6 (continued)

% Gains						% Losses					
Regions	GC (%)	$n=18$	Non-GC (%)	$n=22$	p	Regions	GC (%)	$n=18$	Non-GC (%)	$n=22$	p
11q23.3	50	(9/18)	41	(9/22)	0.8	21q21.1-q21.3	22	(4/18)	5	(1/22)	0.2
12q13	44	(8/18)	14	(3/22)	0.08						
12q24	33	(6/18)	14	(3/22)	0.1						
14q32.12	17	(3/18)	5	(1/22)	0.4						
16p11.2-cent	22	(4/18)	9	(2/22)	0.4						
16q23-qter	33	(6/18)	18	(4/22)	0.5						
17p13.2	33	(6/18)	9	(2/22)	0.1						
17p13.1	17	(3/18)	14	(1/22)	0.5						
20qcent-q13	39	(7/18)	9	(2/22)	0.07						
22q	28	(5/18)	9	(2/22)	0.2						

*Statistically significant

We have confirmed that the regions more frequently affected in patients with DLBCL are 7p22, 11q23, and 3p25, all of which show gains, as well regions 1p31, 6q, and 17p13 which show losses (35, 37). By contrast, in the present study we have observed that the presence of losses in 5q21, 8q23, 9p21, and 11q14 as well as gains on chromosome 18q are common in DBLCL, more than usually reported (35, 37, 38, 69). These differences could be related to the high risk characteristics of the patients in our series. To date, array-CGH studies on NHL have focused mainly on target gene identification rather than upon the clinical correlations (35, 37). Furthermore, the previous series did not included patients who were prospectively and homogeneously treated.

Our study confirmed the previous results from our group (41). Thus we demonstrated that the presence of genomic imbalances correlate with the prognosis in DBLCL: the presence of gains in 2p16 were most frequent in patients failing to achieve a CR. Interestingly, a gain in chromosome 2p has been identified as a recurrent alteration in 20% of DLBCL (36, 70–76), and is the most frequent CGH alteration in classical Hodgkin's lymphoma found in 50% of cases (77, 78). The common recurrent region expands 4.2 Mb in 2p15-16 (77, 78). By using BAC array-CGH, we were able to restrict the minimal common region to 700 kb in length. Two oncogenes, both involved in lymphomagenesis, *REL* and *BCL11A*, are located in this region (36, 79).In fact, overexpression of both genes has been found in DLBCL cases with 2p amplification by means of RQ-PCR (36). In the present study, we were able to identify the minimally targeted genomic regions of 2p16.1 amplification, where *REL*, and not *BCL11A*, was located as the only candidate oncogene. Similar data has been recently reported by array-CGH with the conclusion that the *REL* gene may play a more important role than *BCL11A* in the pathogenesis of DLBCL (80). Furthermore, it has been reported that a high copy number of *REL* correlates with extranodal disease in DLBCL (70). In our multivariate analysis, only the presence of gains in 2p16.1 was associated with MegaCHOP resistant disease, suggesting that this genetic abnormality might be included as a prognostic factor in DLBCL.

Regarding the correlation between genomic changes and outcome, the univariate analysis showed that losses of chromosomes 10q23.31 (9%; $p=0.03$) containing *PTEN* gene, 17p13.1 (18%; $p=0.03$) covering *TP53* gene, and the advanced age ($p=0.02$) were associated with inferior survival. Indeed, in the Cox's proportional hazard model, losses in 10q23.23 ($p=0.03$) and age ($p=0.01$) remained as independent variables influencing overall survival. Recently, it has been reported that 37% of DLBCL cases have reduced *PTEN* expression, and this feature has been associated with short overall survival in a retrospective series of DLBCL (81).

In addition, the loss of nuclear *PTEN* correlated with poor survival germinal center NHL (GC) (82). This data could implicate the PTEN/PI3K/ATK pathway in the development and progression of DLBCL (81, 83). In addition, losses in region 17p13.1 were also associated with a lack of complete response to escalated CHOP and a trend toward a short survival. These results are in agreement with previous studies showing that the inactivation of *TP53* gene is associated with clinical progression, poor prognosis, drug resistance, and low rates of complete remission in patients with aggressive B-cell lymphoma (84).

In our series of 45 DLBCL distributed in GCB and non-GCB, the gains on 2p16 (*BCL11A* and *REL* genes) were associated with GCB DLBCL ($p=0.004$). These findings are in agreement with previous studies (75, 80). By contrast, losses on 10q23.3 (*PTEN* gene) were less frequent in GCB subgroup (6%) than in non-GCB subgroup (14%), although these differences were no significant. In addition, other genomic alterations were more frequent in GCB cases than in non-GCB patients, as gains on 1p36, 6p21.1, 7p22.1, 12q14.1, 12q24, 17p13.2, 20qcent-q13, 22q and losses on 2q36.2, 3q13.31-q13.32, and 21q21.1-21.3, but the differences between both subgroups did reach statistical significance. It should be noted that gains in the 17p13.2 region were associated with GCB DLBCL (6 of 18 patients) while losses in the same region were more frequent in non-GCB DLBCL (6 out of 22). In accordance with recent studies, gains on chromosomes 1q23.3, 7q22, and 12q are found more frequently (>35%) in GCB DLBCL subgroup as well as losses on the long arm of chromosome 6 and gains on 9p in the non-GCB group (18%) (75, 85). All these results suggest a relationship between the presence of genomic imbalances and the gene expression profile that should be analyzed in large series.

To summarize, our results show that the application of high-resolution scanning of the genome has contributed in identifying new regions associated with poor outcome in a group of patients with high-risk DLBCL, and may help in the stratification of this group of aggressive lymphomas. Accordingly, the study of 2p, 10q, and 17p regions should be performed in these patients in the clinical setting.

6. Notes

1. BAC array-CGH enables the identification of chromosomal copy number changes (41, 43, 44).
2. The analysis of the whole genome could be performed in a single experiment. In addition BAC array-CGH allow selected an interest genomic region and focused analysis (45).

3. For the hybridization BAC array-CGH do not need cell culture. With this methodology archival material, such as tissue embedded paraffin, can be used. This DNA not needs restriction enzyme digestion (8, 49).
4. The infiltration of tumor tissue is crucial in the analysis of the BAC array-CGH to avoid obtaining false results (50).
5. The quality and among of DNA is an important factor in the hybridization of BAC array-CGH. To obtain a good hybridization is necessary that the DNA would be optimal to avoid confusing the results with background noise (49, 50).
6. The resolution of the BAC array-CGH could be increase to 50–200 kb by adding more probes (42).
7. There are a variety of types of slides. The type of slide chosen depends on the material spotted and the optimal use of the slide (46).
8. The triplicate deposit of clone in different areas of slide is a critical step in order to avoid, in case of happen spotted errors, losing the fewest possible spots. Another problem to avoid is the appearance of bubbles. These points must be removed from the analysis, without losing data, because a clone must have at least two of the values for the triple spots to be analyzed (47, 48).
9. The normalization of the data is a critical issue that might be assessed with different methods. The method that should be used depends on the sample type (47, 48, 51–62).
10. The identification and the reporting of the CNV is an essential first step toward proper interpretation of array CGH data (50).

Acknowledgments

Partially supported by grants from Spanish Fondo de Investigaciones Sanitarias 09/01543 and Fondo Caja de Burgos de Investigación Clínica, Proyectos de investigación del SACYL 106/A/06 and by the "Acción Transversal del Cáncer" project, through an agreement between Instituto de Salud Carlos III (ISCIII), Spanish Ministry of Science and Innovation, and the Cancer Research Foundation of Salamanca University and Redes de Investigación RTIIC (FIS). We thank Dr Nigel Carter and Dr Heike Fiegler from the Sanger Center (Cambridge, UK) for providing the BAC library; and Eva García, Eva Lumbreras, Teresa Prieto, Irene Rodríguez, and Sara González from "Centro de Investigación del Cáncer, Salamanca" for their technical assistance.

References

1. The Non-Hodgkin's Lymphoma Classification Project (1997) A clinical evaluation of the International Lymphoma Study Group classification of non-Hodgkin's lymphoma. Blood 11:3909–3918
2. Harris NL et al (1994) A revised European-American classification of lymphoid neoplasms: a proposal from the International Lymphoma Study Group. Blood 5:1361–1392
3. Montes-Moreno S et al (2011) MicroRNA expression in diffuse large B-cell lymphoma treated with chemoimmunotherapy. Blood 118(4):1034–1040
4. Swerdlow SH et al (2008) WHO Classifica tion of Tumours of Haematopoietic and Lymphoid Tissues, Fourth Edition. WHO Classification of Tumours, Volume 2. IARC WHO Classification of Tumours, No 2
5. Lossos IS, Morgensztern D (2006) Prognostic biomarkers in diffuse large B-cell lymphoma. J Clin Oncol 6:995–1007
6. Shipp MA, The International Non-Hodgkin's Lymphoma Prognostic Factors Project (1993) A predictive model for aggressive non-Hodgkin's lymphoma. N Engl J Med 14:987–994
7. van Imhoff GW et al (2006) Prognostic impact of germinal center-associated proteins and chromosomal breakpoints in poor-risk diffuse large B-cell lymphoma. J Clin Oncol 25:4135–4142
8. Kluin P, Schuuring E (2011) Molecular cytogenetics of lymphoma: where do we stand in 2010? Histopathology 1:128–144
9. Chaganti RS et al (2000) Recurring chromosomal abnormalities in non-Hodgkin's lymphoma: biologic and clinical significance. Semin Hematol 4:396–411
10. Kameoka Y et al (2004) Contig array CGH at 3p14.2 points to the FRA3B/FHIT common fragile region as the target gene in diffuse large B-cell lymphoma. Oncogene 56:9148–9154
11. Skinnider BF et al (1999) Bcl-6 and Bcl-2 protein expression in diffuse large B-cell lymphoma and follicular lymphoma: correlation with 3q27 and 18q21 chromosomal abnormalities. Hum Pathol 7:803–808
12. Barrans S et al (2010) Rearrangement of MYC is associated with poor prognosis in patients with diffuse large B-cell lymphoma treated in the era of rituximab. J Clin Oncol 20:3360–3365
13. Carbone A et al (2010) B-cell lymphomas with features intermediate between distinct pathologic entities. From pathogenesis to pathology. Hum Pathol 5:621–631
14. Hasserjian RP et al (2009) Commentary on the WHO classification of tumors of lymphoid tissues (2008): "Gray zone" lymphomas overlapping with Burkitt lymphoma or classical Hodgkin lymphoma. J Hematop 2(2):89–95
15. Johnson NA et al (2009) Lymphomas with concurrent BCL2 and MYC translocations: the critical factors associated with survival. Blood 11:2273–2279
16. Tomita N et al (2009) Clinicopathological features of lymphoma/leukemia patients carrying both BCL2 and MYC translocations. Haematologica 7:935–943
17. Le GS et al (2007) The clinical presentation and prognosis of diffuse large B-cell lymphoma with t(14;18) and 8q24/c-MYC rearrangement. Haematologica 10:1335–1342
18. Ventura RA et al (2006) FISH analysis for the detection of lymphoma-associated chromosomal abnormalities in routine paraffin-embedded tissue. J Mol Diagn 2:141–151
19. Boerma EG et al (2009) Translocations involving 8q24 in Burkitt lymphoma and other malignant lymphomas: a historical review of cytogenetics in the light of today's knowledge. Leukemia 2:225–234
20. Hummel M et al (2006) A biologic definition of Burkitt's lymphoma from transcriptional and genomic profiling. N Engl J Med 23:2419–2430
21. Salaverria I et al (2008) Chromosomal alterations detected by comparative genomic hybridization in subgroups of gene expression-defined Burkitt's lymphoma. Haematologica 9:1327–1334
22. Seegmiller AC et al (2010) Simple karyotype and bcl-6 expression predict a diagnosis of Burkitt lymphoma and better survival in IG-MYC rearranged high-grade B-cell lymphomas. Mod Pathol 7:909–920
23. Albertson DG et al (2000) Quantitative mapping of amplicon structure by array CGH identifies CYP24 as a candidate oncogene. Nat Genet 2:144–146
24. Ota A et al (2004) Identification and characterization of a novel gene, C13orf25, as a target for 13q31-q32 amplification in malignant lymphoma. Cancer Res 9:3087–3095
25. Ray ME et al (2004) Genomic and expression analysis of the 8p11-12 amplicon in human breast cancer cell lines. Cancer Res 1:40–47
26. Tagawa H et al (2004) Genome-wide array-based comparative genomic hybridization of diffuse large B-cell lymphoma: comparison between CD5-positive and CD5-negative cases. Cancer Res 17:5948–5955

27. Wessendorf S et al (2003) Hidden gene amplifications in aggressive B-cell non-Hodgkin lymphomas detected by microarray-based comparative genomic hybridization. Oncogene 9:1425–1429
28. Booman M et al (2008) Genomic alterations and gene expression in primary diffuse large B-cell lymphomas of immune-privileged sites: the importance of apoptosis and immunomodulatory pathways. J Pathol 2:209–217
29. de Leeuw RJ et al (2004) Comprehensive whole genome array CGH profiling of mantle cell lymphoma model genomes. Hum Mol Genet 17:1827–1837
30. Schraders M et al (2008) Integrated genomic and expression profiling in mantle cell lymphoma: identification of gene-dosage regulated candidate genes. Br J Haematol 2:210–221
31. Seto M et al (2010) Diversity of genome profiles in malignant lymphoma. Cancer Sci 3:573–578
32. Takeuchi I et al (2009) The potential of copy number gains and losses, detected by array-based comparative genomic hybridization, for computational differential diagnosis of B-cell lymphomas and genetic regions involved in lymphomagenesis. Haematologica 1:61–69
33. Thorns C et al (2007) Chromosomal aberrations in angioimmunoblastic T-cell lymphoma and peripheral T-cell lymphoma unspecified: a matrix-based CGH approach. Genes Chromosomes Cancer 1:37–44
34. Wessendorf S et al (2007) Further delineation of chromosomal consensus regions in primary mediastinal B-cell lymphomas: an analysis of 37 tumor samples using high-resolution genomic profiling (array-CGH). Leukemia 12:2463–2469
35. Tagawa H et al (2005) Comparison of genome profiles for identification of distinct subgroups of diffuse large B-cell lymphoma. Blood 5:1770–1777
36. Bea S et al (2004) Clinicopathologic significance and prognostic value of chromosomal imbalances in diffuse large B-cell lymphomas. J Clin Oncol 17:3498–3506
37. Chen W et al (2006) Array comparative genomic hybridization reveals genomic copy number changes associated with outcome in diffuse large B-cell lymphomas. Blood 6:2477–2485
38. Dierlamm J et al (2008) Gain of chromosome region 18q21 including the MALT1 gene is associated with the activated B-cell-like gene expression subtype and increased BCL2 gene dosage and protein expression in diffuse large B-cell lymphoma. Haematologica 5:688–696
39. Kasugai Y et al (2005) Identification of CCND3 and BYSL as candidate targets for the 6p21 amplification in diffuse large B-cell lymphoma. Clin Cancer Res 23:8265–8272
40. Hernandez JM et al (2001) Novel genomic imbalances in B-cell splenic marginal zone lymphomas revealed by comparative genomic hybridization and cytogenetics. Am J Pathol 5:1843–1850
41. Robledo C et al (2009) Array comparative genomic hybridization identifies genetic regions associated with outcome in aggressive diffuse large B-cell lymphomas. Cancer 16:3728–3737
42. Bredel M et al (2005) Amplification of whole tumor genomes and gene-by-gene mapping of genomic aberrations from limited sources of fresh-frozen and paraffin-embedded DNA. J Mol Diagn 2:171–182
43. Lopez-Villar O et al (2009) Both expanded and uncultured mesenchymal stem cells from MDS patients are genomically abnormal, showing a specific genetic profile for the 5q-syndrome. Leukemia 4:664–672
44. Fiegler H et al (2003) DNA microarrays for comparative genomic hybridization based on DOP-PCR amplification of BAC and PAC clones. Genes Chromosomes Cancer 4:361–374
45. Fiegler H et al (2007) Construction and use of spotted large-insert clone DNA microarrays for the detection of genomic copy number changes. Nat Protoc 3:577–587
46. Affara NA (2003) Resource and hardware options for microarray-based experimentation. Brief Funct Genomic Proteomic 1:7–20
47. Camps J et al (2006) Genomic imbalances and patterns of karyotypic variability in mantle-cell lymphoma cell lines. Leuk Res 8:923–934
48. Rubio-Moscardo F et al (2005) Mantle-cell lymphoma genotypes identified with CGH to BAC microarrays define a leukemic subgroup of disease and predict patient outcome. Blood 11:4445–4454
49. van Beers EH et al (2006) A multiplex PCR predictor for aCGH success of FFPE samples. Br J Cancer 2:333–337
50. Vermeesch JR et al (2005) Molecular karyotyping: array CGH quality criteria for constitutional genetic diagnosis. J Histochem Cytochem 3:413–422
51. Conde L et al (2007) ISACGH: a web-based environment for the analysis of array CGH and gene expression which includes functional profiling. Nucleic Acids Res 35(Web Server issue):W81–W85

52. Herrero J et al (2003) GEPAS: a web-based resource for microarray gene expression data analysis. Nucleic Acids Res 13:3461–3467
53. Montaner D et al (2006) Next station in microarray data analysis: GEPAS. Nucleic Acids Res 34(Web Server issue):W486–W491
54. Chen Y et al (1997) Ratio-base decisions and the quantitative analysis of cDNA microarrays images. J Biomed Opt 2:364–374
55. Kelmansky DM (2006) Normalización. Análisis Exploratorio y Confirmatorio de Datos de Experimentos de Microarrays. 2006:98–108
56. Lapuk A et al (2004) Computational BAC clone contig assembly for comprehensive genome analysis. Genes Chromosomes Cancer 1:66–71
57. Yang YH et al (2002) Normalization for cDNA microarray data: a robust composite method addressing single and multiple slide systematic variation. Nucleic Acids Res 4:e15
58. Bernardini M et al (2005) High-resolution mapping of genomic imbalance and identification of gene expression profiles associated with differential chemotherapy response in serous epithelial ovarian cancer. Neoplasia 6:603–613
59. Lambros MB et al (2005) Analysis of ovarian cancer cell lines using array-based comparative genomic hybridization. J Pathol 1:29–40
60. Pollack JR et al (1999) Genome-wide analysis of DNA copy-number changes using cDNA microarrays. Nat Genet 1:41–46
61. Sebat J et al (2004) Large-scale copy number polymorphism in the human genome. Science 5683:525–528
62. Iafrate AJ et al (2004) Detection of large-scale variation in the human genome. Nat Genet 9:949–951
63. Kaplan E, Meier P (1958) Nonparametric estimation from incomplete observations. J Am Stat Assoc 53:457–481
64. Peto R et al (1977) Desing and analysis of randomized clinical trials requiring prolonged observation of each patient. II. Analysis and Examples. Br J Cancer 35(1):1–39
65. Cox DR (1972) Regression models and life-tables. J Roy Statist Soc 34:187
66. Hans CP et al (2004) Confirmation of the molecular classification of diffuse large B-cell lymphoma by immunohistochemistry using a tissue microarray. Blood 1:275–282
67. Rodriguez J et al (2004) Autologous stem-cell transplantation in diffuse large B-cell non-Hodgkin's lymphoma not achieving complete response after induction chemotherapy: the GEL/TAMO experience. Ann Oncol 10:1504–1509
68. Arranz R et al (2008) Dose-escalated CHOP and tailored intensification with IFE according to early response and followed by BEAM/autologous stem-cell transplantation in poor-risk aggressive B-cell lymphoma: a prospective study from the GEL-TAMO Study Group. Eur J Haematol 80(3):227–235
69. Bentink S et al (2008) Pathway activation patterns in diffuse large B-cell lymphomas. Leukemia 22(9):1746–1754
70. Houldsworth J et al (1996) REL proto-oncogene is frequently amplified in extranodal diffuse large cell lymphoma. Blood 1:25–29
71. Rao PH et al (1998) Chromosomal and gene amplification in diffuse large B-cell lymphoma. Blood 1:234–240
72. Palanisamy N et al (2002) Similar patterns of genomic alterations characterize primary mediastinal large-B-cell lymphoma and diffuse large-B-cell lymphoma. Genes Chromosomes Cancer 2:114–122
73. Houldsworth J et al (2004) Relationship between REL amplification, REL function, and clinical and biologic features in diffuse large B-cell lymphomas. Blood 5:1862–1868
74. Feuerhake F et al (2005) NFkappaB activity, function, and target-gene signatures in primary mediastinal large B-cell lymphoma and diffuse large B-cell lymphoma subtypes. Blood 4:1392–1399
75. Bea S et al (2005) Diffuse large B-cell lymphoma subgroups have distinct genetic profiles that influence tumor biology and improve gene-expression-based survival prediction. Blood 9:3183–3190
76. Joos S et al (1996) Primary mediastinal (thymic) B-cell lymphoma is characterized by gains of chromosomal material including 9p and amplification of the REL gene. Blood 4:1571–1578
77. Barth TF et al (2003) Gains of 2p involving the REL locus correlate with nuclear c-Rel protein accumulation in neoplastic cells of classical Hodgkin lymphoma. Blood 9:3681–3686
78. Joos S et al (2002) Classical Hodgkin lymphoma is characterized by recurrent copy number gains of the short arm of chromosome 2. Blood 4:1381–1387
79. Satterwhite E et al (2001) The BCL11 gene family: involvement of BCL11A in lymphoid malignancies. Blood 12:3413–3420
80. Fukuhara N et al (2006) Characterization of target genes at the 2p15-16 amplicon in diffuse large B-cell lymphoma. Cancer Sci 6:499–504

81. Abubaker J et al (2007) PIK3CA mutations are mutually exclusive with PTEN loss in diffuse large B-cell lymphoma. Leukemia 11:2368–2370
82. Fridberg M et al (2007) Protein expression and cellular localization in two prognostic subgroups of diffuse large B-cell lymphoma: higher expression of ZAP70 and PKC-beta II in the non-germinal center group and poor survival in patients deficient in nuclear PTEN. Leuk Lymphoma 11:2221–2232
83. Uddin S et al (2006) Role of phosphatidylinositol 3 -kinase/AKT pathway in diffuse large B-cell lymphoma survival. Blood 13:4178–4186
84. Ichikawa A et al (1997) Mutations of the p53 gene as a prognostic factor in aggressive B-cell lymphoma. N Engl J Med 8:529–534
85. Campo E (2009) Insights from expression arrays and genomic profiling of aggressive B-cell lymphoma. Personal Comunication

Chapter 9

Genomic Profiling of Mantle Cell Lymphoma

Melissa Rieger Menanteau and Jose A. Martinez-Climent

Abstract

Genomic profiling of mantle cell lymphoma (MCL) cells has enabled a better understanding of the complex mechanisms underlying the pathogenesis of disease. Besides the t(11;14)(q13;q32) leading to cyclin D1 overexpression, MCL exhibits a characteristic pattern of DNA copy number aberrations that differs from those detected in other B-cell lymphomas. These genomic changes disrupt selected oncogenes and suppressor genes that are required for lymphoma development and progression, many of which are components of cell cycle, DNA damage response and repair, apoptosis, and cell-signaling pathways. Additionally, some of them may represent effective therapeutic targets. A number of genomic and molecular abnormalities have been correlated with the clinical outcome of patients with MCL and are considered prognostic factors. However, only a few genomic markers have been shown to predict the response to current or novel targeted therapies. One representative example is the high-level amplification of the *BCL2* gene, which predicts a good response to pro-apoptotic BH3 mimetic drugs. In summary, genomic analyses have contributed to the substantial advances made in the comprehension of the pathogenesis of MCL, providing a solid basis for the identification of optimal therapeutic targets and for the design of new molecular therapies aiming to cure this fatal disease.

Key words: Mantle cell lymphoma, Genomic abnormalities, Comparative genomic hybridization, Cyclin D1, Prognostic factors, Predictive factors of therapeutic response

1. Introduction

Mantle cell lymphoma (MCL) is a distinct B-cell non-Hodgkin's lymphoma that accounts for 6–8% of all human B-cell lymphomas. It is thought to derive from naïve pre-germinal center B lymphocytes localized in primary follicles or in the mantle region of secondary follicles (1). The disease is typically disseminated at presentation, with a leukemic component in the majority of patients. Histologically, two major MCL subsets can be distinguished that share a characteristic CD19+CD5+CD23- immunophenotype: the classical variant,

Diponkar Banerjee and Sohrab P. Shah (eds.), *Array Comparative Genomic Hybridization: Protocols and Applications*, Methods in Molecular Biology vol. 973, DOI 10.1007/978-1-62703-281-0_9, © Springer Science+Business Media, LLC 2013

composed of a monotonous proliferation of small- to medium-sized lymphocytes with irregular nuclei and relatively low proliferation activity, and the blastoid/pleomorphic variant with a more aggressive biological and clinical behavior. Additionally, some patients may follow an indolent clinical evolution (2–4). Nevertheless, MCL is considered one of the most aggressive lymphoid neoplasms with poor responses to conventional therapies including combined immunochemotherapy and radiotherapy and/or intensive high-dose chemotherapy with stem cell transplantation (5–8).

The genetic hallmark that characterizes MCL is the chromosomal translocation t(11;14)(q13;q32) (9). This translocation juxtaposes the *CCND1* gene at 11q13 with enhancers in the gene encoding for the immunoglobulin heavy chain (*IGH*) on chromosome 14, leading to the constitutive overexpression of *CCND1*, which codifies for cyclin D1, a D-type cyclin that is not expressed in normal B lymphocytes (10, 11). In over 95% of MCL cases, the t(11;14)(q13;q32) can be identified by G-banding cytogenetics or fluorescence in situ hybridization (FISH) analysis using probes for *CCND1* and *IGH* gene loci. Alternatively, MCL cells can show variant translocations such as t(2;11)(p11;13) and t(11;22)(q13;11) that fuse the *CCND1* gene with the kappa or lambda immunoglobulin chain light genes (*IGK* and *IGL*), respectively (12–15). Around 15–20% of MCL cases display complex chromosomal rearrangement including the amplification of the *CCND1-IGH* gene fusion (16). Tetraploid karyotypes are detected frequently in pleomorphic and blastoid variants of MCL (17). These cases contain a greater number of centrosome anomalies and overexpress centrosome-associated genes (18, 19). Besides the primary chromosomal rearrangements affecting the D-type cyclins, only a few secondary translocations have been identified in MCL. The most frequent involves *MYC*, which has been consistently involved in different rearrangements with various partner genes, and is usually observed in patients with complex karyotypes and poor prognosis (20–24). However, although cyclin D1 overexpression has been consistently associated with MCL, there are a minority of cyclin D1-negative cases that show similar typical cell morphology and gene expression profile. These cases are often characterized by overexpressing other cyclins such as cyclin D2 or D3 (25).

One of the major consequences of the constitutive activation of cyclin D1 in B cells is the maintenance of the retinoblastoma (RB1) protein in a phosphorylated/activated state, which positively regulates cell cycling and promotes cell proliferation, probably initiating tumorigenesis. However, experiments with transgenic mice overexpressing cyclin D1 in early B lymphocytes demonstrated that cyclin D1 alone cannot induce the malignant transformation (26–28). Therefore, other acquired secondary genetic alterations are required for the development of this lymphoma. Gladden et al. generated a transgenic mouse model with constitutive expression of a nuclear mutant cyclin D1 in lymphocytes that developed mature B-cell lymphomas carrying alterations similar to those of MCL, such as deregulation of the ARF/MDM2/p53 pathway and BCL2 overexpression (29). Similarly, mice in which cyclin D1 expression in B lymphocytes could be externally regulated did not develop tumors, but the addition of secondary changes such as the overexpression of *BCL2* or *MYC* oncogenes induced human-like, blastoid MCL (30). Notably, in vivo inhibition of cyclin D1 expression was not associated with tumor regression, suggesting that this protein is essential for tumor initiation but may be dispensable for MCL maintenance (30). These data indicate that human MCL is generated by a multistep process initiated by *CCND1*/cyclin D1 activation, followed by the accumulation of selected genetic changes that are necessary for tumor development.

Molecular-cytogenetic methods such as conventional comparative genomic hybridization (CGH), CGH to bacterial artificial chromosome (BAC) microarrays or to smaller cDNA probes (aCGH), and oligonucleotide microarrays of single-nucleotide polymorphisms (SNPs) have enabled the study of unbalanced chromosomal abnormalities in tumor cells (31–37). Thus, small chromosomal regions with subtle abnormalities can be detected with an unprecedented accuracy, making it much more feasible for cancer-associated gene localization (38, 39). Application of such technologies to MCL cells has revealed recurrent aberrations that include the loss, gain, and genomic amplification of certain chromosomal regions containing oncogenes or tumor suppressor genes that are implicated in tumor development and progression (7, 40–43). Most of these deregulated genes are components of cancer-related pathways regulating cell cycle, DNA damage response and repair, apoptosis, and cell signaling (41). Some of them have been correlated with the clinical outcome of patients with MCL and are considered prognostic factors, while others can predict the response to selected therapies or may represent valid therapeutic targets (7, 42, 43).

This chapter summarizes our current knowledge on the genetic and molecular changes present in MCL cells that have been detected by using CGH-based genome-wide analysis, with special focus on those genetic abnormalities that play a role in the neoplastic transformation. This review will also highlight how the genomic profiling of MCL cells has been critical for a better understanding of the disease pathogenesis that is paving the way for translating these findings to the clinical setting.

2. Recurrent Genomic Aberrations in MCL

The portrait of genomic aberrations that accompanies the t(11;14)(q13;q32) has been extensively characterized by applying conventional and high-resolution CGH-based methods to primary MCL patient samples and human MCL-derived cell lines (42). Initial studies using conventional CGH to metaphase chromosomes reported that MCL cells show a number of genomic areas recurrently affected that differed from those observed in other B-cell lymphoma subtypes (44). Subsequent studies using similar techniques revealed that the most common DNA copy number variations found in MCL cells were the gain and amplification of chromosomes 3q, 7p, 8q, 10p, 12q, 15q, and 18q, whereas recurrent deletions included chromosomes 1p, 6q, 8p, 9p, 9q, 11q, 13q, and 17p (44–50). The detailed study of such genomic changes by FISH, Southern blot, DNA sequencing, and gene or protein expression measurement techniques allowed the identification of several genes targeted by the genomic abnormalities, such as the *BMI1* in the 10p12.2 amplification (51), *CDK4* in the amplification of 12q14 (45, 52), *P16/INK4a* (*CDKN2A*) inactivation by homozygous deletion in chromosome 9p21.3 (44, 45, 47, 49, 50, 53, 54), and *ATM* in the deletion of chromosome 11q22.3 (44, 45, 47, 49, 50, 55–57). In addition, high-level copy number gains of the *BCL2* gene locus at 18q21.3 were also observed in MCL cells (45). In conjunction with these copy number alterations, mutations in *ATM* and *P53* genes were identified in the MCL cases with 11q and 17p losses, respectively (45, 55, 58, 59). Moreover, tumors that exhibited inactivating mutations of *P53* gene and deletion of the second allele in chromosome 17p13 were commonly associated with poor prognosis (60).

Although conventional CGH has been a very useful approach for the global genomic profiling of the disease, the identification of genes that are targeted by specific genomic rearrangements was hampered by its limited resolution. To overcome this problem, array-based CGH (aCGH) methods in which tumoral and normal DNA were differentially labeled and hybridized onto glass slides containing spotted DNA segments representing distinct genomic loci distributed across the genome were developed, improving the resolution to detect copy number changes (34). The first microarrays containing from 2,460 to 4,100 BAC clones distributed at intervals of ~1 Mb with a resolution of 50–200 kb across the genome provided a more detailed picture of the genomic changes present in MCL cells (3, 61–64). Subsequently, a high-resolution tiling-set array CGH with over 32,000 BAC clones spanning the whole genome led to the identification of previously undetected single copy number changes in MCL samples (33, 65). Additional studies have used microarrays consisting of short synthetic oligonucleotides or SNP microarrays that not only provide higher resolution but also allow the simultaneous detection of areas of loss of heterozygosity (LOH) without changes in the DNA copy number (58, 66–71).

The introduction of these higher resolution techniques has enabled the identification of many putative target genes in MCL, especially in areas of homozygous loss. These include the bi-allelic deletion of chromosomes 1p33 involving *FAF1* and *CDKN2C* genes (69, 72); 2q13 targeting *BCL2L11*, which codifies for BIM protein (63, 69, 72); 2q37 (associated with downregulation of *SP100* gene) (69); and 19p13.3 including the *TNF* superfamily genes 7, 9, and 14 (70). Using a different strategy based on the integrative analysis of high-resolution RNA expression and genomic microarray data, Hartmann et al. found genes with downregulated expression in areas of heterozygous loss, including *PROX1* (1p32), *MCPH1* (8p21.3), *ING1*, and *CUL4A* (13q34). These studies also suggested that the deregulation of the Hippo pathway may have a pathogenetic role in MCL because decreased expression of several members (*MOBKL2A*, *MOBKL2B*, and *LATS2*) showed decreased expression in MCL cells (58). Similarly, a different study reported a decreased expression of the genes encoding the TRAIL receptors DR4 and DR5 in MCL cells with deletion of chromosome 8p21 (73). One report identified monoallelic deletion of *RB1* gene at 13q14.2 with the presence of inactivating mutations in the remaining allele in MCL cells (74). More frequently, homo- and hemizygous deletion of the *TNFAIP3* gene at 6q23.3, often in combination with inactivating mutations or promoter hypermethylation, has been found in MCL and in other B-cell lymphomas (75, 76). Finally, several putative oncogenes targeted by high-level chromosomal amplification in MCL cells have been identified, including *MAP6* in 11q13 (68), *CENTG1* in 12q14 (58), and the *miR-17-92* cluster in 13q31 (69, 77). However, only a minority of these genes (*TNFAIP3*, *CDKN2C*, *BCL2L11*, *miR-17-92*) have been functionally validated as bona fide targeted genes in MCL cells, whereas most others need further evaluation in future studies.

3. Partial Uniparental Disomy in MCL Pathogenesis

In tumor cells, uniparental disomy (UPD), also termed copy number neutral LOH, can be biologically equivalent to the second hit of the Knudson hypothesis and represents 20–80% of the LOH seen in human tumors, including hematologic neoplasia and solid tumors (78–81). Nielander et al. reported that UPD was a common event in human MCL cell lines, involving areas that were frequently deleted in MCL, such as chromosomes 1p, 8p, 11q, and 17p (66). By losing part of one parental chromosome and gaining the homologue from the other parent, UPD (or partial UPD, pUPD) results in LOH without chromosomal deletion. As the gene dosage is not altered, these abnormalities cannot be detected by conventional aCGH (66, 68). Subsequently, other studies using SNP microarray have confirmed that UPD is a frequent phenomenon in MCL cells, being detected in 30–50% of the cases. Interestingly, most regions affected by UPD involve chromosomal segments (pUPD) rather than complete chromosomes, indicating that they probably result from one or multiple mitotic recombination events (58). The most frequent recurrent areas of UPD found in MCL are located in chromosomes 6p (58, 68); 9p, 9q (58, 70); 11q (58, 68, 70); 17p (4, 58, 68–70); 19p (58, 70, 71); and 20q (4, 58, 69). These results suggest that UPD is an alternative mechanism to homozygous deletion that may result in the inactivation of tumor suppressor genes. Thus, one of the most commonly detected UPDs involves chromosomal band 17p13, leading to P53 inactivation. Alternatively, UPD can cause the duplication of an activating somatic mutation (66, 67, 69, 82). However, most of the putative genes targeted by UPD in MCL remain unknown.

The study of MCL genomes with microarray techniques has provided important information to understand different patterns of chromosomal instability and the generation of secondary rearrangements in these tumors. In MCL, several chromosomes show complex patterns with losses and gains all along a single chromosome without diploid dosage. This phenomenon could be a result of chromothripsis in which 10–100 genomic rearrangements, even between chromosomes, are acquired from a single catastrophic event affecting a whole chromosome or a chromosomal arm (83, 84). It is not rare to detect high-level DNA amplifications, especially in MCL blastoid variants. In fact, it is possible to determine complex patterns of amplified regions alternated with diploid or deleted regions that are the result of breakage-fusion-bridge cycles repeated in several cell divisions. This process in time increases genomic instability and leads to amplifications and losses of regions harboring oncogenes and tumor suppressor genes, respectively (85).

4. Cyclin D1-Negative MCL

The t(11;14)(q13,q32) is considered the genetic hallmark of MCL. However, there is a small subset of lymphomas that resemble both morphologically and phenotypically to conventional MCL but lack the t(11;14) translocation and cyclin D1 overexpression (25). Two studies evaluated whether these cyclin D1-negative cases could be classified as MCL or represent distinct lymphoma subtype. The first reported similar morphological, pathological, clinical, and molecular features between six cyclin D1-negative lymphoma cases and typical MCL (86). A second study disclosed that both cyclin D1-negative and cyclin D1-positive cases shared the same profile of secondary genomic aberrations that differed from those of other B-cell neoplasms (56). These results suggest that cyclin D1-negative lymphomas should be considered as molecular variants of MCL.

Interestingly, cyclin D1-negative MCL frequently exhibited overexpression of other cyclins such as cyclin D2 and cyclin D3. High cyclin D2 expression levels have been detected in lymphomas with the chromosomal translocations t(2;12)(p11;p13), t(12;22) (p13;q21), and t(12;14)(p13;q32), which result in the fusion of the *CCND2* gene (which codifies for cyclin D2) in 12p13 to the *IGK* (87), *IGL* (88), or *IGH* (89) gene loci, respectively. On the other hand, rare chromosomal translocations involving *CCND3* (which codifies for cyclin D3) and *IGH* genes have been reported in B-cell malignancies other than MCL (90–92). These findings suggest that the deregulation of other cyclins in the absence of cyclin D1 expression may be an alternative mechanism in cyclin D1-negative MCL pathogenesis (86).

The clinical diagnosis of cyclin D1-negative MCL is difficult because it may resemble phenotypically other small cell B-cell lymphomas. In an aim to identify a reliable biomarker for cyclin D1-negative MCL, Mozos et al. conducted a study of 54 cases of conventional MCL, 12 cyclin D1-negative MCL, and 209 additional lymphoid neoplasms. They reported that the nuclear transcription factor SOX11 was highly expressed in 93% of the conventional MCL cases and in all of the cyclin D1-negative cases studied, but was virtually absent in other mature B-cell lymphomas (93). Similarly, Ek et al. found overexpression of SOX11 mRNA and nuclear protein in both cyclin D1-positive and cyclin D1-negative MCL (94). These studies suggest that the nuclear expression of SOX11 protein may be considered a specific marker for typical MCL and a reliable indicator for detecting the cyclin D1-negative MCL subtype (93, 94).

5. Indolent MCL

Recent observations have suggested that the clinical behavior of MCL may not be as homogeneous as initially thought. Thus, although most patients follow a very rapid progression and die of the disease, others have an indolent clinical course and can survive more than 10 years, even without receiving immunochemotherapy (2, 3, 95). Orchard et al. reported a subset of patients with MCL with leukemic non-nodal disease, typical phenotype, and mutated IgV_H genes that had an indolent clinical presentation with a prolonged survival, even without receiving immunochemotherapy (2). Subsequent studies have revealed that these indolent MCL cases have a biased usage of V_H34-9 and display stable genomes with rare deletion or mutation of *P53, CDKNA2*, and *ATM* genes (2, 3). Despite the differences in the genetic and clinical characteristics between conventional MCL and the indolent form, both show the t(11;14)(q13;q32) translocation with cyclin D1 overexpression and share a similar global gene expression profile (96). However, indolent and conventional MCL differ in the expression of a small signature of genes, including *SOX11*, which is usually not expressed in indolent MCL but in all conventional MCL cases (4, 93, 94, 96).

The identification of clinically aggressive and indolent MCL variants suggests the use of different treatment approaches in these two subgroups. Conventional MCL is characterized by short remissions and high rates of relapses after intensive therapy. Given its poor prognosis aggressive immunochemotherapy is mandatory at diagnosis, but whether similar treatment strategies are necessary in indolent MCL is unknown. Two retrospective studies have reported the outcome of patients with MCL in whom initial chemotherapy was deferred (95, 97). Interestingly, time to treatment did not influence overall survival in newly diagnosed MCL, suggesting that in selected patients with asymptomatic indolent MCL, deferred initial treatment ("watch and wait") is an acceptable management approach.

6. The MCL Genomic Signature

Despite many genomic aberrations are common to different B-cell lymphomas, MCL shows has a distinctive genomic profile. Many of the genes targeted by genomic changes are involved in common cancer pathways and represent candidates that may cooperate with cyclin D1 in MCL development (42) (Table 1). In addition to cyclin D1 overexpression, which promotes cell proliferation through the constitutive phosphorylation of the RB protein, genomic changes of genes regulating the cell cycle are common in MCL, preferentially the homozygous deletion of *CDKN2A* gene in 9p21.3 (53, 54). *CDKN2A* encodes for two different proteins: INK4A, a cyclin-dependent kinase (CDK) inhibitor, and ARF, an activator of P53. Thus, disruption of the *CDKN2A* locus causes alterations in two major cell-cycle regulatory pathways: INK4A/CDK4/RB1 and ARF/MDM2/P53 (98, 99). An alternative mechanism to INK4A inactivation is the genomic amplification of the gene encoding the polycomb protein BMI1 in chromosome 10p12.2, which is a transcriptional repressor of *CDKN2A* (51, 100). Other cell-cycle abnormalities are caused by the genetic inactivation of a different CDK inhibitor, *CDKN2C*, in 1p32.3 (72) by deletion and mutation of *RB1* gene in 13q14 (74) and by *CDK4* amplification in 12q14 (52). *CDKN2A* homozygous deletions also involve the ARF protein, which negatively regulates MDM2. Thus, ARF inactivation results in MDM2 overexpression which in turn promotes P53 degradation (98). Similar tumorigenic effects are induced by the amplifications of *MDM2* in 12q15 (52) and by inactivating mutations or deletions of *P53* in chromosome 17p13, which are frequently observed in MCL (60).

Alterations of the DNA damage response machinery are also frequently observed in MCL, including deletion and mutation of the *ATM* gene in 11q22.3 (55, 101). Downstream ATM targets

Table 1
Recurrent genomic abnormalities in MCL

Chromosome region	DNA copy number change	Target gene	Function	Prognostic factor	Predictor of therapeutic response
1p21	Loss	Unknown		No	No
1p32.3	Loss	*CDKN2C*	CDK inhibitor	No	No
2q13	Loss	*BCL2L11*	Encodes pro-apoptotic BH3-only BIM protein	No	No
3q27-q29	Gain, amplification	Unknown		Short survival	No
6q23.3	Loss	*TNFAIP3*	NF-kB inhibitor	No	NF-kB inhibitors?
8p21	Loss	Unknown		Short survival	No
8q24.21	Gain, amplification	*MYC*	Multi-functional: Regulates proliferation and apoptosis	Short survival	No
9p21.3	Loss	*CDKN2A*	Encodes INK4a and ARF proteins	Short survival	No
9q21-q22	Loss	Unknown		Short survival	No
10p12.2	Gain, amplification	*BMI1*	Transcriptional inhibitor of *CDKN2A;* regulates cell self-renewal	No	No
11q13.3	Amplification	*CCND1-IGH*	Cyclin D1 overexpression	Unknown	No
11q22.3	Loss	*ATM*	Sensor of DNA damage and regulator of DNA repair	No	No
12q14.1	Gain, amplification	*CDK4*	Complex with D-type cyclins to promote cell cycle	Short survival	PD-0332991?
12q15	Gain, amplification	*MDM2*	Ubiquitin E3 ligase that targets P53 for degradation	Short survival	Nutlin-3?
13q14.2	Loss	*RB1*	Negatively regulates cell cycle	Unknown	No
13q14.2	Loss	*miR-15a/miR-16-1*	Inhibit coding genes including *CCND1*	Short survival	No
13q31.3	Gain, amplification	*miR-17-92*	Inhibit coding genes *BCL2L11*	No	No
17p13.1	Loss	*P53*	Multi-functional: induces cell cycle arrest and apoptosis	Short survival	No
18q21.3	Gain, amplification	*BCL2*	Anti-apoptotic regulator	No	Yes: BH3 mimetics

such as *CHK1* and *CHK2* can also be deregulated in MCL, although rarely (102, 103). Notably, germ-line mutations in *ATM* and *CHK2* genes have been identified in patients with MCL, suggesting that may have a role in the genetic predisposition to these tumors (59, 103). Disruption of gene components of the apoptosis pathways in MCL includes the genomic amplification of the anti-apoptotic *BCL2* gene (3, 61, 63, 65, 68) and the homozygous deletion of the *BCL2L11* gene that encodes the pro-apoptotic protein BIM (63).

Other genomic lesions frequently detected in MCL include the gain or genomic amplification of chromosome 8q24 involving *MYC* gene, which is frequently observed in highly proliferative tumors (104), and the hemi- and homozygous deletion of *TNFAIP3* in 6q23.3, which is an inhibitor of the NF-κB pathway (75, 76).

In addition, noncoding microRNAs are recurrently targeted by genomic abnormalities in MCL cells (77). Hemi- and homozygous deletion of 13q14 induces *miR-15a* and *miR-16-1* downregulation. The *miR-16-1* regulates *CCND1* gene at the posttranscriptional level by binding to its 3 untranslated region (UTR), thus blocking its translation or initiating its degradation (105). Consequently, deletion of *miR-16-1* increases cyclin D1 expression levels. A second common abnormality in MCL cells is the genomic amplification of 13q31.3 involving the *miR-17-92* cluster, whose overexpression was associated with high *MYC* mRNA levels and a clinical aggressive behavior (77, 106).

7. Prognostic Factors and Predictive Markers of Therapeutic Response

In combination with other morphological and clinical parameters, the profile and number of genomic imbalances are correlated with the outcome of patients with MCL. Using conventional CGH analysis, Bea et al. reported that the number of chromosomal gains, losses, and DNA amplifications was significantly higher in the MCL blastoid variants, particularly the gains of chromosomes 3q and 12q and losses of 9p and 17p (45). The gains or amplification of chromosomes 3q27-q29 and of 12q14-q15 (involving *CDK4* and *MDM2* genes) have been consistently associated with shorter survival (15, 45, 56, 58, 64). In addition, a number of chromosomal losses such as those 9p21 (targeting *CDKN2A*), 17p13 (*P53*), 8p21, 13q14 (*miR-15a/miR-16*), and 9q21-q22 were associated with poor prognosis in different studies (3, 15, 45, 49, 50, 56, 58, 107, 108). Likewise, the gain of chromosome 8q involving *MYC* gene locus, although uncommon, has been detected in MCL with a very rapid clinical evolution. Accordingly, *MYC* rearrangements, either at diagnosis or at relapse, represent a negative prognostic factor (42, 109). Notably, deletions or mutations in *ATM* locus occur both in classical and blastoid variants and are not related to prognosis (3, 45, 55, 59, 110), suggesting that this alteration is an initial event in lymphomagenesis. However, in several common genomic abnormalities correlated with poor prognosis, such as the amplification of 3q27-q29 and the deletions of 8p21 and 9q21-q22, the target genes have not yet been identified.

Despite the number of prognostic factors has substantially increased in patients with MCL, very few genomic abnormalities that can predict the response to current or, even more important, to novel therapies have been found. One example is the high-level amplification of the *BCL2* gene, which indicates a good response to small BH3 mimetic molecules targeting BCL2 (30). Other putative predictive markers of response to selected drugs are (1) the genomic amplification and overexpression of *MDM2*, which may correlate with response to the MDM2-antagonist nutlin-3 (111); the genomic amplification of *CDK4*, which might select cases sensitive to PD-0332991, an oral cyclin-dependent kinase 4/6 inhibitor (112); and the inactivation of the NF-κB inhibitor *TNFAIP3* gene, which induces NF-κB signaling and may predict response to NF-κB inhibitor drugs (75, 76). More recently, Kridel et al. applied whole transcriptome sequencing to MCL cells and identified recurrent mutations in *NOTCH1* in 12% of primary samples (113). Interestingly, inhibition of the NOTCH pathway reduced proliferation and induced apoptosis in MCL cell lines, suggesting that these mutations may sensitize the cells to NOTCH inhibitor drugs (113).

8. Conclusion

Human MCL is generated by a multistep process initiated by *CCND1*/cyclin D1 activation followed by selected gene mutations that are necessary for tumor development, many of which have been identified through the application of array-based CGH methods. Most of these deregulated genes are components of cancer-related pathways regulating cell-cycle, DNA damage response and repair, apoptosis, and cell signaling. Some of them have been correlated with the clinical outcome of patients with MCL and are considered prognostic factors, while others may represent valid therapeutic targets. However, very few specific markers that can adequately predict the response to current or, even more important, new targeted therapies have been found. In any case, the genomic analysis has contributed to the substantial advances made in the comprehension of the pathogenesis of MCL, giving new perspectives for the identification of the optimal therapeutic targets and for the design of new molecular therapies that can ultimate cure this fatal disease.

References

1. Swerlow SH, Berger F, Isaacson P, Muller-Hermelink HK, Nathwani BN, Piris MA, Harris NL (2001) Mantle cell lymphoma. World Health Organization Classification of tumors pathology and genetics of tumours of haematopoietic and lymphoid tissues:168–170
2. Orchard J, Garand R, Davis Z et al (2003) A subset of t(11;14) lymphoma with mantle cell features displays mutated igvh genes and includes patients with good prognosis, non-nodal disease. Blood 101(12):4975–4981. doi:10.1182/blood-2002-06-1864 2002-06-1864 [pii]
3. Rubio-Moscardo F, Climent J, Siebert R et al (2005) Mantle-cell lymphoma genotypes identified with cgh to bac microarrays define a leukemic subgroup of disease and predict patient outcome. Blood 105(11):4445–4454. doi:2004-10-3907 [pii] 10.1182/blood-2004-10-3907
4. Fernandez V, Salamero O, Espinet B et al (2010) Genomic and gene expression profiling defines indolent forms of mantle cell lymphoma. Cancer Res 70(4):1408–1418. doi:0008-5472.CAN-09-3419 [pii] 10.1158/0008-5472.CAN-09-3419
5. Dreyling M, Hiddemann W (2008) Dose-intense treatment of mantle cell lymphoma: can durable remission be achieved? Curr Opin Oncol 20(5):487–494. doi:10.1097/CCO.0b013e32830b61c2 00001622-200809000-00002 [pii]
6. Witzig TE (2005) Current treatment approaches for mantle-cell lymphoma. J Clin Oncol 23(26):6409–6414. doi:23/26/6409 [pii] 10.1200/JCO.2005.55.017
7. O'connor OA (2007) Mantle cell lymphoma: identifying novel molecular targets in growth and survival pathways. Hematology Am Soc Hematol Educ Program:270–276. doi: 2007/1/270 [pii] 10.1182/asheducation-2007.1.270
8. Leonard JP, Williams ME, Goy A et al (2009) Mantle cell lymphoma: biological insights and treatment advances. Clin Lymphoma Myeloma 9(4):267–277. doi:11G5T6V781XQ3127 [pii] 10.3816/CLM.2009.n.055
9. Campo E, Raffeld M, Jaffe ES (1999) Mantle-cell lymphoma. Semin Hematol 36(2):115–127
10. Rosenberg CL, Wong E, Petty EM et al (1991) Pradl, a candidate bcll oncogene: mapping and expression in centrocytic lymphoma. Proc Natl Acad Sci USA 88(21):9638–9642
11. Bosch F, Jares P, Campo E et al (1994) Prad-1/cyclin d1 gene overexpression in chronic lymphoproliferative disorders: a highly specific marker of mantle cell lymphoma. Blood 84(8):2726–2732
12. Wlodarska I, Meeus P, Stul M et al (2004) Variant t(2;11)(p11;q13) associated with the igk-ccnd1 rearrangement is a recurrent translocation in leukemic small-cell b-non-hodgkin lymphoma. Leukemia 18(10):1705–1710. doi:10.1038/sj.leu.2403459 2403459 [pii]
13. Komatsu H, Iida S, Yamamoto K et al (1994) A variant chromosome translocation at 11q13 identifying prad1/cyclin d1 as the bcl-1 gene. Blood 84(4):1226–1231
14. Rocha CK, Praulich I, Gehrke I et al (2011) A rare case of t(11;22) in a mantle cell lymphoma like b-cell neoplasia resulting in a fusion of igl and ccnd1: case report. Mol Cytogenet 4(1):8. doi:1755-8166-4-8 [pii] 10.1186/1755-8166-4-8
15. Espinet B, Salaverria I, Bea S et al (2010) Incidence and prognostic impact of secondary cytogenetic aberrations in a series of 145 patients with mantle cell lymphoma. Genes Chromosomes Cancer 49(5):439–451. doi:10.1002/gcc.20754
16. Gruszka-Westwood AM, Atkinson S, Summersgill BM et al (2002) Unusual case of leukemic mantle cell lymphoma with amplified ccnd1/igh fusion gene. Genes Chromosomes Cancer 33(2):206–212. doi:10.1002/gcc.1216 [pii]
17. Ott G, Kalla J, Ott MM et al (1997) Blastoid variants of mantle cell lymphoma: frequent bcl-1 rearrangements at the major translocation cluster region and tetraploid chromosome clones. Blood 89(4):1421–1429
18. Kramer A, Schweizer S, Neben K et al (2003) Centrosome aberrations as a possible mechanism for chromosomal instability in non-hodgkin's lymphoma. Leukemia 17(11):2207–2213. doi:10.1038/sj.leu.2403142 2403142 [pii]
19. Neben K, Ott G, Schweizer S et al (2007) Expression of centrosome-associated gene products is linked to tetraploidization in mantle cell lymphoma. Int J Cancer 120(8):1669–1677. doi:10.1002/ijc.22404
20. Tirier C, Zhang Y, Plendl H et al (1996) Simultaneous presence of t(11;14) and a variant burkitt's translocation in the terminal phase of a mantle cell lymphoma. Leukemia 10(2):346–350
21. Vaishampayan UN, Mohamed AN, Dugan MC et al (2001) Blastic mantle cell lymphoma associated with burkitt-type translocation and hypodiploidy. Br J Haematol 115(1):66–68. doi:3056 [pii]
22. Hao S, Sanger W, Onciu M et al (2002) Mantle cell lymphoma with 8q24 chromosomal abnor-

malities: a report of 5 cases with blastoid features. Mod Pathol 15(12):1266–1272. doi:10.1097/01.MP.0000037310.82136.99
23. Michaux L, Wlodarska I, Theate I et al (2004) Coexistence of bcl1/ccnd1 and cmyc aberrations in blastoid mantle cell lymphoma: a rare finding associated with very poor outcome. Ann Hematol 83(9):578–583. doi:10.1007/s00277-004-0879-2
24. Oliveira FM, Tone LG, Simoes BP et al (2007) Blastoid mantle cell lymphoma with t(2;8) (p12;q24). Leuk Lymphoma 48(10):2079–2082. doi:782725470 [pii] 10.1080/10428190701606834
25. Rosenwald A, Wright G, Wiestner A et al (2003) The proliferation gene expression signature is a quantitative integrator of oncogenic events that predicts survival in mantle cell lymphoma. Cancer Cell 3(2):185–197. doi:S153561080300028X [pii]
26. Hinds PW, Dowdy SF, Eaton EN et al (1994) Function of a human cyclin gene as an oncogene. Proc Natl Acad Sci USA 91(2):709–713
27. Lovec H, Grzeschiczek A, Kowalski MB et al (1994) Cyclin d1/bcl-1 cooperates with myc genes in the generation of b-cell lymphoma in transgenic mice. EMBO J 13(15):3487–3495
28. Bodrug SE, Warner BJ, Bath ML et al (1994) Cyclin d1 transgene impedes lymphocyte maturation and collaborates in lymphomagenesis with the myc gene. EMBO J 13(9):2124–2130
29. Gladden AB, Woolery R, Aggarwal P et al (2006) Expression of constitutively nuclear cyclin d1 in murine lymphocytes induces b-cell lymphoma. Oncogene 25(7):998–1007. doi:1209147 [pii] 10.1038/sj.onc.1209147
30. Beltran E, Fresquet V, Martinez-Useros J et al (2011) A cyclin-d1 interaction with bax underlies its oncogenic role and potential as a therapeutic target in mantle cell lymphoma. Proc Natl Acad Sci USA 108(30):12461–12466. doi:1018941108 [pii] 10.1073/pnas.1018941108
31. Wong KK, Tsang YT, Shen J et al (2004) Allelic imbalance analysis by high-density single-nucleotide polymorphic allele (snp) array with whole genome amplified DNA. Nucleic Acids Res 32(9):e69. doi:10.1093/nar/gnh072 32/9/e69 [pii]
32. Cutler DJ, Zwick ME, Carrasquillo MM et al (2001) High-throughput variation detection and genotyping using microarrays. Genome Res 11(11):1913–1925. doi:10.1101/gr.197201
33. Ishkanian AS, Malloff CA, Watson SK et al (2004) A tiling resolution DNA microarray with complete coverage of the human genome. Nat Genet 36(3):299–303. doi:10.1038/ng1307 ng1307 [pii]
34. Snijders AM, Nowak N, Segraves R et al (2001) Assembly of microarrays for genome-wide measurement of DNA copy number. Nat Genet 29(3):263–264. doi:10.1038/ng754 ng754 [pii]
35. Pollack JR, Perou CM, Alizadeh AA et al (1999) Genome-wide analysis of DNA copy-number changes using cdna microarrays. Nat Genet 23(1):41–46. doi:10.1038/12640
36. Kallioniemi A, Kallioniemi OP, Sudar D et al (1992) Comparative genomic hybridization for molecular cytogenetic analysis of solid tumors. Science 258(5083):818–821
37. Pinkel D, Segraves R, Sudar D et al (1998) High resolution analysis of DNA copy number variation using comparative genomic hybridization to microarrays. Nat Genet 20(2):207–211. doi:10.1038/2524
38. Pinkel D, Albertson DG (2005) Comparative genomic hybridization. Annu Rev Genomics Hum Genet 6:331–354. doi:10.1146/annurev.genom.6.080604.162140
39. Pinkel D, Albertson DG (2005) Array comparative genomic hybridization and its applications in cancer. Nat Genet 37(Suppl):S11–S17. doi:ng1569 [pii] 10.1038/ng1569
40. Royo C, Salaverria I, Hartmann EM et al (2011) The complex landscape of genetic alterations in mantle cell lymphoma. Semin Cancer Biol 21(5):322–334. doi:S1044-579X(11)00056-3 [pii] 10.1016/j.semcancer.2011.09.007
41. Fernandez V, Hartmann E, Ott G et al (2005) Pathogenesis of mantle-cell lymphoma: all oncogenic roads lead to dysregulation of cell cycle and DNA damage response pathways. J Clin Oncol 23(26):6364–6369. doi:23/26/6364 [pii] 10.1200/JCO.2005.05.019
42. Jares P, Colomer D, Campo E (2007) Genetic and molecular pathogenesis of mantle cell lymphoma: perspectives for new targeted therapeutics. Nat Rev Cancer 7(10):750–762. doi:nrc2230 [pii] 10.1038/nrc2230
43. Perez-Galan P, Dreyling M, Wiestner A (2010) Mantle cell lymphoma: biology, pathogenesis, and the molecular basis of treatment in the genomic era. Blood 117(1):26–38. doi:blood-2010-04-189977 [pii] 10.1182/blood-2010-04-189977
44. Monni O, Oinonen R, Elonen E et al (1998) Gain of 3q and deletion of 11q22 are frequent aberrations in mantle cell lymphoma. Genes Chromosomes Cancer 21(4):298–307. doi:10.1002/ (SICI)1098-2264(199804) 21:4<298: : AID-GCC3>3.0.CO;2-U [pii]

45. Bea S, Ribas M, Hernandez JM et al (1999) Increased number of chromosomal imbalances and high-level DNA amplifications in mantle cell lymphoma are associated with blastoid variants. Blood 93(12):4365–4374
46. Cuneo A, Bigoni R, Rigolin GM et al (1999) Cytogenetic profile of lymphoma of follicle mantle lineage: correlation with clinicobiologic features. Blood 93(4):1372–1380
47. Bentz M, Plesch A, Bullinger L et al (2000) T(11;14)-positive mantle cell lymphomas exhibit complex karyotypes and share similarities with b-cell chronic lymphocytic leukemia. Genes Chromosomes Cancer 27(3):285–294. doi:10.1002/ (SICI)1098-2264(200003) 27:3<285::AID-GCC9>3.0.CO;2-M [pii]
48. Bigoni R, Cuneo A, Milani R et al (2001) Secondary chromosome changes in mantle cell lymphoma: cytogenetic and fluorescence in situ hybridization studies. Leuk Lymphoma 40(5–6):581–590. doi:10.3109/10428190109097656
49. Martinez-Climent JA, Vizcarra E, Sanchez D et al (2001) Loss of a novel tumor suppressor gene locus at chromosome 8p is associated with leukemic mantle cell lymphoma. Blood 98(12):3479–3482
50. Allen JE, Hough RE, Goepel JR et al (2002) Identification of novel regions of amplification and deletion within mantle cell lymphoma DNA by comparative genomic hybridization. Br J Haematol 116(2):291–298. doi:3260 [pii]
51. Bea S, Tort F, Pinyol M et al (2001) Bmi-1 gene amplification and overexpression in hematological malignancies occur mainly in mantle cell lymphomas. Cancer Res 61(6):2409–2412
52. Hernandez L, Bea S, Pinyol M et al (2005) Cdk4 and mdm2 gene alterations mainly occur in highly proliferative and aggressive mantle cell lymphomas with wild-type ink4a/arf locus. Cancer Res 65(6):2199–2206. doi:65/6/2199 [pii] 10.1158/0008-5472.CAN-04-1526
53. Pinyol M, Hernandez L, Cazorla M et al (1997) Deletions and loss of expression of p16ink4a and p21waf1 genes are associated with aggressive variants of mantle cell lymphomas. Blood 89(1):272–280
54. Dreyling MH, Bullinger L, Ott G et al (1997) Alterations of the cyclin d1/p16-prb pathway in mantle cell lymphoma. Cancer Res 57(20):4608–4614
55. Greiner TC, Dasgupta C, Ho VV et al (2006) Mutation and genomic deletion status of ataxia telangiectasia mutated (atm) and p53 confer specific gene expression profiles in mantle cell lymphoma. Proc Natl Acad Sci USA 103(7):2352–2357. doi:0510441103 [pii] 10.1073/pnas.0510441103
56. Salaverria I, Zettl A, Bea S et al (2007) Specific secondary genetic alterations in mantle cell lymphoma provide prognostic information independent of the gene expression-based proliferation signature. J Clin Oncol 25(10):1216–1222. doi:JCO.2006.08.4251 [pii] 10.1200/JCO.2006.08.4251
57. Stilgenbauer S, Winkler D, Ott G et al (1999) Molecular characterization of 11q deletions points to a pathogenic role of the atm gene in mantle cell lymphoma. Blood 94(9):3262–3264
58. Hartmann EM, Campo E, Wright G et al (2010) Pathway discovery in mantle cell lymphoma by integrated analysis of high-resolution gene expression and copy number profiling. Blood 116(6):953–961. doi:blood-2010-01-263806 [pii] 10.1182/blood-2010-01-263806
59. Camacho E, Hernandez L, Hernandez S et al (2002) Atm gene inactivation in mantle cell lymphoma mainly occurs by truncating mutations and missense mutations involving the phosphatidylinositol-3 kinase domain and is associated with increasing numbers of chromosomal imbalances. Blood 99(1):238–244
60. Greiner TC, Moynihan MJ, Chan WC et al (1996) P53 mutations in mantle cell lymphoma are associated with variant cytology and predict a poor prognosis. Blood 87(10):4302–4310
61. Kohlhammer H, Schwaenen C, Wessendorf S et al (2004) Genomic DNA-chip hybridization in t(11;14)-positive mantle cell lymphomas shows a high frequency of aberrations and allows a refined characterization of consensus regions. Blood 104(3):795–801. doi:10.1182/blood-2003-12-4175 2003-12-4175 [pii]
62. Schrader C, Janssen D, Klapper W et al (2005) Minichromosome maintenance protein 6, a proliferation marker superior to ki-67 and independent predictor of survival in patients with mantle cell lymphoma. Br J Cancer 93(8):939–945. doi:6602795 [pii] 10.1038/sj.bjc.6602795
63. Tagawa H, Karnan S, Suzuki R et al (2005) Genome-wide array-based cgh for mantle cell lymphoma: identification of homozygous deletions of the proapoptotic gene bim. Oncogene 24(8):1348–1358. doi:1208300 [pii] 10.1038/sj.onc.1208300
64. Flordal Thelander E, Ichimura K, Collins VP et al (2007) Detailed assessment of copy number alterations revealing homozygous deletions in 1p and 13q in mantle cell lymphoma. Leuk Res 31(9):1219–1230.

doi:S0145-2126(06)00415-2 [pii] 10.1016/j.leukres.2006.10.022

65. De Leeuw RJ, Davies JJ, Rosenwald A et al (2004) Comprehensive whole genome array cgh profiling of mantle cell lymphoma model genomes. Hum Mol Genet 13(17):1827–1837. doi:10.1093/hmg/ddh195 ddh195 [pii]
66. Nielander I, Martin-Subero JI, Wagner F et al (2006) Partial uniparental disomy: a recurrent genetic mechanism alternative to chromosomal deletion in malignant lymphoma. Leukemia 20(5):904–905. doi:2404173 [pii] 10.1038/sj.leu.2404173
67. Rinaldi A, Kwee I, Taborelli M et al (2006) Genomic and expression profiling identifies the b-cell associated tyrosine kinase syk as a possible therapeutic target in mantle cell lymphoma. Br J Haematol 132(3):303–316. doi:BJH5883 [pii] 10.1111/j.1365-2141.2005.05883.x
68. Vater I, Wagner F, Kreuz M et al (2009) Genechip analyses point to novel pathogenetic mechanisms in mantle cell lymphoma. Br J Haematol 144(3):317–331. doi:BJH7443 [pii] 10.1111/j.1365-2141.2008.07443.x
69. Bea S, Salaverria I, Armengol L et al (2009) Uniparental disomies, homozygous deletions, amplifications, and target genes in mantle cell lymphoma revealed by integrative high-resolution whole-genome profiling. Blood 113(13):3059–3069. doi:blood-2008-07-170183 [pii] 10.1182/blood-2008-07-170183
70. Kawamata N, Ogawa S, Gueller S et al (2009) Identified hidden genomic changes in mantle cell lymphoma using high-resolution single nucleotide polymorphism genomic array. Exp Hematol 37(8):937–946. doi:S0301-472X(09)00175-1 [pii] 10.1016/j.exphem. 2009.04.012
71. Halldorsdottir AM, Sander B, Goransson H et al (2011) High-resolution genomic screening in mantle cell lymphoma–specific changes correlate with genomic complexity, the proliferation signature and survival. Genes Chromosomes Cancer 50(2):113–121. doi:10.1002/gcc.20836
72. Mestre-Escorihuela C, Rubio-Moscardo F, Richter JA et al (2007) Homozygous deletions localize novel tumor suppressor genes in b-cell lymphomas. Blood 109(1):271–280. doi:blood-2006-06-026500 [pii] 10.1182/blood-2006-06-026500
73. Rubio-Moscardo F, Blesa D, Mestre C et al (2005) Characterization of 8p21.3 chromosomal deletions in b-cell lymphoma: trail-r1 and trail-r2 as candidate dosage-dependent tumor suppressor genes. Blood 106(9):3214–3222. doi:2005-05-2013 [pii] 10.1182/blood-2005-05-2013
74. Pinyol M, Bea S, Pla L et al (2007) Inactivation of rb1 in mantle-cell lymphoma detected by nonsense-mediated mrna decay pathway inhibition and microarray analysis. Blood 109(12):5422–5429. doi:blood-2006-11-057208 [pii] 10.1182/blood-2006-11-057208
75. Kato M, Sanada M, Kato I et al (2009) Frequent inactivation of a20 in b-cell lymphomas. Nature 459(7247):712–716. doi:nature07969 [pii] 10.1038/nature07969
76. Honma K, Tsuzuki S, Nakagawa M et al (2009) Tnfaip3/a20 functions as a novel tumor suppressor gene in several subtypes of non-hodgkin lymphomas. Blood 114(12):2467–2475. doi:blood-2008-12-194852 [pii] 10.1182/blood-2008-12-194852
77. Navarro A, Bea S, Fernandez V et al (2009) Microrna expression, chromosomal alterations, and immunoglobulin variable heavy chain hypermutations in mantle cell lymphomas. Cancer Res 69(17):7071–7078. doi:0008-5472.CAN-09-1095 [pii] 10.1158/0008-5472.CAN-09-1095
78. Mao X, Young BD, Lu YJ (2007) The application of single nucleotide polymorphism microarrays in cancer research. Curr Genomics 8(4):219–228
79. Gondek LP, Tiu R, O'keefe CL et al (2008) Chromosomal lesions and uniparental disomy detected by snp arrays in mds, mds/mpd, and mds-derived aml. Blood 111(3):1534–1542. doi:blood-2007-05-092304 [pii] 10.1182/blood-2007-05-092304
80. Beroukhim R, Lin M, Park Y et al (2006) Inferring loss-of-heterozygosity from unpaired tumors using high-density oligonucleotide snp arrays. PLoS Comput Biol 2(5):e41. doi:10.1371/journal.pcbi.0020041
81. Lo KC, Bailey D, Burkhardt T et al (2008) Comprehensive analysis of loss of heterozygosity events in glioblastoma using the 100 k snp mapping arrays and comparison with copy number abnormalities defined by bac array comparative genomic hybridization. Genes Chromosomes Cancer 47(3):221–237. doi:10.1002/gcc.20524
82. Fitzgibbon J, Smith LL, Raghavan M et al (2005) Association between acquired uniparental disomy and homozygous gene mutation in acute myeloid leukemias. Cancer Res 65(20):9152–9154. doi:65/20/9152 [pii] 10.1158/0008-5472.CAN-05-2017
83. Stephens PJ, Mcbride DJ, Lin ML et al (2009) Complex landscapes of somatic rearrangement

in human breast cancer genomes. Nature 462(7276):1005–1010. doi:nature08645 [pii] 10.1038/nature08645

84. Stephens PJ, Greenman CD, Fu B et al (2011) Massive genomic rearrangement acquired in a single catastrophic event during cancer development. Cell 144(1):27–40. doi:S0092-8674(10)01377-2 [pii] 10.1016/j.cell.2010.11.055
85. Coquelle A, Pipiras E, Toledo F et al (1997) Expression of fragile sites triggers intrachromosomal mammalian gene amplification and sets boundaries to early amplicons. Cell 89(2):215–225. doi:S0092-8674(00)80201-9 [pii]
86. Fu K, Weisenburger DD, Greiner TC et al (2005) Cyclin d1-negative mantle cell lymphoma: a clinicopathologic study based on gene expression profiling. Blood 106(13):4315–4321. doi:2005-04-1753 [pii] 10.1182/blood-2005-04-1753
87. Gesk S, Klapper W, Martin-Subero JI et al (2006) A chromosomal translocation in cyclin d1-negative/cyclin d2-positive mantle cell lymphoma fuses the ccnd2 gene to the igk locus. Blood 108(3):1109–1110. doi:108/3/1109 [pii] 10.1182/blood-2006-01-0015
88. Shiller SM, Zieske A, Holmes H 3rd et al (2011) Cd5-positive, cyclind1-negative mantle cell lymphoma with a translocation involving the ccnd2 gene and the igl locus. Cancer Genet 204(3):162–164. doi:S0165-4608(10)00479-6 [pii] 10.1016/j.cancergencyto.2010.08.016
89. Herens C, Lambert F, Quintanilla-Martinez L et al (2008) Cyclin d1-negative mantle cell lymphoma with cryptic t(12;14)(p13;q32) and cyclin d2 overexpression. Blood 111(3):1745–1746. doi:111/3/1745 [pii] 10.1182/blood-2007-10-120824
90. Shaughnessy J Jr, Gabrea A, Qi Y et al (2001) Cyclin d3 at 6p21 is dysregulated by recurrent chromosomal translocations to immunoglobulin loci in multiple myeloma. Blood 98(1):217–223
91. Sonoki T, Harder L, Horsman DE et al (2001) Cyclin d3 is a target gene of t(6;14)(p21.1;q32.3) of mature b-cell malignancies. Blood 98(9):2837–2844
92. Wlodarska I, Dierickx D, Vanhentenrijk V et al (2008) Translocations targeting ccnd2, ccnd3, and mycn do occur in t(11;14)-negative mantle cell lymphomas. Blood 111(12):5683–5690. doi:blood-2007-10-118794 [pii] 10.1182/blood-2007-10-118794
93. Mozos A, Royo C, Hartmann E et al (2009) Sox11 expression is highly specific for mantle cell lymphoma and identifies the cyclin d1-negative subtype. Haematologica 94(11):1555–1562. doi:94/11/1555 [pii] 10.3324/haematol.2009.010264
94. Ek S, Dictor M, Jerkeman M et al (2008) Nuclear expression of the non b-cell lineage sox11 transcription factor identifies mantle cell lymphoma. Blood 111(2):800–805. doi:blood-2007-06-093401 [pii] 10.1182/blood-2007-06-093401
95. Martin P, Chadburn A, Christos P et al (2009) Outcome of deferred initial therapy in mantle-cell lymphoma. J Clin Oncol 27(8):1209–1213. doi:JCO.2008.19.6121 [pii] 10.1200/JCO.2008.19.6121
96. Ondrejka SL, Lai R, Smith SD et al (2011) Indolent mantle cell leukemia: a clinicopathological variant characterized by isolated lymphocytosis, interstitial bone marrow involvement, kappa light chain restriction, and good prognosis. Haematologica 96(8):1121–1127. doi:haematol.2010.036277 [pii] 10.3324/haematol.2010.036277
97. Eve HE, Furtado MV, Hamon MD et al (2009) Time to treatment does not influence overall survival in newly diagnosed mantle-cell lymphoma. J Clin Oncol 27(32):e189–e190. doi:JCO.2009.23.9731 [pii] 10.1200/JCO.2009.23.9731, author reply e191
98. Zhang Y, Xiong Y, Yarbrough WG (1998) Arf promotes mdm2 degradation and stabilizes p53: Arf-ink4a locus deletion impairs both the rb and p53 tumor suppression pathways. Cell 92(6):725–734. doi:S0092-8674(00)81401-4 [pii]
99. Serrano M, Hannon GJ, Beach D (1993) A new regulatory motif in cell-cycle control causing specific inhibition of cyclin d/cdk4. Nature 366(6456):704–707. doi:10.1038/366704a0
100. Jacobs JJ, Scheijen B, Voncken JW et al (1999) Bmi-1 collaborates with c-myc in tumorigenesis by inhibiting c-myc-induced apoptosis via ink4a/arf. Genes Dev 13(20):2678–2690
101. Salaverria I, Espinet B, Carrio A et al (2008) Multiple recurrent chromosomal breakpoints in mantle cell lymphoma revealed by a combination of molecular cytogenetic techniques. Genes Chromosomes Cancer 47(12):1086–1097. doi:10.1002/gcc.20609
102. Tort F, Hernandez S, Bea S et al (2005) Checkpoint kinase 1 (chk1) protein and mrna expression is downregulated in aggressive variants of human lymphoid neoplasms. Leukemia 19(1):112–117. doi:2403571 [pii] 10.1038/sj.leu.2403571
103. Tort F, Hernandez S, Bea S et al (2002) Chk2-decreased protein expression and infrequent genetic alterations mainly occur in

aggressive types of non-hodgkin lymphomas. Blood 100(13):4602–4608. doi:10.1182/blood-2002-04-1078 2002-04-1078 [pii]
104. Hernandez L, Hernandez S, Bea S et al (1999) C-myc mrna expression and genomic alterations in mantle cell lymphomas and other nodal non-Hodgkin's lymphomas. Leukemia 13(12):2087–2093
105. Chen RW, Bemis LT, Amato CM et al (2008) Truncation in ccnd1 mrna alters mir-16-1 regulation in mantle cell lymphoma. Blood 112(3):822–829. doi:blood-2008-03-142182 [pii] 10.1182/blood-2008-03-142182
106. He L, Thomson JM, Hemann MT et al (2005) A microrna polycistron as a potential human oncogene. Nature 435(7043):828–833. doi:nature03552 [pii] 10.1038/nature03552
107. Parry-Jones N, Matutes E, Morilla R et al (2007) Cytogenetic abnormalities additional to t(11;14) correlate with clinical features in leukaemic presentation of mantle cell lymphoma, and may influence prognosis: a study of 60 cases by fish. Br J Haematol 137(2):117–124. doi:BJH6526 [pii] 10.1111/j.1365-2141.2007.06526.x
108. Katzenberger T, Kienle D, Stilgenbauer S et al (2008) Delineation of distinct tumour profiles in mantle cell lymphoma by detailed cytogenetic, interphase genetic and morphological analysis. Br J Haematol 142(4):538–550. doi:BJH7199 [pii] 10.1111/j.1365-2141.2008.07199.x
109. Nagy B, Lundan T, Larramendy ML et al (2003) Abnormal expression of apoptosis-related genes in haematological malignancies: overexpression of myc is poor prognostic sign in mantle cell lymphoma. Br J Haematol 120(3):434–441. doi:4121 [pii]
110. Schaffner C, Idler I, Stilgenbauer S et al (2000) Mantle cell lymphoma is characterized by inactivation of the atm gene. Proc Natl Acad Sci USA 97(6):2773–2778. doi:10.1073/pnas.050400997 050400997 [pii]
111. Tabe Y, Sebasigari D, Jin L et al (2009) Mdm2 antagonist nutlin-3 displays antiproliferative and proapoptotic activity in mantle cell lymphoma. Clin Cancer Res 15(3):933–942. doi:15/3/933 [pii] 10.1158/1078-0432.CCR-08-0399
112. Marzec M, Kasprzycka M, Lai R et al (2006) Mantle cell lymphoma cells express predominantly cyclin d1a isoform and are highly sensitive to selective inhibition of cdk4 kinase activity. Blood 108(5):1744–1750. doi:blood-2006-04-016634 [pii] 10.1182/blood-2006-04-016634
113. Kridel R, Meissner B, Rogic S et al (2012) Whole transcriptome sequencing reveals recurrent notch1 mutations in mantle cell lymphoma. Blood. doi:blood-2011-11-391474 [pii] 10.1182/blood-2011-11-391474

Chapter 10

Copy Number Analysis in EBV-Positive Nodal Peripheral T Cell Lymphoma, Unspecified

Kyusam Choi, Hyunjeong Ju, and Young Hyeh Ko

Abstract

Copy number alteration was investigated in three cases of EBV+nodal peripheral T cell lymphoma of cytotoxic phenotype using Agilent array comparative genomic hybridization platform. While no recurrent abnormality among three cases was identified, genomic abnormalities involving 9p23.3, 16p13.3, and 7q34 were overlapped with changes reported in other type of NK or T cell lymphomas previously. Chromosomal region 9p23.3 contains CDKN2A and CDKN2B which are frequently mutated or deleted in various types of cancer and may be a target to find genes associated with the pathogenesis of EBV-positive nodal peripheral T cell lymphoma.

Key words: Array comparative genomic hybridization, EBV, T cell, Lymphoma

1. Introduction

Epstein–Barr virus (EBV) is a ubiquitous human herpes virus and infects more than 90% of individuals. In normal hosts, EBV has been shown to be latent in B cells for life after primary infection. Uncommonly, EBV can infect T or natural killer (NK) cells to cause unique systemic lymphoproliferative diseases (LPDs) which are much more common in Asia, especially Japan, Korea, and China. Such epidemiologic data strongly suggest that common genetic factor predisposes to development of this disease in the Orientals.

EBV-positive systemic T cell LPD of childhood, extranodal NK/T cell lymphoma of nasal type, and aggressive NK cell leukemia are well-recognized EBV+LPDs listed in the 2008 WHO classification of the tumor of lymphoid cells (1). In addition to these diseases, EBV is also detected in subsets of peripheral T cell lymphoma. Frequently EBV is detected only in a proportion of

Diponkar Banerjee and Sohrab P. Shah (eds.), *Array Comparative Genomic Hybridization: Protocols and Applications*, Methods in Molecular Biology vol. 973, DOI 10.1007/978-1-62703-281-0_10, © Springer Science+Business Media, LLC 2013

the lymphoid cells, usually in nonneoplastic cells; however, rare cases show EBV infected in the vast majority of tumor cells. Such peripheral T cell lymphoma with high viral burden has been described previously in Asian patients (2, 3) and usually of cytotoxic T cell phenotype and pursue very aggressive clinical course (4, 5).

Genetic changes in EBV-associated LPD are well investigated in extranodal NK/T cell lymphoma. The commonest chromosomal changes detected by aCGH are 6q−, 2q+, 8p−, 11q−, 12q−, 13q−, 15q+, 17q+, and 22q+. Specific chromosomal translocations have not been identified, but translocations involving the 8p23 breakpoint are reported in a number of cases (6, 7).

Because the cases are very uncommon, genetic alteration of EBV+peripheral T cell lymphoma of cytotoxic phenotype has not been studied. Here we investigated copy number change of three cases of EBV-positive peripheral T cell lymphomas.

2. Materials

Fresh frozen tissue from lymph node biopsy was used for study. The quality of frozen samples was assessed by H&E stain. All samples contained EBV in more than 50% of atypical lymphocytes.

3. Methods

3.1. Extraction and Purification of Genomic DNA gDNA with QIAGEN QIAmp DNA Mini Kit (Cat. No. 51306)

1. Put the tissue (0.2–0.3 cm) in Petri dish with 1× PBS.
2. Remove OCT media by pipetting.
3. Add 1× PBS and centrifuge at 6,000 × *g* for 30 s.
4. Repeat step 3.
5. Chop the tissue using the surgical knife.
6. Add 180 μl of ATL buffer and add 20 μl of proteinase K, mix by vortexing, and overnight incubate in rocking heat block at 56 °C.
7. Centrifuge at 6,000 × *g* for 30 s.
8. Add 4 μl RNaseA (100 mg/ml), vortex, and incubate for 2 min at RT.
9. Briefly centrifuge, add 200 μl buffer AL, mix, and incubate at 70 °C for 10 min.
10. Briefly centrifuge, add 200 μl 100% ethanol, and centrifuge at 6,000 × *g* for 30 s.
11. Transfer the sample mixture onto a DNeasy Mini spin column and centrifuge at 6,000 × *g* for 1 min.

12. Add 500 μl buffer AW1 to QIAamp Mini spin column. Centrifuge at 6,000 × *g* for 1 min. Discard the flow-through.
13. Add 500 μl of buffer AW2 to QIAamp Mini spin column. Centrifuge at 20,000 × *g* for 3 min. Discard the flow-through.
14. Place the column in a new 2 ml collection tube. Centrifuge at 20,000 × *g* for 3 min.
15. Place the column in a clean microcentrifuge tube. Add 200 μl of nuclease-free water. Incubate at RT for 1 min, and centrifuge at 6,000 × *g* for 1 min.
16. Repeat step 15. Combine the duplicate samples in one microcentrifuge tube for a final volume of 400 μl.
17. Use the NanoDrop ND-1000 UV–VIS spectrophotometer to assess gDNA concentration and purity. High-quality gDNA samples should have an *A260*/*A280* ratio of 1.8–2.0 and *A260*/*A230* ratio of >2.0.

3.2. Restriction Digestion

Use the following procedure to aCGH in a 4× microarrays format.

1. Make the gDNA mixture (test genomic DNA and reference genomic DNA).

Genomic DNA (option: heat the DNA at 55/60 °C for 10 min before using)	1.5 μg
Nuclease-free water up to volume	20.2 μl

2. Preparation of digestion master mix (for 4 × 44 K microarrays).

Component per reaction	(μl)
10× buffer C[a]	2.6
Acetylated BSA (10 μg/μl)[a]	0.2
*Alu*I (10 U/μl)	0.5
*Rsa*I (10 U/μl)	0.5
Nuclease-free water	2.0
Final volume	5.8

[a]Supplied with the restriction enzyme *Rsa*I

3. Add 5.8 μl of digestion master mix to the genomic DNA and make a total volume of 26 ml.
4. Volume of 26 μl mix well by pipetting up and down.
5. Incubate at 37 °C for 2 h.
6. Incubate at 65 °C for 20 min to inactivate the enzymes.
7. Move the sample tubes to ice.
8. Spin the samples in a centrifuge for 1 min at 6,000 × *g*.
9. Add 5 μl of random primers (random primers supplied with the Agilent Genomic DNA enzymatic labeling kit).

10. Incubate at 95 °C for 3 min then move to ice.
11. Centrifuge for 1 min at 6,000×g.
12. Prepare labeling master mix

Component per reaction	(μl)
5× buffer	10.0
10× dNTP	5.0
Exo-Klenow fragment	1.0
Cyanine 3-dUTP(1.0 mM) or Cyanine 5-dUTP(1.0 mM)	3.0
Final volume	19.0

13. Add 19 μl of labeling master mix to 31 μl of digested gDNA, mix well.
14. Gently pipette up and down.
15. Incubate at 37 °C for 2 h.
16. Incubate at 65 °C for 10 min to inactivate the enzyme, then move to ice.

3.3. Cleanup of Labeled Genomic DNA

1. Spin the samples in a centrifuge for 1 min at 6,000×g.
2. Add 430 μl of 1× TE (pH 8.0).
3. Load each labeled gDNA into the Amicon 30 kDa filter.
4. Centrifuge for 10 min at 14,000×g and discard the flow-through.
5. Add 480 μl of 1× TE (pH 8.0) to filter. Spin for 10 min at 14,000×g.
6. Invert the filter into a fresh tube and centrifuge for 1 min at 1,000×g.
7. Spin until each sample volume is 21 μl by using SpeedVac.
8. Bring total sample volume to 21 μl with 1× TE (pH 8.0).
9. Take 1.5 μl of each sample to determine the yield and specific activity by using the NanoDrop ND-1000 UV–VIS spectrophotometer.
10. Expected yield (5–7 μg) and specific activity(Cy3: 25 to 40 pmol/mg, Cy5: 20 to 35 pmol/mg) after labeling and clean up.
11. Combine test and reference sample using the cyanine 5-labeled sample and cyanine 3-labeled sample for a total mixture volume 39 ml.

3.4. Preparation of Labeled Genomic DNA for Hybridization

1. Preparation of hybridization master mix:

Component	Volume (μl) per hybridization
Cyanine 5- and cyanine 3-labeled gDNA mixture	39

Human Cot-1 DNA (1.0 mg/ml)	5
Agilent 10× blocking agent	11
Agilent 2× hybridization buffer	55
Final hybridization sample volume	110

2. Mix the sample by pipetting up and down and incubate at 95 °C for 3 min.
3. Incubate at 37 °C for 30 min.
4. Spin the samples in a centrifuge for 1 min at 6,000 × *g*.

3.5. Hybridization Assembly

1. Load a clean gasket slide into the Agilent SureHyb chamber base with the gasket label facing up and aligned with the rectangular section of the chamber base.
2. Slowly dispense 110 μl of hybridization sample mixture onto the gasket well in a "drag and dispense" manner.
3. Place a microarray "active side" down onto the SureHyb gasket slide, so the numeric barcode side is facing up and the "Agilent"-labeled barcode is facing down. Assess that the sandwich pair is properly aligned.
4. Place the SureHyb chamber cover onto the sandwiched slides and slide the clamp assembly onto both pieces.
5. Hand tighten the clamp onto the chamber.
6. Vertically rotate the assembled chamber to wet the slides and assess the mobility of the bubbles.
7. Place assembled slide chamber in the rotator rack in a hybridization oven set to 65 °C. Set your hybridization rotator to rotate at 20 rpm.
8. Hybridize at 65 °C for 24 h.

3.6. Wash Preparation

1. Prewarm oligo aCGH Wash Buffer 2 (overnight).
2. Put a clean magnetic stir bar into the stabilization and drying solution bottle and recap.
3. Partially fill a plastic bucket with hot water at approximately 40–45 °C.

3.7. Microarray Washing

1. Completely fill slide-staining dish #1 with oligo aCGH Wash Buffer 1 at room temperature.
2. Place a slide rack into slide-staining dish #2. Add a magnetic stir bar and fill with aCGH Wash Buffer 1 at room temperature.
3. Put the prewarmed 1.5 l glass dish filled with water and containing slide-staining dish #3 on a magnetic stir plate with heating element and fill with aCGH Wash Buffer 2 warmed to 37 °C.

4. Turn on the heating element and maintain temperature of aCGH Wash Buffer 2 at 37 °C; monitor using a thermometer.
5. Remove one hybridization chamber from incubator.
6. Place the hybridization chamber assembly on a flat surface and loosen the thumbscrew, turning counterclockwise.
7. Slide off the clamp assembly and remove the chamber cover.
8. With gloved fingers, remove the array-gasket sandwich from the chamber base by grabbing the slides from their ends. Keep the microarray slide numeric barcode facing up as you quickly transfer the sandwich to slide-staining dish #1.
9. Without letting go of the slides, submerge the array-gasket sandwich into slide-staining dish #1 containing aCGH Wash Buffer 1.
10. With the sandwich completely submerged in oligo aCGH Wash Buffer 1, pry the sandwich open from the barcode end only.
11. Do this by slipping one of the blunt ends of the forceps between the slides and then gently turn the forceps upwards or downwards to separate the slides.
12. Let the gasket slide drop to the bottom of the staining dish.
13. Remove the microarray slide and place into slide rack in the slide-staining dish #2.
14. Stir using setting 4 for 5 min.
15. Transfer slide rack to slide-staining dish #3 containing aCGH Wash Buffer 2 at 37 °C and stir for 1 min.
16. Slowly remove the slide rack trying to minimize droplets on the slides. It should take 5–10 s to remove the slide rack.
17. Put the slides with Agilent barcode facing up in a slide holder.

3.8. Microarray Scanning Using Agilent

1. Agilent scanner settings

   Agilent scanner control software v7.0 is recommended for 5 μm scan of 44 K microarray.

   (a) Assemble slides into appropriate slide holders such that the numeric barcode (not the "Agilent"-labeled barcode) is visible.

   (b) Place assembled slide holders into scanner carousel.

   (c) Verify default scan settings.

   - Scan region is set to scan area (61 × 21.6 mm).
   - Scan resolution (μm) is set to 5 for 44 K microarrays.
   - Dye channel is set to red and green.
   - Green PMT is set to 100%.
   - Red PMT is set to 100%.

(d) Select settings for the automatic file naming.
  - Prefix1 is set to instrument serial number.
  - Prefix2 is set to array barcode.

(e) Verify that the scanner status in the main window says scanner ready.

(f) Start reading.

3.9. Data Analysis

1. CGH data ware extracted from scanned images (TIFF files) using Feature Extraction software (version 10.3 Agilent Technologies).
2. Feature Extraction employs a process by which data are extracted from the scanned microarray image and translated into log2 ratios of the Cy3 (green)-labeled normal DNA and Cy5 (red)-labeled tumor DNA signals in each of 60 K probes, allowing us to measure DNA copy number aberrations in our experiments in conjunction with Agilent Genomic Workbench v6.5 software (Agilent Technologies).
3. Raw data text files of the log2 ratios were transferred to the Genomic Workbench v6.5 software (Agilent) for further analysis.
4. The raw data was visualized as a graphical figure within the Genomic Workbench v6.5 software.
5. Aberrations were detected with the aberration detection method 2 (ADM-2), whose threshold was 6.0, and filtering options of the Genomic Workbench v6.5 software.

   The ADM-2 algorithm identifies all aberrant intervals in a given sample with consistently high or low log ratios based on the statistical score.
6. We defined gains and losses over a continuous minimum number of absolution 3 probe dataset as a linear log2 ratio average of ≥ 0.25 and ≤ -0.25.
7. The linear average of the log2 ratio refers to the non-weighted average of the absolute log2 ratio.
8. Finally, we validated significant differences in the log2 ratio averages of sequential probe datasets of each aberration region and those of normal regions by the Welch test with $p \leq 0.05$.

4. Result

4.1. Clinical and Pathologic Findings

Three cases of peripheral T cell lymphoma were diagnosed based on the 2008 WHO classification (1). The patients were 49, 58, and 65 years old and comprised of two men and one woman. One patient had HCV hepatitis and another patient was HBV carrier.

The patients presented with systemic lymphadenopathy and died of disease in 4 days, 24 days, and 7 months, respectively, after diagnosis. The cause of death was septic shock in two patients and disease progression in one patient. Immunohistochemical stain of the lymph node was performed using paraffin embedded tissues and showed CD3 + CD4−CD8 + CD56-TIA1 + immunophenotype. EBV in situ hybridization was performed using EBER fluorescein conjugated oligonucleotide probe (BOND™ ISH probe). EBER positive cells were more than 50, 80, and 90% of all lymphocytes of tumor tissue in each case.

4.2. Copy Number Alteration in EBV + Peripheral T Cell Lymphoma

Case 1 showed gains of 8p11.22 and loss of 3q26.1, 7q34, 14q11.2, 15q11.1–q11.2, and 19q13.33 with heterozygous deletion of 4q34.1 which contains GALNTL6. Case 2 showed loss of 2p24.1, 7q21.13, and 9p21.31. Case 3 showed loss of 9q34.3 and 16p13.3 and gain of 1p36.33–p36.31, 2q13, 3p25.2, 3q22.1, 4p16.3,5p15.33, 5q35.2, 6p22.1, 6q25.1, 7p22.3, 9q34.2-q34.3, 10q26.3, 11q13.1–q13.4, 16p13.3, 16q22.1, 16q24.2-q24.3, 17q25.3, 19q13.32, and 20q13.33 with high-level amplification of 7q22.1, 8q24.3, 9q34.2, 11p15.5, 11q23.3, 14q32.33, 16q24.3, 19p13.3–p13.11, 19p13.3, 19p13.12, 20p11.22, 21q22.3, and 22q13.33. There was no recurrent copy number alteration among three cases.

4.3. Comparison with the Data from Other T or NK Cell Lymphoma

aCGH data of our cases was compared with the data reported in other T or NK cell lymphomas (8). Two cases showed altered region involving 7q34 and 16p13.3, respectively, of which change was reported in nasal-type NK/T cell lymphoma. There was gain of 7q34 which contains PRSS1, TRY6, and PRSS2 genes. Chromosome region 16p13.3 contains FBXL16, METRN, and FAM173A. Loss of 16p13.3 was present in our case, whereas NK/T cell lymphoma reported previously showed gain of 16p13.3. One case showed loss of 9p21.3 which was described in peripheral T cell lymphoma, unspecified with aggressive clinical course (9).

5. Discussion

In this study, we demonstrated genetic alterations in EBV-positive peripheral T cell lymphoma through genome-wide aCGH. Unfortunately we failed to find recurrent copy number alteration in EBV-positive peripheral T cell lymphoma which may ascribe to small number of cases or genetic heterogeneity of EBV-positive peripheral T cell lymphoma. Nevertheless, two altered regions in EBV-positive peripheral T cell lymphoma overlapped with those described in EBV-associated NK/T cell lymphoma which is a prototype of EBV-associated lymphoma. The Epstein–Barr virus (EBV) genome is maintained as an extrachromosomal episome during latent infection of B lymphocytes (10). Latent infected B cells express viral antigens including LMP1, LMP2, and EBNA2 which play a role in the lymphomagenesis. In malignant lymphomas, EBV exists in episomal form but can integrate into the host genome (11). So far distinct genetic abnormality associated with EBV-positive malignant T or NK cell lymphoma is not known.

Nasal-type NK/T cell lymphoma is the most common form of EBV-associated non-B cell malignant lymphoma which occurs predominantly in Asian population. According to the literature, 6q21 loss was the most frequent aberration. The gains of 1q31.2–44, 7q11.22–36.3, 16p13.3, and 11p15.5, and the losses of 6q16.1–27 and 7p15.3–22.2, were found in more than 20% of cases (8). Although alterations involving similar region were observed in EBV-positive peripheral T cell lymphoma in our case, the significance and relation with EBV infection are not known.

It is intriguing that loss of 9p21.3 observed in our case was also reported in peripheral T cell lymphoma with poor prognosis. Nakagawa et al. analyzed genomic profile using BAC array and found that frequent genomic imbalances including homozygous loss of 9q21.2 and high copy number gain at 14q32.2. Gains of 7p and 7q and loss of 9p21.3 were significantly associated with poor prognosis (9). In their BAC array, 9p21.3 contains MLLT3, IFNA5, KLHL9, IFNA17, MTAP, CDKN2A, and CDKN2B. In our platform, 9p21.3 contains IFNA5, KLHL9, IFNA6, CDKN2A, and CDKN2B. *CDKN2A* gene encodes cyclin-dependent kinase inhibitor 2A, (CDKN2A, $p16^{Ink4A}$) also known as multiple tumor suppressor 1 (MTS-1) (12). P16 plays an important role in regulating the cell cycle, and mutations or deletion in p16 increases the risk of developing a variety of cancers including malignant lymphoma. CDKN2B lies adjacent to the tumor suppressor gene CDKN2A in a region that is frequently mutated and deleted in a wide variety of tumors. This gene encodes a cyclin-dependent kinase inhibitor, also known as $p15^{Ink4b}$ protein, which forms a complex with CDK4 or CDK6 and prevents the activation of the CDK kinases by cyclin D, thus the encoded protein functions as a cell growth regulator that inhibits cell cycle G1 progression (13). Homozygous loss of 9p21.3 in peripheral T cell lymphoma with poor prognosis suggests that CDKN2A and CDKN2B genes located in this region play an important role in the biology of peripheral T cell lymphoma. No viral study was performed in the study by Nakagawa et al. (9).

In summary, we analyzed EBV-positive peripheral T cell lymphomas with Agilent aCGH platform. There was no recurrent copy number alteration; however, 9p23.3, 16p13.3, and 7q34 may be targets to find candidate genes associated with the pathogenesis of EBV-positive T cell lymphomas.

Acknowledgment

This study is supported by Samsung Biomedical Research Institute grant (C-B0-202-2) and a grant of the Korea Healthcare technology R&D Project, Ministry for Health and Welfare Affairs, Republic of Korea (A092255).

References

1. Steven H, Swerdlow EC, Harris NL et al (2008) WHO classification of tumours of haematopoietic and lymphoid tissues, 4th edn. IARC, Lyon
2. Asano N, Suzuki R, Kagami Y et al (2005) Clinicopathologic and prognostic significance of cytotoxic molecule expression in nodal peripheral T-cell lymphoma, unspecified. Am J Surg Pathol 29:1284–1293
3. Kagami Y, Sobue R, Ito N et al (1999) Cytotoxic large T-cell lymphoma with fulminant clinical course, CD8+ and CD56- phenotype, and its relation to Epstein–Barr virus: a report of two cases. Int J Hematol 70:105–111
4. Dupuis J, Emile JF, Mounier N et al (2006) Prognostic significance of Epstein–Barr virus in nodal peripheral T-cell lymphoma, unspecified: a Groupe d'Etude des Lymphomes de l'Adulte (GELA) study. Blood 108:4163–4169
5. Langer R, Geissinger E, Rüdiger T et al (2010) Peripheral T-cell lymphoma with progression to a clonally related, Epstein–Barr virus+, cytotoxic aggressive T-cell lymphoma: evidence for secondary EBV infection of an established malignant T-cell clone. Am J Surg Pathol 34:1382–1387
6. Wong N, Wong KF, Chan JK et al (2000) Chromosomal translocations are common in natural killer-cell lymphoma/leukemia as shown by spectral karyotyping. Hum Pathol 31:771–774
7. MacLeod RA, Nagel S, Kaufmann M et al (2002) Multicolor-FISH analysis of a natural killer cell line (NK-92). Leuk Res 26:1027–1033
8. Karube K, Nakagawa M, Tsuzuki S et al (2011) Identification of FOXO3 and PRDM1 as tumor-suppressor gene candidates in NK-cell neoplasms by genomic and functional analyses. Blood 118:3195–3204
9. Nakagawa M, Nakagawa-Oshiro A, Karnan S et al (2009) Array comparative genomic hybridization analysis of PTCL-U reveals a distinct subgroup with genetic alterations similar to lymphoma-type adult T-cell leukemia/lymphoma. Clin Cancer Res 15:30–38
10. Dheekollu J, Lieberman PM (2011) The replisome pausing factor Timeless is required for episomal maintenance of latent Epstein–Barr virus. J Virol 85:5853–5863
11. Ohshima K, Suzumiya J, Kanda M (1998) Integrated and episomal forms of Epstein–Barr virus (EBV) in EBV associated disease. Cancer Lett 122:43–50
12. Nobori T, Miura K, Wu DJ et al (1994) Deletions of the cyclin-dependent kinase-4 inhibitor gene in multiple human cancers. Nature 368:753–756
13. Hannon GJ, Beach D (1994) p15INK4B is a potential effector of TGF-beta-induced cell cycle arrest. Nature 371:257–261

Chapter 11

Mycosis Fungoides and Sézary Syndrome

Blanca Espinet and Rocío Salgado

Abstract

The development of array comparative genomic hybridization (aCGH) techniques has allowed to characterize more precisely several human neoplasms with the aim of providing prognostic markers and targets for directed therapeutic intervention. Recently, several studies applying aCGH technique have been reported in which an exhaustive genetic characterization of mycosis fungoides (MF) and Sézary syndrome (SS) has been performed. Regarding MF, a genomic profile characterized by the gains of 7q, 17q, and 8q and losses in 9p, 13q, 17p, and 10q has been described. In SS, the most common abnormalities are gains in 8q and 17q and losses at 17p and 10q. One of the main contributions of the aCGH studies in MF and SS has been the description of genetic markers associated with a poor prognosis. In MF, three specific chromosomal regions, 9p21.3 (*CDKN2A*, *CDKN2B*, and *MTAP*), 8q24.21 (*MYC*), and 10q26qter (*MGMT* and *EBF3*) have been defined as prognostic markers exhibiting a significant correlation with overall survival ($P=0.042$, $P=0.017$, and $P=0.022$, respectively). Moreover, two MF genomic subgroups have been described, distinguishing a stable group (0–5 DNA aberrations) and an unstable group (>5 DNA aberrations), showing that the genomic unstable group had a shorter overall survival ($P=0.05$).

Key words: aCGH, Primary cutaneous T-cell lymphomas, Mycosis fungoides, Sézary syndrome, Conventional cytogenetics, Multicolor FISH, Genomic profile

1. Introduction

1.1. Primary Cutaneous T-Cell Lymphomas

Primary cutaneous lymphoproliferative disorders represent approximately 30% of all lymphomas. They are the second most common type of extranodal non-Hodgkin lymphomas, after the gastrointestinal tract, with an estimated annual incidence of 1:100,000 (1). Primary cutaneous lymphomas account for a heterogeneous group of neoplasms that present in the skin with no evidence of extracutaneous disease at the time of diagnosis. These entities often show clinical, histological, immunophenotypic, and disease progression characteristics that differ from those of their nodal counterparts. Primary cutaneous lymphomas are divided mainly into two types depending on the origin of the neoplastic cells: T-cell and NK

Diponkar Banerjee and Sohrab P. Shah (eds.), *Array Comparative Genomic Hybridization: Protocols and Applications*, Methods in Molecular Biology vol. 973, DOI 10.1007/978-1-62703-281-0_11, © Springer Science+Business Media, LLC 2013

cutaneous lymphomas and B-cell cutaneous lymphomas (2, 3). Unlike nodal lymphomas, of which the B-cell lymphoproliferative syndromes represent the majority, cutaneous T-cell lymphomas constitute the most frequent type, being approximately 77% of primary cutaneous lymphomas.

Primary cutaneous T-cell lymphomas (PCTCL) are characterized by a clonal proliferation of lymphoid T cells with a tendency to infiltrate the epidermis (epidermotropism). The clinical presentation is highly variable, with macules or erythematous, infiltrated scaly plaques observed in the early stages of the disease, whereas in advanced stages, erythematous violaceous nodules with a tendency to ulceration are seen, as well as erythroderma in the leukemic variants of the disease (Sézary syndrome [SS]) (4, 5). The most frequent type of PCTCL is mycosis fungoides (60%), followed by primary cutaneous CD30-positive T-cell lymphoproliferative disorders (25%), including lymphomatoid papulosis and primary cutaneous anaplastic large cell lymphomas (PCALCL).

1.2. Mycosis Fungoides and Sézary Syndrome

Mycosis fungoides (MF) is a primary cutaneous lymphoma of mature CD4+ T-helper cells showing epidermotropism. It is considered an indolent lymphoma, with a clinical course that can take years or even decades, characterized by the sequential development of early lesions as macules that may evolve into infiltrated plaques and occasionally to tumors. As the disease progresses, the cancer cells lose their affinity for the skin and tend to develop a nodal extracutaneous involvement, including hematologic or visceral. MF is the most common of all PCTCL with an annual incidence of 0.9 per 100,000 in the United States (1).

Sézary syndrome (SS) is considered the leukemic variant of MF. Its clinical manifestation is characterized by the presence of erythroderma, lymphadenopathy, and more than 1,000 cells per mm^3 circulating atypical lymphoid cells (Sézary cells/mm^3). In the WHO/EORTC cutaneous T-cell lymphoma classification, SS is considered as an aggressive lymphoma, as it progresses faster and it is more resistant to treatment than conventional MF. SS is a rare entity with an annual incidence of 0.3 cases per 1,000,000 people in the United States (1).

1.3. Chromosomal Abnormalities of Mycosis Fungoides

MF is a relatively little-studied tumor with conventional cytogenetic techniques because of the difficulty in obtaining dividing cells from the skin tumor (6). The same limitation is found when multicolor fluorescence in situ hybridization techniques (M-FISH and Spectral Karyotyping or SKY) are applied, due to the need of metaphases for the interpretation. However, there are several publications in which tumor cells are analyzed in peripheral blood, from which it is easier to obtain metaphases and to perform the conventional karyotypes and/or M-FISH (7, 8). The most prominent studies in which M-FISH is applied are those published by Batista et al. (8) and

Padilla-Nash et al. (9). In the first study, Batista et al. applied the SKY technique using peripheral blood samples from eight patients with advanced-stage MF. Only two of these patients had cytogenetic abnormalities, which could not be considered recurrent among different patients (8). Later, Padilla-Nash et al. studied four patients with MF in early stages of which four cell lines from skin biopsies were established. Of these, chromosome abnormalities were analyzed by using the SKY technique, in which each chromosome is identified by a different color. Of the four patients, two had a normal karyotype while the remaining two patients had isolated abnormal clones, being trisomy 21, del(5)(p15.1), del(17)(p12), and del(18)(p11.2) the most frequent alterations. With respect to structural alterations detected, none of them might be considered recurrent (9).

Regarding comparative genomic hybridization (CGH), it is a molecular cytogenetic technique that allows the study of chromosomal gains and losses in the whole genome. This technique has two main advantages: (1) it provides an overview of the genome in a single experiment and (2) it does not require dividing cells as with conventional cytogenetics or M-FISH methodologies. Due to the difficulty of obtaining dividing cells from tumoral skin samples, the CGH technique has been the most widely used for genetic characterization of MF before the development of aCGH platforms. The most frequently affected chromosomes in MF are 1, 8, 9, 10, 11, 12, and 17 as described in different series (10–12). Although there are many published studies describing recurrent genetic alterations in MF, some of which have been associated with a poor prognosis, it seems that the relevance of these cytogenetic markers is limited because of the low number of patients included in those series as well as the heterogeneity of patients selection (some series include a mixture of MF, SS, and primary cutaneous ALCL). Recently, Prochazkova et al. (13) analyzed a homogeneous series of patients diagnosed with advanced-stage MF by conventional CGH. The recurrent changes identified were losses of 17p, 10q, 2q36qter, and 9p21 and gains of chromosome 7, 17q, 9q34qter, and 8q (13).

1.4. Cytogenetic Abnormalities of Sézary Syndrome

In contrast to MF, in SS there are numerous studies that have applied both conventional cytogenetics and M-FISH techniques, due to the availability of tumoral cells in peripheral blood (8, 14–17). In the review published by Mao et al. (16), they compared their results with data from other published studies using conventional cytogenetics and/or M-FISH. Mao et al. (16) concluded that the most frequently reported changes in the SS were losses of chromosomes 10 and 1p, the presence of isochromosome 17q, and translocations affecting 1p, 10q, and 14q.

With respect to conventional CGH, there are few published studies in SS (10, 11, 18). The most relevant was carried out by

Mao et al. (11), in which a total of 20 SS were analyzed. The most frequently observed changes were losses at 1p, 10, 10q, 17p, 19 and gains at 17q and 18.

2. Materials

In the first publications using the conventional CGH technique, the most frequently used samples were from peripheral blood (10, 11, 16, 19). Despite several chromosomal abnormalities which have been described using this source, blood involvement in mycosis fungoides is variable and could be very low in early stages. In fact, the results among these different studies are very heterogeneous, and one of the reasons is the degree of involvement.

The first homogeneous series studied by conventional CGH technique was described by Prochazkova et al. (13), which included 11 patients diagnosed with tumor-stage MF (MFt). In this study, genomic DNA was extracted from fresh frozen tissue which is the most appropriate tissue sample. The analysis of an advanced stage of MF using fresh frozen tissue will ensure the presence of a high proportion of tumoral cells and also good DNA quality which will ensure high-quality aCGH results.

3. Methods

The most important step to ensure a good result of the aCGH technique is the starting material. It is well known that genomic DNA extracted from fresh or frozen tissue is intact and will probably lead to high-quality hybridization.

One advantage of frozen tissue is that a hematoxylin–eosin staining of a tissue section can be assessed prior to DNA extraction in order to check the content of tumoral cells in each sample. It is recommended that hematoxylin–eosin-stained sections of all frozen samples be reviewed in order to verify the presence of a minimum of 70% of atypical lymphocytes infiltration.

3.1. Tissue Preservation

To ensure the high quality of DNA samples, it is advisable to embed the skin biopsies in Tissue-Tek® OCT™ Compound (Sakura Finetek Tissue-Tek OCT Compound, Zoeterwoude, Netherlands) and store them at −80 °C. This initial step will allow good preservation of DNA in the samples.

3.2. DNA Extraction

A minimum of 14 × 10 μm snap frozen sections from skin lesions are necessary to obtain enough DNA. DNA extraction can be

Table 1
Different aCGH platforms used for the study of MF and SS genomic profiles

References	Type of CTCL	aCGH platform	Characteristics
Mao et al. (16)	MF SS	BAC aCGH	*AmpliOnc I DNA array* (Vysis): 59 clones which cover 57 known oncogenes
Vermeer et al. (23)	SS	BAC aCGH	Homemade aCGH platform that contains 3,500 BAC clones
Van Doorn et al. (20)	MF	BAC aCGH	Homemade aCGH platform that contains 3,500 BAC clones
Caprini et al. (24)	SS	SNP arrays	GeneChip human mapping 10 K array (Affymetrix) containing 10,000 tiled SNPs
Salgado et al. (21)	MF	Oligonucleotide aCGH	44 K human CGH microarray (Agilent Technologies): 44,000 probes that cover all the genome with a resolution of 75–100 kb
Laharanne et al. (22)	MF SS	BAC aCGH	3 K and 5 K IntegraChip (IntegraGen), platforms that contains 3,113 and 5,600 BAC clones, respectively

performed using a classical Phenol:Chloroform extraction or commercial kits. Of the commercial kits, one of the most commonly used one is the DNeasy Blood and Tissue Kit (Qiagen, Hilden, Germany).

The quality of DNA extraction is dependent on the optimum preservation conditions and the sample size.

3.3. Quality Control

Two parameters can be checked in order to know the integrity of the DNA samples. Firstly, to quantify the purity of genomic DNA obtained from the extraction process, the sample is analyzed using a spectrophotometer (Nanodrop, Wilmington, DE, USA). Secondly, to analyze the integrity of the DNA, a 1% agarose gel is performed to check that the DNA is intact.

3.4. Array Comparative Genomic Hybridization

The vast majority of aCGH protocols are based on a DNA digestion step followed by a labeling process, a hybridization procedure, and finally washings and scanning. Nowadays, there are several commercial platforms with standard protocols that are consistently robust. All technical protocols can be downloaded from the Web sites of each company. The detailed technical protocols of laboratory-developed aCGH platforms can be obtained from the original papers. aCGH platforms used in the analysis of the genomic profile of SS and MF, as found in the literature, are summarized in Table 1.

3.4.1. Data Analysis and Interpretation

For the interpretation and visualization of results, each commercial platform has its own analysis software. Basically, all these versions of software allow the evaluation of the quality of the hybridization process with a specific parameter (e.g., Agilent Technologies use the derivative log ratio (DLR) spread which indicates the degree of dispersion between two consecutive clones of the array). Later, all these types of software provide a tool to analyze the results at the genomic, gene, and chromosome level to visualize each region.

4. Clinical and Research Applications

4.1. MFt Genomic Profile

Numerous cytogenetic techniques have been used to characterize the MF. As discussed in the introduction, conventional cytogenetic analyses are limited due to the difficulty in obtaining dividing cells (6). Thus, molecular cytogenetic techniques, principally CGH and later genomic microarrays, have allowed the description of the genetic alterations of MF.

In the first publications performed with the conventional CGH technique, the most frequently altered chromosomes were chromosomes 1, 8, 9, 10, 11, 12, and 17 (10, 11, 19). However, these studies had some limitations: (1) the analysis of a heterogeneous series which included different types of cutaneous T cell at different stages and, therefore, with a variable clinical course and (2) the limited number of patients.

The first homogeneous series studied by conventional CGH technique was described by Prochazkova et al. (13), which included only 11 patients diagnosed with MFt. In this paper, the most frequent alterations were gains of chromosome 7 and 17q regions, 9q34qter, and 8q and losses of 17p, 10q, and 9p21 2q36qter.

Recently, three studies applying aCGH technique have been reported in which an exhaustive genetic characterization of MFt has been performed. The first one was described by a Dutch group, in which the technique was applied in 22 patients (20). This study used a "noncommercial" BAC aCGH platform containing about 3,500 probes. The main objective of this study was to identify genomic differences between MFt and SS. Gains in different regions of chromosome 7 (7q36, 7q32q35, 7q21q22, 7q11.2, 7p13p14, 7p21p22) 1q31q32, 1p36.2, 8q24, and 17q22 and losses on chromosomes 5q13, 9p21, and 13q14q31 were the most frequent alterations in MFt.

The second study (21) included 41 MFt patients, which is the largest series studied so far. It should be noted that 10 of the 41

Table 2
Recurrent alterations in MFt

References	*n*	Type of aCGH platform	Recurrent alterations Gains	Losses
Van Doorn et al. (20)	22	BAC aCGH	+7q36	-5q13
			+7q21q22	-9p21
			+1p36	-13q14q31
			+17q21	
			+8q24	
Salgado et al. (21)	41	Oligo aCGH	+7q33q35	-9p21.3
			+17q21.1	-9q31.2
			+8q24.21	-17p13.1
			+9q34qter	-13q14.11
			+10p14	
Laharanne et al. (22)	24	BAC aCGH	+7q21	-10q26
		Oligo aCGH	+7q31	-13q34
			+7p22p11.2	-9q21q22
			+1q25q31	-16q23q24
			+17q12	-9p21

patients included were from the Leiden Medical Center (Netherlands) and had been previously analyzed using a different aCGH platform (20). The most frequent anomalies found were, in order of frequency, gains of 7q33.3q35, 17q21.1, 8q24.21, 9q34, and 10p14 and losses of 9p21.3, 9q31.2, 17p13.1, 13q14.11, 6q21.3, 10p11.22, 16q23.2, and 16q24.3.

Subsequently, Laharanne et al. (22) analyzed 58 patients diagnosed with PCTCL, among which 24 were MFt. In this study, there were recurring genetic abnormalities in different regions of chromosome 7 (7p22p11.2, 7q21, 7q31) and 17q12 1q25q31 as well as losses of 9p21, 10p11.2, 10q26, and 17p.

When comparing the results obtained in the publications mentioned above, many of the changes are common (Table 2). This leads us to suggest that MFt has a genomic profile characterized by the gains of 7q, 17q, and 8q and losses in 9p, 13q, 17p, and 10q. However, discrepancies exist between studies, as with 1q gains or losses on 5q, which are more frequently altered in studies conducted by Dutch and French groups, respectively (20, 22). These discrepancies may be due to the different platforms used, the different number of patients included, and the clinical characteristics of these patients (clinical stages).

4.2. SS Genomic Profile

Unlike MF, SS is the most studied CTCL subtype by conventional cytogenetic techniques due to the pathological cells, CD4+ lymphocytes, circulating in peripheral blood. Conventional cytogenetic studies show that over 50% of SS patients have many abnormalities, mainly complex karyotypes (8, 14–17). The most frequent alterations are losses of chromosome 10, deletions of 1p, isochromosome 17q, and translocations involving 1p, 10q, and 14q (16, 17).

Recently, Vermeer et al. (23) reported an aCGH study of 20 SS using a homemade platform of 3,500 BACs. The most frequent genetic abnormalities found were 17q23q25 and 8q24.1q24.3 gains and losses at 17p13.1 and 10q25. Subsequently, Caprini et al. analyzed 28 SS patients using two commercial microarrays: GeneChip Human Mapping 10 K (Affymetrix) and Microarray Kit Human Genome CGH 44B (Agilent Technologies) (24). Both platforms had a higher resolution than the array used previously (23). In this study, Caprini et al. observed many changes similar to those detected by Vermeer et al. as gains that affect 8q and 17q regions and losses at 17p and 10q. However, they also have found differences not only in specific chromosomal regions affecting the chromosome 10q and 17q but also in the frequencies of different genetic changes. Such discrepancies between studies could be due to the different platforms used in each study. Finally, a third study published by Laharanne et al. (22), which included 16 SS patients, reproduced the results found in the aforementioned studies suggesting that the SS is characterized by a high chromosomal instability, with recurring gains and losses. The most frequent chromosomal aberrations and the platforms used are summarized in Table 3.

4.3. Differences and Similarities Between MF and SS Profiles

With regard to the MFt in the work published by van Doorn et al. (20) whose aim was to compare the genetic profiles of MFt and SS, they concluded that the two entities have very different profiles and postulated that they are separate entities. With respect to the results obtained by Salgado et al. (21), among all the detected alterations, many of them (gains of 17q21.1 and 8q24.21 and losses of 17p13.1) were similar to those found in SS (23). Interestingly, these alterations were observed in three patients with involvement of the peripheral blood. Therefore, although the MFt genomic profile is different to that described for SS, similar alterations were detected in those patients with peripheral blood infiltration indicating that there may be a relationship between them. Future studies aimed at analyzing and comparing the genomic profiles of these two types of pathologies (de novo SS vs. SS developed in patients with a previous MF) will identify the differences and similarities in these entities.

Table 3
Recurrent chromosomal aberrations in SS

References	*n*	Type of aCGH platform	Recurrent alterations	
			Gains	Losses
Vermeer et al. (23)	20	BAC aCGH	+17q23	−17p13.1p13.3
			+17q24q25	−10q24q26
			+8q24.1q24.3	−10q25
			+17q25	−16p11.2
			+8q22q23	−5q14.3
Caprini et al. (24)	28	Oligo aCGH/SNP arrays	+17p11.2q25.3	−17p13.2p11.2
			+8p23.3q24.3	−10p12.1q26.3
			+10p15.3p12.2	-9q13q21.33
Laharanne et al. (22)	16	BAC aCGH	+8q23q24.3	-9p21
		Oligo aCGH	+17q23q24	−10p12.11.2
				−10q22q24
				−10q25q26
				−17p13q11.1

4.4. Genomic Instability Profile

It has been suggested that alterations in those processes that maintain the integrity of the genome may contribute to the emergence of different types of genetic aberrations (25, 26). These alterations are total or partial chromosome gains, the presence of amplified regions, the presence of homozygous deletions, and the number of chromosomal breaks. Genomic instability has been analyzed in different types of human tumors such as bladder cancer, breast cancer, neuroblastoma, Ewing's sarcoma, and B-cell lymphomas (26–29). These studies have allowed establishing different groups based on genetic and genomic instability related to poor prognosis.

In the Salgado et al. series (21), the application of oligo aCGH technique helped to define the profile of genomic instability in MFt. In this sense, the conjunction of different bioinformatics tools used to quantify these changes allowed the distinction of two groups of patients: one characterized by having few chromosomal abnormalities (0–5) which was called genetically stable and a second group called unstable, which had a high number of chromosomal alterations (>5). It was also observed that the unstable group had a worse overall survival than the group which was genetically stable. These data are similar to those proposed by Fischer et al. (12). In Fischer's study, the main aim was to analyze genetic aberrations by conventional CGH technique and to describe poor prognosis genetic factors. One of the findings of this study was the association between the presence of five or more genetic alterations and poorer 5-year overall survival. However, the Fischer study included different types of PCTCL (MF, ALCL CD30+, SS, PCTCL-CD8+), while Salgado et al. focused exclusively on MFt cases. The Spanish group also described the association between gains of 7q and a greater genomic instability (21). It is remarkable that both the gain of chromosome 7 and the presence of isochromosome 7q are common abnormalities in different types of aggressive T-cell lymphomas such as hepatosplenic T-cell lymphoma (30, 31), extranodal NK/T-cell lymphoma, nasal type, and ALK-negative anaplastic large cell lymphoma (32). The presence of numerous genes in 7q involved in the pathogenesis of different types of human tumors (*BRAF* and *EZH2*, among others) and the association with a more unstable genomic profile suggest a pathogenic role of this chromosomal region in MFt.

5. Discussion Related to Clinically Significant Abnormalities and the Impact of the Results on Patient Management, Prognosis, or Prediction of Therapeutic Outcomes

The discovery of recurrent chromosomal abnormalities in several types of cancer has allowed a better classification of the different entities, as well as a more accurate prediction of the disease aggressiveness and the therapeutic outcome.

One of the main contributions of aCGH studies in MF and SS has been the description of genetic markers associated with a poor prognosis.

In MF, previous conventional CGH studies (12) pointed to specific genetic alterations associated with poorer survival in a series of PCTCLs. Specifically, losses on chromosomes 6q and 13q and gains in 8q regions and the presence of more than five genetic alterations were related to a shorter 5-year overall survival. However, information obtained from this work is limited because, as mentioned above, the series included various types of cutaneous T-cell lymphomas (12).

Recently, van Doorn et al. (20) reported that gains in 8q24.21 and 1q21q22 as well as losses of 9p21.3 were associated with a decreased overall survival. A later study by Salgado et al. (21), including a larger series of uniformly selected tumoral MF patients, described three chromosomal aberrations associated with a worse overall survival of patients: deletions of 9p21.3 and 10q26qter and 8q24.21 gain. Moreover, in another study by Laharanne et al. (22), 8q24 gains and losses of 9p21 were significantly associated with poorer survival. Therefore, these three regions have been explored to find possible candidate genes related with the pathogenesis of MFt.

Deletion of the Region 9p21.3. The minimal deleted region on 9p21.3 found in the Salgado et al. study included only 200 kb and involved mainly the tumor suppressor genes *CDKN2A*, *CDKN2B*, and *MTAP*. Unlike in van Doorn et al. (20) series, which defined a 2 Mb minimal altered region, the application of oligonucleotide aCGH platforms and bioinformatics tools allowed a more precise delineation of the minimal altered region. In addition, Salgado et al. (21) described that among all patients who had the 9p21 region affected ($n = 17$), six showed an homozygous deletion, a genetic abnormality not previously described, and which was subsequently confirmed by Laharanne et al. (22, 33). *CDKN2A* and *CDKN2B* genes encode for p16INK4a and p15INK4B proteins, which are capable of inducing cell cycle arrest in G1 phase by inhibiting the phosphorylation of Rb. It has been reported that *CDKN2A* and *CDKN2B* genes are implicated in the pathogenesis of MF. Navas et al. (34, 35) showed that the more common genetic phenomena affecting the function of these genes were hypermethylation of the promoter region, mutations, or loss of heterozygosity. Recently, Laharanne et al. have studied the genetic and methylation status of both genes in a series of 24 tumoral MF, 16 SS, and 18 PCALCL by FISH techniques, aCGH, quantitative PCR, and bisulfite PCR (33). This study has shown that 9p21 losses, either homozygous or heterozygous, which are frequently observed in MFt and SS, correlate with a worse overall survival as previously described (20, 21). On the other hand, the MTAP gene encodes for an enzyme essential for the adenine and methionine synthesis. It has been postulated that this gene acts as a tumor suppressor gene in various types of cancer (36–42), and it is inactivated due to loss of the region 9p21.3. *MTAP*'s proximity to the well-defined tumor suppressor genes *CDKN2A* and *CDKN2B* led to the suggestion that the loss of this gene is incidental. However, it was recently demonstrated in a knockout mouse model for *MTAP* that mice with homozygous genotype acquired an embryonic lethal phenotype while heterozygote mice developed primary T-cell lymphomas (43). These findings support not only the *MTAP* gene function as a tumor suppressor gene but also suggest a possible pathogenic role in the development of the tumoral MF.

Gain of 8q24.21 Region. Complete or partial gains of chromosome 8 have been described in the MFt (13, 20–22). Salgado et al. defined a minimal amplification of the altered region 8q24.21 (750 Mb) involving the *MYC* oncogene. MYC is implicated in the regulation of proliferation, growth, differentiation, and apoptosis (44, 45) and plays an important role in the etiology of many hematological malignancies (46, 47). Recently, the altered function of MYC has also been described in SS (23). Vermeer et al. studied the genetic profile of 20 SS by microarray techniques (gene expression arrays and BAC aCGH). In this study, a frequent gain of 8q24 region was detected, as well as an increased MYC expression and a decreased expression of its antagonists: MXI1 and MNT. Recent studies in murine models showed that loss of Mnt and MXI1 is associated with a greater proliferative capacity and therefore facilitate tumorigenesis, leading to abnormal development of T cells and finally the development of a T-cell lymphoma (48). These data therefore suggest an important role not only in the pathogenesis of MFt but also of other aggressive PCTCL as the SS.

Deletion of the Region 10q26qter. Alterations in chromosome 10 have been described in MFt and SS techniques of conventional cytogenetics, M-FISH, CGH, and microsatellite markers (7, 10–13, 16, 17, 49–52). With the application of oligonucleotide aCGH techniques, Salgado et al. have defined the 10q26qter region as the most frequently deleted in the MFt (21). The size of this deletion is about 700 kb and it contains 31 genes. Among them, there are two tumor suppressor genes: *MGMT* and *EBF3*. *MGMT* gene silencing by methylation mechanisms has been studied in many human cancers including PCTCL (53, 54). *MGMT* gene encodes a DNA repair enzyme that protects cells from alkylating agents and has been suggested to participate in the protection of carcinogenesis. With respect to MFt, although it was initially observed that the *MGMT* methylation was a frequent feature of this disease (53), in a second study by the Dutch group (54), they observed that the *MGMT* promoter methylation is a phenomenon that occurs not only in PCTCL but also in control samples. Therefore, the methylation status of this gene has been discarded as a marker of malignancy. Moreover, the recent description of gene *EBF3* as a tumor suppressor gene implicated in the disruption of cell cycle and apoptosis suggests a possible pathogenic role in MFt.

References

1. Criscione VD, Weinstock MA (2007) Incidence of cutaneous T-cell lymphoma in the United States, 1973–2002. Arch Dermatol 1437:854–859
2. Willemze R, Jaffe ES, Burg G et al (2005) WHO-EORTC classification for cutaneous lymphomas. Blood 10510:3768–3785
3. Swerdlow SH, Campo E, Harris NL, Jaffe ES, Pileri SA, Stein H et al eds (2008) World Health Organization classification of tumours of haematopoietic and lymphoid tissues. IARC Press: Lyon
4. Kuzel TM, Roenigk HH Jr, Rosen ST (1991) Mycosis fungoides and the Sézary syndrome: a review of pathogenesis, diagnosis, and therapy. J Clin Oncol 97:1298–1313
5. Diamandidou E, Cohen PR, Kurzrock R (1996) Mycosis fungoides and Sézary syndrome. Blood 887:2385–2409
6. Harwix S, Gunzl HJ, Blaschke V, Zachmann K, Neumann C (2001) Inability to culture the dominant T-cell clone from the skin of primary cutaneous T-cell lymphoma as proven by TCR gamma-chain gene sequencing. Arch Dermatol Res 2933:139–146
7. Karenko L, Hyytinen E, Sarna S, Ranki A (1997) Chromosomal abnormalities in cutaneous T-cell lymphoma and in its premalignant conditions as detected by G-banding and interphase cytogenetic methods. J Invest Dermatol 1081:22–29
8. Batista DA, Vonderheid EC, Hawkins A et al (2006) Multicolor fluorescence in situ hybridization (SKY) in mycosis fungoides and Sézary syndrome: search for recurrent chromosome abnormalities. Genes Chromosomes Cancer 454:383–391
9. Padilla-Nash HM, Wu K, Just H, Ried T, Thestrup-Pedersen K (2007) Spectral karyotyp-

ing demonstrates genetically unstable skin-homing T lymphocytes in cutaneous T-cell lymphoma. Exp Dermatol 162:98–103

10. Karenko L, Kahkonen M, Hyytinen ER, Lindlof M, Ranki A (1999) Notable losses at specific regions of chromosomes 10q and 13q in the Sézary syndrome detected by comparative genomic hybridization. J Invest Dermatol 1123:392–395
11. Mao X, Lillington D, Scarisbrick JJ et al (2002) Molecular cytogenetic analysis of cutaneous T-cell lymphomas: identification of common genetic alterations in Sézary syndrome and mycosis fungoides. Br J Dermatol 1473:464–475
12. Fischer TC, Gellrich S, Muche JM et al (2004) Genomic aberrations and survival in cutaneous T cell lymphomas. J Invest Dermatol 1223:579–586
13. Prochazkova M, Chevret E, Mainhaguiet G et al (2007) Common chromosomal abnormalities in mycosis fungoides transformation. Genes Chromosomes Cancer 469:828–838
14. Sole F, Woessner S, Vallespi T et al (1995) Cytogenetic abnormalities in seven patients with the Sézary syndrome. Sangre (Barc) 405:431–433
15. Thangavelu M, Finn WG, Yelavarthi KK et al (1997) Recurring structural chromosome abnormalities in peripheral blood lymphocytes of patients with mycosis fungoides/Sézary syndrome. Blood 899:3371–3377
16. Mao X, Lillington DM, Czepulkowski B, Russell-Jones R, Young BD, Whittaker S (2003) Molecular cytogenetic characterization of Sézary syndrome. Genes Chromosomes Cancer 363:250–260
17. Espinet B, Salido M, Pujol RM et al (2004) Genetic characterization of Sézary's syndrome by conventional cytogenetics and cross-species color banding fluorescent in situ hybridization. Haematologica 892:165–173
18. Barba G, Matteucci C, Girolomoni G et al (2008) Comparative genomic hybridization identifies 17q11.2 approximately q12 duplication as an early event in cutaneous T-cell lymphomas. Cancer Genet Cytogenet 1841:48–51
19. Karenko L, Sarna S, Kahkonen M, Ranki A (2003) Chromosomal abnormalities in relation to clinical disease in patients with cutaneous T-cell lymphoma: a 5-year follow-up study. Br J Dermatol 1481:55–64
20. van Doorn R, van Kester MS, Dijkman R et al (2009) Oncogenomic analysis of mycosis fungoides reveals major differences with Sézary syndrome. Blood 1131:127–136
21. Salgado R, Servitje O, Gallardo F et al (2010) Oligonucleotide array-CGH identifies genomic subgroups and prognostic markers for tumor stage mycosis fungoides. J Invest Dermatol 1304:1126–1135
22. Laharanne E, Oumouhou N, Bonnet F et al (2010) Genome-wide analysis of cutaneous T-cell lymphomas identifies three clinically relevant classes. J Invest Dermatol 1306:1707–1718
23. Vermeer MH, van Doorn R, Dijkman R et al (2008) Novel and highly recurrent chromosomal alterations in Sézary syndrome. Cancer Res 688:2689–2698
24. Caprini E, Cristofoletti C, Arcelli D et al (2009) Identification of key regions and genes important in the pathogenesis of Sézary syndrome by combining genomic and expression microarrays. Cancer Res 6921:8438–8446
25. Lengauer C, Kinzler KW, Vogelstein B (1998) Genetic instabilities in human cancers. Nature 3966712:643–649
26. Blaveri E, Brewer JL, Roydasgupta R et al (2005) Bladder cancer stage and outcome by array-based comparative genomic hybridization. Clin Cancer Res 1119(Pt 1):7012–7022
27. Fridlyand J, Snijders AM, Ylstra B et al (2006) Breast tumor copy number aberration phenotypes and genomic instability. BMC Cancer 6:96
28. Ferreira BI, Alonso J, Carrillo J et al (2008) Array CGH and gene-expression profiling reveals distinct genomic instability patterns associated with DNA repair and cell-cycle checkpoint pathways in Ewing's sarcoma. Oncogene 2714:2084–2090
29. Ferreira BI, Garcia JF, Suela J et al (2008) Comparative genome profiling across subtypes of low-grade B-cell lymphoma identifies type-specific and common aberrations that target genes with a role in B-cell neoplasia. Haematologica 935:670–679
30. Wlodarska I, Martin-Garcia N, Achten R et al (2002) Fluorescence in situ hybridization study of chromosome 7 aberrations in hepatosplenic T-cell lymphoma: Isochromosome 7q as a common abnormality accumulating in forms with features of cytologic progression. Genes Chromosomes Cancer 333:243–251
31. Tamaska J, Adam E, Kozma A et al (2006) Hepatosplenic gamma delta T-cell lymphoma with ring chromosome 7, an isochromosome 7q equivalent clonal chromosomal aberration. Virchows Arch 4494:479–483
32. Feldman AL, Law M, Grogg KL et al (2008) Incidence of TCR and TCL1 gene translocations and isochromosome 7q in peripheral T-cell lymphomas using fluorescence in situ hybridization. Am J Clin Pathol 1302:178–185

33. Laharanne E, Chevret E, Idrissi Y et al (2010) CDKN2A-CDKN2B deletion defines an aggressive subset of cutaneous T-cell lymphoma. Mod Pathol 234:547–558
34. Navas IC, Algara P, Mateo M et al (2002) p16(INK4a) is selectively silenced in the tumoral progression of mycosis fungoides. Lab Invest 822:123–132
35. Navas IC, Ortiz-Romero PL, Villuendas R et al (2000) p16(INK4a) gene alterations are frequent in lesions of mycosis fungoides. Am J Pathol 1565:1565–1572
36. Nobori T, Takabayashi K, Tran P et al (1996) Genomic cloning of methylthioadenosine phosphorylase: a purine metabolic enzyme deficient in multiple different cancers. Proc Natl Acad Sci U S A 9312:6203–6208
37. Dreyling MH, Roulston D, Bohlander SK, Vardiman J, Olopade OI (1998) Codeletion of CDKN2 and MTAP genes in a subset of non-Hodgkin's lymphoma may be associated with histologic transformation from low-grade to diffuse large-cell lymphoma. Genes Chromosomes Cancer 221:72–78
38. Christopher SA, Diegelman P, Porter CW, Kruger WD (2002) Methylthioadenosine phosphorylase, a gene frequently codeleted with p16(cdkN2a/ARF), acts as a tumor suppressor in a breast cancer cell line. Cancer Res 6222:6639–6644
39. Subhi AL, Tang B, Balsara BR et al (2004) Loss of methylthioadenosine phosphorylase and elevated ornithine decarboxylase is common in pancreatic cancer. Clin Cancer Res 1021:7290–7296
40. Marce S, Balague O, Colomo L et al (2006) Lack of methylthioadenosine phosphorylase expression in mantle cell lymphoma is associated with shorter survival: Implications for a potential targeted therapy. Clin Cancer Res 1212:3754–3761
41. Mirebeau D, Acquaviva C, Suciu S et al (2006) The prognostic significance of CDKN2A, CDKN2B and MTAP inactivation in B-lineage acute lymphoblastic leukemia of childhood. Results of the EORTC studies 58881 and 58951. Haematologica 917:881–885
42. Worsham MJ, Chen KM, Tiwari N et al (2006) Fine-mapping loss of gene architecture at the CDKN2B (p15INK4b), CDKN2A (p14ARF, p16INK4a), and MTAP genes in head and neck squamous cell carcinoma. Arch Otolaryngol Head Neck Surg 1324:409–415
43. Kadariya Y, Yin B, Tang B et al (2009) Mice heterozygous for germ-line mutations in methylthioadenosine phosphorylase (MTAP) die prematurely of T-cell lymphoma. Cancer Res 6914:5961–5969
44. Meyer N, Kim SS, Penn LZ (2006) The Oscar-worthy role of myc in apoptosis. Semin Cancer Biol 164:275–287
45. Vita M, Henriksson M (2006) The myc oncoprotein as a therapeutic target for human cancer. Semin Cancer Biol 164:318–330
46. Dalla-Favera R, Bregni M, Erikson J, Patterson D, Gallo RC, Croce CM (1982) Human c-myc oncogene is located on the region of chromosome 8 that is translocated in Burkitt lymphoma cells. Proc Natl Acad Sci USA 7924:7824–7827
47. Klapproth K, Wirth T (2010) Advances in the understanding of MYC-induced lymphomagenesis. Br J Haematol 1494:484–497
48. Dezfouli S, Bakke A, Huang J, Wynshaw-Boris A, Hurlin PJ (2006) Inflammatory disease and lymphomagenesis caused by deletion of the myc antagonist mnt in T cells. Mol Cell Biol 266:2080–2092
49. Limon J, Nedoszytko B, Brozek I et al (1995) Chromosome aberrations, spontaneous SCE, and growth kinetics in PHA-stimulated lymphocytes of five cases with Sézary syndrome. Cancer Genet Cytogenet 831:75–81
50. Scarisbrick JJ, Woolford AJ, Russell-Jones R, Whittaker SJ (2000) Loss of heterozygosity on 10q and microsatellite instability in advanced stages of primary cutaneous T-cell lymphoma and possible association with homozygous deletion of PTEN. Blood 959:2937–2942
51. Scarisbrick JJ, Woolford AJ, Russell-Jones R, Whittaker SJ (2001) Allelotyping in mycosis fungoides and Sézary syndrome: common regions of allelic loss identified on 9p, 10q, and 17p. J Invest Dermatol 1173:663–670
52. Wain EM, Mitchell TJ, Russell-Jones R, Whittaker SJ (2005) Fine mapping of chromosome 10q deletions in mycosis fungoides and Sézary syndrome: identification of two discrete regions of deletion at 10q23.33-24.1 and 10q24.33-25.1. Genes Chromosomes Cancer 422:184–192
53. Gallardo F, Esteller M, Pujol RM, Costa C, Estrach T, Servitje O (2004) Methylation status of the p15, p16 and MGMT promoter genes in primary cutaneous T-cell lymphomas. Haematologica 8911:1401–1403
54. van Doorn R, Zoutman WH, Dijkman R et al (2005) Epigenetic profiling of cutaneous T-cell lymphoma: promoter hypermethylation of multiple tumor suppressor genes including BCL7a, PTPRG, and p73. J Clin Oncol 2317:3886–3896

Chapter 12

Array CGH Reveals Clonal Evolution of Adult T-Cell Leukemia/Lymphoma

Akira Umino and Masao Seto

Abstract

Adult T-cell leukemia/lymphoma (ATLL) is the neoplasm caused by human T-cell leukemia virus type 1 (HTLV-1). We performed oligoarray comparative genomic hybridization (CGH) against paired samples comprising peripheral blood (PB) and lymph node (LN) samples from patients with acute-type ATLL. Results disproved the established theory that true monoclonal proliferation, such as identical clonal expansion in all respects, occurred in acute-type ATLL, and our findings revealed that acute-type ATLL contains multiple subclones with differing genomic aberrations. Oligoarray CGH technology has been developed not only for high-resolution application but also for use in various analyses. Our original analysis is a method of identifying two or more clones with different chromosomal aberrations in one sample. Herein, we describe the analysis and clonal evolution of acute-type ATLL.

Key words: ATL, Array CGH, Clonal evolution, HTLV-1, Log2 ratio imbalance

1. Introduction

Adult T-cell leukemia/lymphoma (ATLL) is a neoplasm caused by human T-cell leukemia virus type 1 (HTLV-1) (1, 2). It is also known that HTLV-1 infection alone does not trigger progression of infected CD4+ T cells to fully malignant ATLL cells. Therefore, the search for genes involved in ATLL development is underway, and specific genes for each ATLL type have been actively pursued, albeit with limited success. It is often cited in regard to causality that ATLL does not have specific chromosomal aberrations or common provirus integration sites. High-resolution oligoarray CGH for comprehensive chromosome analysis is a powerful tool to search for different genomic aberrations in each ATLL case, whereas investigations relying on G-band (3) and fluorescence in situ hybridization analyses (4) have not been able to reveal detailed

Diponkar Banerjee and Sohrab P. Shah (eds.), *Array Comparative Genomic Hybridization: Protocols and Applications*, Methods in Molecular Biology vol. 973, DOI 10.1007/978-1-62703-281-0_12, © Springer Science+Business Media, LLC 2013

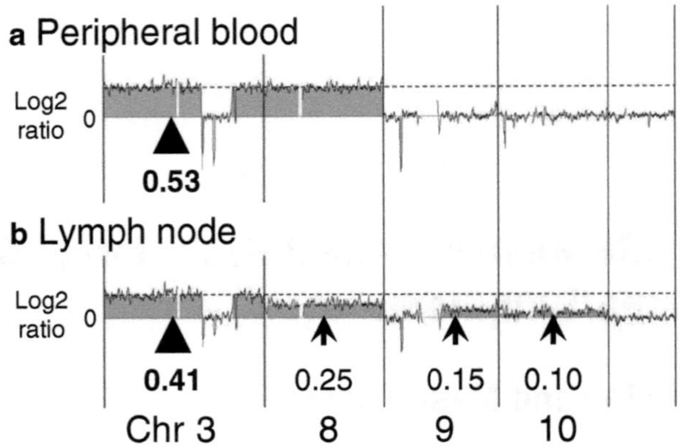

Fig. 1. Log2 ratio imbalance in an ATLL patient. Representative log2 imbalances in an ATLL patient are shown. Log2 ratios of gain regions were 0.53 in PB [(**a**), *arrow head*]. In LN, regions of gain were detected on chromosomes 3, 8, 9, and 10 as shown by the log2 ratio in parentheses: chromosomes 3 (0.41), chromosome 8 (0.25), chromosome 9 (0.15), and chromosome 10 (0.10) [(**b**), *arrow head* and *arrows*].

genomic aberrations. Shimoyama et al. classified ATLL into four subtypes: smoldering, chronic, lymphoma, and acute (5). In particular, the lymphoma and acute types of ATLL, which are aggressive, have a poor prognosis. Even now, it is difficult to cure the lymphoma and acute types of ATLL due to drug resistance. Lymphoma and acute-type ATLL differ in the presence of flower cells in peripheral blood (PB). A case with more than 2% flower cells in PB is classified as the acute type. Therefore, we speculated that acute-type ATLL contains overt leukemic cases of lymphoma-type ATLL. We thus set out to analyze the genomic aberrations of acute-type ATLL with paired PB and lymph node (LN) samples in more detail using oligoarray CGH (6). The analysis showed that LN samples had a log2 ratio imbalance among chromosomes. The log2 ratio imbalance indicated the existence of multiple clones with different genomic profiles. In theory, identical copy number regions have a fixed log2 ratio revealed by array CGH. Although log2 ratios of aberration regions of the PB sample were constant, those of the LN samples exhibited various values in each patient (Fig. 1). The numerical formula is theoretically expressed as "$\Upsilon = \log_2(2 + XS)/2$," where Υ represents the log2 ratio, X the copy number change (e.g., 1-copy gain present: $X = 1$; 1-copy loss: $X = -1$), and S the apparent tumor content. This formula indicates that all log2 ratios of each different copy number should be determined if a log2 ratio of an aberrant copy number is obtained. However, log2 ratio imbalances of LN samples did not follow the numerical formula, whereas those of PB conformed to the rule. These findings demonstrated the occurrence of log2 ratio imbalance in LN samples. We prepared two cell lines, SP-49 and HANK1, which possess different genomic aberrations. The genomic DNA of SP-49 was mixed with that of HANK1. When we conducted

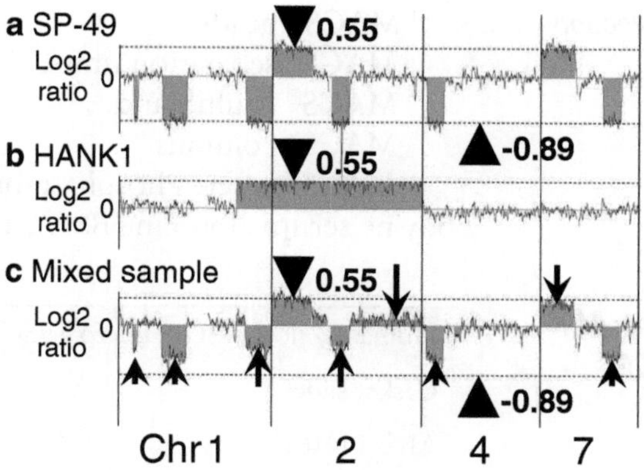

Fig. 2. Log2 imbalance resulting from a mixture of two cell lines. A log2 ratio imbalance among chromosomes can be made by mixing two kinds of cell lines. These data provide confirmation. We used the cell lines SP-49 and HANK1. Each cell line is monoclonal. Log2 ratios of 1-copy gain regions in SP-49 were 0.55, and those of 1-copy loss regions were −0.89 (**a**). Log2 ratios of 1-copy gain regions in HANK1 were 0.55 (**b**). Using the mixed sample at a ratio of 7:3 (SP-49:HANK1), the log2 ratio imbalance was reproduced [(**c**), *arrows*].

oligoarray CGH using the mixed-genomic DNA sample, the log2 ratio was calculated and demonstrated imbalance (Fig. 2). Hence, multiple subclones with different genomic aberrations develop in LNs of ATLL. In addition, subclones in PB were identical to LNs in many ATLL cases.

2. Materials

High-molecular-weight DNA was extracted from LNs and CD4+ cells of PB. The quality of DNA affects array CGH data. It is therefore important to extract high-quality DNA.

2.1. Cell Separation

Dulbecco's PBS(−).

Ficoll-Paque PLUS centrifugation (GE Healthcare, Piscataway, NJ).

2.2. DNA Extraction

Phenol/Chloroform

TENS: 10 mM Tris–HCl (pH 7.4), 10 mM EDTA, 150 mM NaCl, 0.4% SDS.

PK: Proteinase K (Wako) is dissolved in H_2O to 20 mg/ml.

TENS/PK: 10 ml TENS supplemented with 500 μl PK.

TE(10/1): 10 mM Tris–HCl (pH 7.4)/1 mM EDTA, pH 8.0.

100% EtOH.

3 M NaOAc pH 7.0.

NanoDrop ND-1000 spectrophotometer (NanoDrop).

2.3. CD4+ Selection

MACS® beads.

MACS® separation units.

MACS® MultiStand.

MACS® columns.

MACS buffer: Phosphate-buffered saline (PBS), pH 7.2, 0.5% bovine serum albumin (BSA), and 2 mM EDTA.

2.4. Agilent 4 × 44 K Oligo Array CGH

Human genome CGH micro array Kit 4×44	Agilent
Gasket slides	Agilent
ALU I (10 U/μl)	Promega
Rsa I (10 U/μl)	Promega
Genomic DNA labeling kit	Agilent
Microcon YM-30 filter units	Millipore
1×TE	Promega
Human cot-1	Invitrogen
Oligo aCGH hybridization kit	Agilent
Agilent oligo aCGH/ChIP wash buffer 1	Agilent
Agilent oligo aCGH/ChIP wash buffer 2	Agilent
Acetonitrile	Wako
Agilent scanner	Agilent
Feature Extraction Software	Agilent
Agilent Genomic Workbench v5.0 Software	Agilent

3. Methods

3.1. Peripheral Mononuclear Cell Separation

1. Transfer 10 ml heparinized blood into a 50 ml Falcon 2070 tube using a disposable pipette.
2. Add 20 ml PBS(−).
3. Carefully layer 10 ml of Ficoll-Paque PLUS under heparinized blood (see Note 1).
4. Centrifuge at 500 × *g* (1,600 rpm) for 30 min at 22°C with ACCEL slow and DECEL slow.
5. Take the buffy coat between the plasma fractionation (upper) and Ficoll-Paque PLUS (lower) using a new disposable pipette (see Note 2).
6. Pour the buffy coat into a new 50 ml Falcon 2070 tube.

7. Add PBS(−) to obtain a total of 40 ml, and mix by pipetting up and down.
8. Centrifuge at 560 × *g* (1,700 rpm) for 10 min at 22°C with ACCEL fast and DECEL fast.
9. Discard supernatant.
10. Tap the pellet.
11. Add PBS(−) to obtain a total of 20 ml.
12. Count the cell number (assess the number of dead cells by trypan blue staining).

3.2. CD4+ Selection

Procedures were performed according to the manufacturer's protocol (Miltenyi Biotec).

1. Resuspend peripheral mononuclear cell (PBMC) in MACS buffer, and add the CD4+ MACS beads.
2. Set the MACS® columns on the MACS® separation units.
3. Apply the labeled cells into the MACS® columns under the magnetic field of the MACS® separation units.
4. Collect the unlabeled cells which pass through, and wash with the buffer. Collect total effluent as the negative fraction.
5. Remove the MACS® columns from the MACS® separation units, place column on a suitable collection tube, pipette buffer onto the columns, and firmly flush out the positive fraction using the plunger.

3.3. DNA Extraction

Prepare CD4+ cells or chopped frozen lymph nodes in a 1.5 ml microtube.

1. Centrifuge at 1,200 × *g* (4,000 rpm) for 2 min, discard supernatant, and tap the pellet.
2. Elute in 400 μl of TENS/PK.
3. Incubate at 37°C for overnight digestion.
4. Add an equal volume of Phenol/Chloroform, and gently turn the tube upside down and upright for 5 min (never shake vigorously).
5. Centrifuge at 11,000 × *g* (12,000 rpm) for 5 min at room temperature.
6. Transfer the supernatant with a pipette to a new 1.5 ml microtube (see Note 3).
7. Add 400 μl of Phenol/Chloroform, and gently turn the tube upside down and upright for 5 min (never shake vigorously).
8. Centrifuge at 11,000 × *g* (12,000 rpm) for 5 min at room temperature.

9. Transfer the supernatant with a pipette to a new 1.5 ml microtube.
10. Add 40 μl of 3 M NaOAc pH 7.0 and 1 ml of 100% EtOH. Turn the tube upside down and upright carefully and slowly (see Note 4).
11. Centrifuge at 15,000 × *g* (14,000 rpm) for 20 min at 4°C.
12. Discard the supernatant.
13. Resuspend (wash) 150 μl of 70% EtOH.
14. Centrifuge at 15,000 × *g* (14,000 rpm) for 5 min at 4°C.
15. Discard the supernatant.
16. Resuspend (wash) 150 μl of 70% EtOH.
17. Centrifuge at 15,000 × *g* (14,000 rpm) for 5 min at 4°C.
18. Discard the supernatant.
19. Air-dry for 10 min.
20. Elute in TE.

3.4. Agilent 44 K Whole Human Genome CGH Arrays

Procedures of Agilent 44 K Whole Human Genome CGH arrays (Agilent Technologies, Palo Alto, CA) containing 44,000 probes were performed according to the manufacturer's protocol (Agilent Technologies).

1. DNA digestion.
2. DNA labeling.
3. Purification of labeled DNA.
4. Check the labeled DNA concentration and quality with the NanoDrop (see Note 5).
5. Hybridization on the array slide.
6. Wash the slide.
7. Scan the washed slide.
8. Data analysis.

3.5. Data Analysis

1. CGH data were extracted from scanned images (TIFF files) using Feature Extraction Software (version 10.3 Agilent Technologies).
2. Feature Extraction is a process by which data are extracted from a scanned microarray image and translated into log2 ratios of the Cy3 (green) labeled normal DNA and Cy5 (red) labeled tumor DNA signals in each of the 44 K probes, allowing us to measure DNA copy number changes in our experiments in conjunction with Agilent Genomic Workbench v5.0 Software (Agilent Technologies).
3. Raw data text files of the log2 ratios were transferred to the Genomic Workbench v5.0 Software for further analysis. The raw data were visualized as a graphical figure within the Genomic Workbench v5.0 Software (see Note 6).

4. Aberrations were detected using the ADM1 algorithm, whose threshold was 2.5, and filtering options of the Genomic Workbench v5.0 Software; the filtering options comprised a minimum number of probes in the region of 15 probes, and the minimum number of absolution log2 ratio average was 0.05. We defined gains and losses over a continuous 15-probe dataset as a log2 ratio average of ≥0.05 or ≤−0.05, respectively (see Note 7).
5. Finally, we validated significant differences in log2 ratio averages of sequential probe datasets of each aberration region and those of normal regions by the Welch test with $p \leq 0.05$ (see Note 8).
6. The theoretical formula is "$Y = \log_2(2 + XS)/2$," where Y represents the log2 ratio, X the copy number change, and S the apparent tumor content (see Note 9).

4. Notes

1. You need to pipette 11 ml of Ficoll-Paque PLUS and place the point of the pipette to the bottom of the 50 ml tube. Ficoll-Paque PLUS is layered until the 10 ml line.
2. Ficoll-Paque PLUS must be used at room temperature. Granulocyte contamination occurs in the buffy coat if Ficoll-Paque is cool. Plasma proteins and platelets are contained in the plasma (just above the buffy coat), and granulocytes are present in the Ficoll-Paque (just under the buffy coat).
3. Be careful not to pull strings from a lower layer. If you pull strings, phenol is contaminated by protein.
4. After the process, the tube can be kept at 4°C, but RNA contamination increases if the temperature falls too much.
5. If you have a low yield, the DNA quality may be poor.
6. Genomic Workbench revises the zero point of the log2 ratio. Check that the centralization mark of the Genomic Workbench is turned on. It is possible that log2 ratios of normal copy number regions are off the zero point in samples with many genomic aberrations.
7. Agilent recommends selecting aberration settings so that they yield a balance between false positives and false negatives. Agilent states that self–self controls are useful for selecting settings that minimize the false positives and that each laboratory needs to adjust threshold and filter options. You must adjust threshold and filter options with self–self controls as determined for your laboratory and define these states.

8. A sequential probe set defined as an aberration region may contain some different copy number regions if it does not belong to a normal distribution.
9. If log2 ratios of aberration regions do not accord with the abovementioned numerical formula, multiple clones with different genomic aberrations must exist in the used sample.

References

1. Uchiyama T, Yodoi J et al (1977) Adult T-cell leukemia: clinical and hematologic features of 16 cases. Blood 50(3):481–492
2. Poiesz BJ, Ruscetti FW et al (1980) Detection and isolation of type C retrovirus particles from fresh and cultured lymphocytes of a patient with cutaneous T-cell lymphoma. Proc Natl Acad Sci U S A 77(12):7415–7419
3. Shimoyama M, Abe T et al (1987) Chromosome aberrations and clinical features of adult T cell leukemia-lymphoma not associated with human T cell leukemia virus type I. Blood 69(4):984–989
4. Maciejewski JP, Tiu RV et al (2009) Application of array-based whole genome scanning technologies as a cytogenetic tool in haematological malignancies. Br J Haematol 146(5):479–488
5. Shimoyama M (1991) Diagnostic criteria and classification of clinical subtypes of adult T-cell leukaemia-lymphoma. A report from the Lymphoma Study Group (1984–87). Br J Haematol 79(3):428–437
6. Umino A, Seto M et al (2011) Clonal evolution of adult T-cell leukemia/lymphoma takes place in the lymph nodes. Blood 117(20):5473–5478

Chapter 13

Array-CGH Analysis of Cutaneous Anaplastic Large Cell Lymphoma

K. Szuhai, R. van Doorn, C.P. Tensen, and Van Kester

Abstract

This chapter describes a study in which the pattern of numerical chromosomal alterations in cutaneous anaplastic large cell lymphoma (C-ALCL) tumor samples was defined using array-based comparative genomic hybridization (CGH). First, the array-based CGH technique applied is outlined in detail. Next, its application in the analysis of C-ALCL tumor specimens is described. This approach resulted in the identification of highly recurrent chromosomal alterations in C-ALCL that include gain of 7q31 and loss on 6q16-6q21 and 13q34, each affecting 45% of the patients. The pattern characteristic of C-ALCL differs markedly from chromosomal alterations observed in other CTCL such as mycosis fungoides and Sézary syndrome and yielded several candidate genes with potential relevance in the pathogenesis of C-ALCL.

Key words: aCGH, CTCL, Cutaneous anaplastic large cell lymphoma

1. Introduction

Primary cutaneous anaplastic large cell lymphoma (C-ALCL) is a non-Hodgkin lymphoma of T cell origin that presents in the skin without evidence of extracutaneous disease at the time of diagnosis. It is characterized by large cells with an anaplastic, pleomorphic, or immunoblastic cytomorphologic presentation and by expression of the CD30 antigen by more than 75% of the tumor cells (1). Previous cytogenetic studies were able to identify gross chromosomal changes but did not reveal sufficient details (2, 3).

In order to obtain a higher resolution and more precise mapping of (genetic) aberrations in C-ALCL, an array CGH (aCGH) approach was developed. In addition, aCGH has the potential to be used for tumor classification and to predict tumor progression and prognosis. This chapter describes in detail the construction, use, and analysis of these arrays and their application for analyzing

Diponkar Banerjee and Sohrab P. Shah (eds.), *Array Comparative Genomic Hybridization: Protocols and Applications*, Methods in Molecular Biology vol. 973, DOI 10.1007/978-1-62703-281-0_13, © Springer Science+Business Media, LLC 2013

the genome of C-ALCL. Using this approach, we were able to identify numerical chromosomal alterations in malignant T cells from C-ALCL samples and defined a pattern of characteristic recurrent chromosomal alterations.

2. Protocols for Genomic Microarray CGH

2.1. Array-Chip Production Steps

The first part of the protocol provides detailed information about the array-chip production, starting from DNA isolation from large genomic insert clones (typically BAC, PAC, fosmid, and cosmid clones). This is followed by a modified DOP-PCR amplification using three primers selected to have the lowest chance to hybridize, consequently amplify to *Escherichia coli* genome (4, 5). The DOP-PCR reaction not only amplifies but also fragments the isolated genomic sequences. The primers have a common tail sequence for the secondary amplification reaction in which the primer contains an aminolinker group at its 5′ ends. This aminolinker group facilitates covalent binding of the spotted DNA sequences to a glass surface with active aldehyde groups. A major part of this protocol has been adapted from the Wellcome Trust Sanger Institute Group of Nigel Carter with some adjustments (6, 7). The large genomic insert clone set consist of ~3,700 clones are part of the so-called 1 Mb clones provided by the Wellcome Trust Sanger Institute. All these clones have been sequence verified at the Wellcome Trust Sanger Institute and displayed at the Ensembl Web Database (http://www.ensembl.org/Homo_sapiens/Info/Index).

2.2. Culturing E. coli Containing Large Genomic Insert Clones

- Prepare LB medium containing the right antibiotics (Chloramphenicol: 25 μg/mL or Kanamycin: 30 μg/mL, depending on the resistance profile of BAC/PAC clones) and fill it out in a 96-deep well plate, 2 mL in each well.
- Inoculate wells in the plate from a 96-well plate containing the glycerol stocks of the clone set, using a 96 pin replicator.
- Seal plates with a gas permeable seal and incubate for 18 h at 37 °C, shaking at ~220 rpm.
- Spin plates for 10 min at 4,000 rpm (1800 rcf) and pour the supernatant off. The pellets can be immediately used for BAC/PAC DNA isolation.

2.3. Isolating BAC and PAC DNA from E. coli

- Isolate DNA using the "Wizard SV Plasmid DNA Purification System" from Promega following recommended protocol and using the Beckman "Biomek 2000" robot.
- Plates containing eluted DNA (100 μL product per well) can be sealed and stored at −20 °C.

2.4. Primary DOP PCR

2.4.1. TAPS2 Salt Solution (Final Volume 96 mL)

250 mM TAPS (Sigma) pH 9.3	6.08 g
166 mM $(NH_4)_2SO_4$	2.20 g
25 mM $MgCl_2$	2.5 mL of 1 M stock solution

TAPS solution has been optimized as PCR buffer for the DOP-PCR reaction using Amplitaq Taq-polymerase. The use of other buffer and/or polymerase requires optimization of the PCR reaction.

Aliquots of UV-sterilized TAPS2 salt solution (960 μL per tube) can be stored at −20 °C. Make sure the pH has been adjusted to pH 9.3.

TAPS	*N*-tris(hydroxymethyl)methyl-3-amino-proanesulphonic acid
	Sigma order. No. T5130 (100 g)
BSA	Bovine serum albumin
	Sigma order. No. A4628, 5% stock solution
W1	Polyoxyethylene ether W1
	Sigma order. No. P7516 (powder)
2 ME	β-mercaptoethanol (BDH)

2.4.2. Prepare TAPS2 Buffer

Prior to use, add BSA (33 μL/mL) and 2 ME (7 μL/mL) to thawed buffer aliquots. Avoid frequent thawing and freezing of BSA and 2 ME containing buffer aliquots.

TAPS2 salt solution	960 μL (=1 tube)
BSA, 5%	33 μL
β-Mercaptoethanol	7 μL

2.4.3. Preparation of PCR Master Mixes (Tables 1 and 2)

Add 41 μL solution to each PCR well using a multichannel pipette. Add 9 μL DNA template from plates containing large genomic insert isolated form BAC/PAC cultures. To avoid contamination,

Table 1
Pipetting scheme for primary DOP PCR master mixes

	1 reaction	1 plate	2 plates
Water	24 μL	2,400 μL	4,800 μL
DOP primer (20 μM)	5 μL	500 μL	1,000 μL
dNTPs (2.5 mM)	4 μL	400 μL	800 μL
W1	2.5 μL	250 μL	500 μL
TAPS2 buffer	5 μL	500 μL	1,000 μL
Amplitaq polymerase	0.5 μL	50 μL	100 μL

Table 2
Sequences of human genome specific primary DOP PCR mixes

DOP 1 primer	CCGACTCGAGNNNNNNCTAGAA
DOP 2 primer	CCGACTCGAGNNNNNNTAGGAG
DOP 3 primer	CCGACTCGAGNNNNNNTTCTAG

use a Biomek 2000 robot for pipetting steps. Seal plates with Bioseal 6 (BIOplastics bv) and place an ABI PRISM Optical Cover Compression Pad (Applied Biosystems) on top to ensure a better sealing and run 1st DOP PCR program on Primus PCR machine.

The PCR program consist of

Step 1	94 °C for 3 min
Step 2	94 °C for 30 s
Step 3	30 °C for 1 s
Step 4	Ramp at 0.1 °C per second to 72 °C
Step 5	72 °C for 3 min
Step 6	Go to step 2 for an additional 29 cycles
Step 7	72 °C for 7 min
Step 8	Cool down to RT
Step 9	End

PCR machines with high ramping speed need adjustment for the ramp time as described in the protocol. Deviation from this may result in PCR failure.

Run 5 μL on a 2.5% agarose gel. A mix of distinct bands and a smear between 200 bp and 2 kb is expected. Always perform negative controls to ensure that your PCR has not been contaminated. Samples with very short (less than 200 bp) may represent concatenated primer byproduct and should be repeated. Randomly select 5–8 position per plate for labeling with nick translation and hybridize to a normal metaphase slide. This quality control step is important to check possible carry-over contamination.

2.5. Secondary DOP PCR

2.5.1. Prepare Aminolinking Buffer

1 M KCl	5 mL
0.1 M $MgCl_2$	2.5 mL
1 M TRIS/HCl; pH 8.5	0.5 mL
H_2O	2 mL

Table 3
Pipetting scheme for secondary PCR using aminolinker-labeled primers

	1 reaction	1 plate	6 plates
Water	42.4 μL	4,240 μL	25,440 μL
Aminolinking buffer	6 μL	600 μL	3,600 μL
dNTPs 2.5 mM	6 μL	600 μL	3,600 μL
Aminoprimer	3 μL	300 μL	1,800 μL
Amplitaq polymerase	0.6 μL	60 μL	360 μL

Filter through a 0.2 μM filter using a syringe. Final concentrations: 500 mM KCl, 25 mM $MgCl_2$, 50 mM Tris, and pH 8.5. This buffer can be stored for about a week at room temperature. The sequence of the aminoprimer: 5′-aminolinker-GGAAACAGCCCG ACTCGAG-3′.

2.5.2. Preparation of PCR Master Mixes (Table 3)

Add 58 μL PCR mix to each well. Use Biomek 2000 robot to pipet template, 2 μL. Seal plates with Bioseal 6 (BIOplastics bv), put a ABI PRISM Optical Cover Compression Pad (Applied Biosystems) on top, and run 2nd DOP PCR program on Primus PCR machine.

The PCR program consists of:

Step 1	95 °C for 10 min
Step 2	95 °C for 1 min
Step 3	60 °C for 1 min 30 s
Step 4	72 °C for 7 min
Step 5	Go to step 2 for 34 more cycles
Step 6	72 °C for 10 min
Step 7	Cool down to RT
Step 8	End

Run 5 μL on a 2.5% agarose gel. Check your negative controls.

2.6. Spotting of DOP PCR Products on Glass Slides

- Using the Biomek 2000 robot, combine the secondary DOP PCR samples for each probe. The resulting ~180 μL DOP-PCR amplified, aminolinker-labeled DNA mix with an estimated concentration of 250 ng/μL can be used for spotting.
- Mix 100 μL aminolinker-labeled solution with 25 μL "4× spotting buffer" (1 M Phosphate buffer, pH 8.5 + 0.001% Sarkosyl, 0.20 μm filter sterilized) using a Biomek 2000 robot and filtered through a 96-well Multiscreen-GV clear plates (Millipore) at 2,000 rpm (450 rcf) for 5 min.

- Generate combined 384-well plates from the 96-well plates using 20 μL solution from each well, using a Biomek 2000 robot. The combined 384-well spotting plates are ready to use or sealed plates can be stored at −20 °C.
- Thaw the 384-well spotting plates at RT and place into a 37 °C for at least 30 min. During the freezing step, precipitates of phosphate buffer can be formed. It is important to solve them completely as they may clog the pins system in the spotter.
- For spotting sue, the Omnigrid microarrayer from Genemachines equipped with SMP3 Stealth Spotting pins (Telechem) in a Stealth 48 pin printhead (SPH48) (http://www.arrayit.com/Products/Printing/Stealth/stealth.html). For optimal spot formation on the CodeLink™ Activated Slides (Amersham Biosciences), set a relative humidity of 42–47% and a temperature of 20–23 °C.
- Triplicated spots (reporters) were generated from each well representing one large genomic insert clone derived-labeled DNA.
- Use the Omnigrid software to compose a Genepix Array List file (GAL file). This reference file identifies spot coordinates and link to reporter (spotted, amplified BAC/PAC DNA) name.

2.7. Blocking Treatment of Spotted Array Slides

- Incubate spotted slides for about 18 h (overnight) at RT, in a chamber with 70–80% relative humidity, containing saturated NaCl solution at the bottom.
- Dry slides in a desiccator at room temperature.
- Then immerse microarray slides into a 1% ammonium-hydroxide for 5 min.
- Transfer slides to a 0.1% SDS solution and incubate for 5 min.
- Denature the slides in water at 95–100 °C for 2 min.
- Rinse the slides in ice-cold water, followed by two changed washes in water, each of 10 min.
- Dry the slides in a centrifuge by spinning at 1,000 rpm (110 rcf) for 5 min.
- Finally, dried slides can be stored at room temperature (RT) in the dark in a desiccator until further use. Slides are stable for 9–12 months. Prepare big batches to minimize batch-to-batch variation.

3. Sample Preparation

3.1. Labeling of Genomic DNA

For labeling reaction in studies before April 2006 random primer labeling kit with separately added Cy3 and Cy5 labeled dCTP nucleotides were used. Samples analyzed later were labeled with a random prime kit containing ready-to-use mixes of "Alexa Fluor 3" and "Alexa Fluor 5." Both labeling protocols are given in this section.

3.2. Labeling of Genomic DNA Using Random Priming with Cy3 and Cy5 Labeled dCTP's

3.2.1. Equipment and Reagents

BioPrime Labeling Kit (Invitrogen 18094-011).

10× dNTP mix (1 mM dCTP, 2 mM dATP, 2 mM dGTP, 2 mM dTTP in H_2O).

1 mM Cy3-dCTP (Amersham).

1 mM Cy5-dCTP (Amersham).

1 M NaCl.

3.2.2. Method 1

1. Label 330 ng test or reference DNA in a final reaction volume of 100 μL by mixing the DNA with:
   40 μL 2.5× Random Primers Solution.
   2 μL of 1 M NaCl and make up to 86.7 μL with water.
2. Denature DNA in a water bath for 5′ at 100 °C and immediately place the tube on ice.
3. Add the following reagents on ice:
   10 μL 10× dNTP mix.
   1.3 μL Cy3 (typically for test sample) or Cy5 (typically for reference sample) labeled dCTP (1 mM).
   2 μL Klenow fragment.
   Mix gently but thoroughly.
4. Incubate the reaction at 37 °C overnight.

Do not stop the labeling reaction with the stop buffer as EDTA will interfere with precipitation (it may form unsolvable precipitates.) Proceed with precipitation of DNA without cleanup.

(*Alternatively, store labeled sample at* –20 °C.)

3.3. Labeling of Genomic DNA Using Random Priming AlexaFluor 3 and AlexaFluor 5 Labeled dCTP's

3.3.1. BioPrime Total Genomic Labeling System (Invitrogen 18097-012, for 30 Reactions, Labeling Only Module)

Method

1. Dilute 300 ng DNA sample for labeling (test or reference) in TE buffer in a 33 μL end volume.
2. Briefly vortex each Alexa Fluor 2× Reaction Mix and centrifuge to collect the contents of the tubes. Place the tubes on ice.
3. Table 4 shows the pipetting scheme for BioPrime labeling reaction.
4. Mix and incubate it at 95 °C protected from light, for 5 min. Place tubes on ice for 5 min.
5. On ice, add 4.5 μL of exo-Klenow fragment to each tube, for a final reaction volume of 75 μL.
6. Vortex tubes briefly and centrifuge to collect the contents and incubate at 37 °C for 2 h protected from light.

Proceed with precipitation of DNA without cleanup.

(*Alternatively, store labeled sample at* –20 °C.)

Table 4
Pipetting scheme for BioPrime labeling reaction

	Tube 1	Tube 2
Alexa Fluor 3 2× Reaction Mix	37.5 μL	
Alexa Fluor 5 2× Reaction Mix		37.5 μL
Genomic DNA sample (test)	33 μL	
Genomic DNA sample (reference)		33 μL
Total volume	70.5 μL	70.5 μL

3.4. Hybridization of Labeled Genomic DNA on CodeLink DNA Array Slides Using a Tecan HS400 or 4,800 Hybridization Station

3.4.1. Equipment and Reagents

- Pre-/hybridization buffer: (50% formamide, 7% dextran sulfate, 0.1% Tween 20, 2× SSC, 10 mM TRIS/HCl (pH 7.4), 25 mM EDTA).
- 3 M NaAc pH 5.2.
- Human Cotl DNA (Invitrogen, test the quality prior to use).
- Sonicated Fish sperm DNA.
- 100% Ethanol.
- 70% Ethanol.

Yeast tRNA (Invitrogen) (100 μg/μL, dissolved in H_2O).

3.4.2. Precipitation

150 μL	Cy3/OR Alexa Fluor 3 labeled DNA
150 μL	Cy5/OR Alexa Fluor 5 labeled DNA
125 μL	Human Cotl DNA
42.5 μL	3 M NaAc pH 5.2
1,000 μL	100% EtOH (cold)

Mix the tubes gently and precipitate at −20 °C for 30 min to overnight.

3.4.3. Preparation of the Slide

Place the slide in the Tecan slide holder.

3.4.4. Preparation of the Probe

1. Warm hybridization buffer up to 37 °C.
2. Spin the labeled, mixed DNA at 13,200 rpm (18500 rcf) in an Eppendorf microcentrifuge for 15 min.
3. Remove supernatant, add 500 μL 70% EtOH, and spin again at 13,000 rpm (17950 rcf) for 1 min.
4. Remove supernatant and spin again at 13,000 rpm (17950 rcf) for 1 min. Collect all fluid from the tube and let the pellet is dry.
5. Resuspend the DNA pellet in mixture containing 68 μL Hybridization buffer and 6 μL yeast tRNA.

The final volume of the probe mix is for a Tecan hybridization station module with dimensions of 21 by 50 mm. Modules with other sizes needs accordingly adjusted volumes.

To ensure that the DNA pellet dissolves completely place the tube for 2–3 h in a 42 °C heat block and Vortex it three to four times.

- Denature the probe mix at 80 °C for 5 min.
- Shortly spin the tube to collect vapor.

3.5. Hybridization and Washing

Before denaturing the probe mixes, make sure all the washing solutions are fresh and available for the hybridization station at the right channels.

Channel 1	50% formamide, 2× SSC, pH 7
Channel 2	40% formamide, 2× SSC, pH 7
Channel 3	2× SSC, 0.1% SDS
Channel 4	0.1× SSC
Channel 6	MilliQ water
N_2 channel	Open

While denaturing the probe, put the slides to be hybridized onto the machine and close the block. Start the program for array-CGH hybridization. Start injection of the probes when the machine asks for it.

3.5.1. TECAN-Array-CGH Hybridization and Wash program

Step 1: Wash step at 37 °C, channel 1, 1 wash run for 30 s, no soak.

Step 2: Probe injection at 37 °C.

Step 4: Hybridization at 37 °C for 23:00:00, Agitation Frequency: medium.

Step 5: Wash step at 47 °C, channel 2, 2 runs, wash time 1:15 min, soak time 1 min.

Step 6: Wash step at 47 °C, channel 3, 4 runs, wash time 1:30 min, soak time 1 min.

Step 7: Wash step at 30 °C, channel 4, 1 wash run for 30 s, no soak.

Step 8: Slide drying step at 30 °C for 3 min, final manifold cleaning using channel 6.

Collect slides dry from the machine when the program is finished.

Slides are ready for scanning.

4. Data Acquisition

4.1. Scanning and Analysis of Array Data

Scanning of the arrays was performed either with a GenePix 4100A scanner (Molecular Devices, Wokingham, Berkshire, UK) or with an Agilent DNA microarray scanner (Agilent Technologies,

Amstelveen, the Netherlands). If the Agilent scanner is used, use the TIFF splitter to split the image into two channels and flip and rotate the image. Spots were identified, and spot intensities were measured for each channel using GenePixPro 4.1 software tool. Spots with measured fluorescence intensities within the reference channel (typically Cy5 or Alexa Flour 5 channel label) with values below five times the average background or presented more than 3% saturated pixels were flagged and excluded from further analysis. These raw data values were exported and used with an in-house developed Microsoft Excel routine. Values from triplicate reporter elements (spots) were averaged, and spot values outside of the 20% confidence interval were excluded. Reporter elements presenting at least two values within the 20% confidence interval of the average were included. Slides with less than 95% retained reporter elements were regarded as bad quality and were excluded. Log 2 ratios of the test (typically the Cy3 or Alexa Fluor 3 labeled sample) and reference (typically the Cy5 or Alexa Fluor 5 sample) fluorescence intensities were calculated. These values were used for global normalization using the median of the overall log2 ratio values. Further use the software to analyze the picture and normalize the data.

Use any platform to further analyze the data and generate a graphical interpretation of the genomic microarray. For the analysis of C-ALCL, BAC clone and oligonucleotide probe positions were established based on Ensembl release 44 (April 2007). Recurrent minimal common regions (MCRs) with copy number alterations (CNA) affecting at least 35% of analyzed samples were computed in CAPWeb (Institute Curie, http://bioinfo-out.curie.fr/actudb/cgh.php) using the algorithm proposed by Rouveirol et al. (8). Only CNAs characterized by gain or loss of at least 2 clones were taken into consideration. The nearby borders of adjacent clones were chosen to delineate MCRs.

5. Research Applications

Genomic microarray CGH analysis was used to gain more insight in the genetic events involved in the pathogenesis of primary C-ALCL. C-ALCL, formerly designated as primary cutaneous CD30-positive large T-cell lymphoma, is a T-cell lymphoma composed of large cells with an anaplastic, pleomorphic, or immunoblastic cytomorphology, which show expression of the CD30 receptor (1). C-ALCL has an indolent clinical behavior and rarely shows extracutaneous dissemination. This cutaneous T-cell lymphoma (CTCL) is commonly treated with radiotherapy and has an excellent prognosis with an estimated 5-year survival exceeding 90% (9–11).

Table 5
Clinical characteristics

No.	Sex	Age (years)	Extent	Therapy	Result	Site of relapse	Current status	Follow-up (months)
1	M	48	LOC	RT	CR	S; LN; LUNG	DOD	231
2	F	62	GEN	Chemo	PR	S; LN	DOD	36
3	F	74	SOL	RT	CR	S	ACR	258
4	F	45	GEN	Chemo	PR	S; LN	DOD	308
5	M	69	LOC	RT	CR	–	ACR	120
6	M	51	GEN	–	–	S	ACR	105
7	M	63	GEN	RT	PR	S	AWD	139
8	M	86	LOC	RT	CR	–	ACR	29
9	M	70	SOL	RT	CR	S	ACR	12
10	F	79	SOL	RT	CR		ACR	13
11	F	44	LOC	EXC	CR	S	AWD	62

ACR alive in complete remission, *AWD* alive with clinical symptoms of disease, *BM* bone marrow; *chemo* chemotherapy, *C-ALCL* cutaneous anaplastic large cell lymphoma, *CNS* central nervous system, *CR* complete remission, *DOD* death by disease, *EXC* excision, *F* female, *GEN* generalized (multifocal) skin disease, *LN* lymph node, *LOC* localized disease, *M* male, *PD* progressive disease, *PR* partial remission, *RT* radiotherapy, *S* skin, *SOL* solitary lesion

Previous cytogenetic analyses of C-ALCL have shown recurrent CNAs of several chromosomal regions, including gains on 6p, 7q, and 19 and losses on 6q, 9, and 18 (2, 3, 12–14).

5.1. Selection of Patient Material

Pretreatment biopsies of 11 patients with C-ALCL were selected for this study (Table 5). In all cases, the biopsies contained more than 75% malignant T cells. All patients had been retrieved from the database of the Dutch Cutaneous Lymphoma Group, reviewed by an expert panel of dermatologists and hematopathologists before entry in this database, and classified using criteria of the WHO-EORTC classification for the primary cutaneous lymphomas (1). In all patients, routine staging procedures including physical examination, complete and differential blood cell counts and serum biochemical analysis, computed tomography scan of the neck, chest, and abdomen, and a bone marrow biopsy had been negative.

6. Results

The overall pattern of chromosomal alterations of C-ALCL is characterized by gains of large regions on chromosome 7q and 17 and losses of regions on chromosome 6q and 13. A total of 30 MCRs present in at least 35% of the patients were identified and are listed in Table 6. Of these recurrent MCRs with CNA, 20 represent gains of chromosomal regions and 10 correspond to losses. The most highly recurrent chromosomal alterations are gain of 7q31 and loss of 13q34 and 6q16-6q21, all affecting 45% of patients. Cross-referencing of the genes residing in these 30 MCRs with the Cancer Gene Census, a list of genes for which mutations have been causally implicated in cancer (15) revealed 26 known oncogenes and tumor suppressor genes. The most highly recurrent MCR with gain in C-ALCL at 7q31 harbors a single oncogene, the MET gene that encodes the hepatocyte growth factor receptor. Three putative tumor suppressor genes, CDC16, CUL4A, and PRDM1, reside in the loci with loss on 13q34 and 6q16-6q21.

7. Discussion

Comprehensive analysis of CNAs in C-ALCL showed that the most highly recurrent MCR with gain was located on 7q31 and with loss on 6q16-6q21 and 13q34, each affecting 45% of the patients. The focal MCR with gain on 7q31 harbors the MET oncogene. MET amplification has previously been shown to result in its increased expression in nodal ALCL (16), and deregulated expression has been noted in adult T-cell leukemia (17, 18). The MCR with loss on 6q16-6q21 contains the PRDM1/BLIMP-1 gene encoding a transcription factor that is implicated in T cell homeostasis and differentiation. In mice lacking Prdm1/Blimp-1 activity, accumulation of CD4+ T cells is observed (19, 20). Other cancer-associated genes located in regions with recurrent CNA with potential relevance in the pathogenesis of C-ALCL include FOXO1A and BRCA2 on 13q12-13q14 (loss in 36%), PRDM16/MEL1 on 1p36 (gain in 27%), and TP53 on 17p13 (loss in 27% of patients). We have identified previously unreported recurrent chromosomal alterations in C-ALCL and confirmed several gains and losses found in the few studies in which conventional CGH was applied. Consistent with our results, Zettl et al. (14) observed gain of regions on 7q and 6p in 2 of 11 C-ALCL samples. Mao et al. (2) described gains of 1p, 5, 6, 7, 8p, and 19, showing partial overlap with our findings. In a group of seven patients with relapsing C-ALCL, Prochazkova et al. (3) observed recurrent gain of regions on chromosome 9 and losses on chromosome 6 and 18. Furthermore in nodal ALCL, recurrent CNAs affecting 7q and 13q have been described (14, 21).

The applied aCGH analysis revealed (novel) detailed information on genomic alterations in cALCL. Furthermore, our analysis also uncovered genomic differences between C-ALCL and two other CTCL entities, mycosis fungoides and Sézary syndrome

Table 6
Minimal common regions affected in C-ALCL

Chrom.	Cytogenetic band	CNA	Position start	Position end	No. of cases affected	Genes
6	6q16–6q21	Loss	104,974,085	107,020,676	5	PRDM1
7	7q31	Gain	110,975,058	118,629,320	5	MET
13	13q34	Loss	110,681,700	114,142,980	5	CDC16 CUL4A
1	1p36.2	Gain	12,462,984	15,425,500	4	
6	6q25–6q27	Loss	156,122,661	170,373,241	4	MLLT4, FGFR1OP
7	7q21	Gain	80,528,144	84,729,683	4	
7	7q21	Gain	85,485,924	91,973,205	4	AKAP9
7	7q21	Gain	92,090,477	97,314,287	4	
7	7q21–7q22	Gain	97,314,794	100,976,355	4	
7	7q32–7q34	Gain	130,792,900	139,186,224	4	TIF1
7	7q34	Gain	140,558,165	142,236,215	4	
7	7q35	Gain	144,233,517	147,084,269	4	
7	7q35–7q36	Gain	147,259,180	149,650,854	4	
8	8p22–8p21	Loss	18,644,291	22,394,183	4	
13	13q12–13q14	Loss	29,008,462	45,700,578	4	FOXO1A, BRCA2, LHFP, LCP1
7	17q12–17q21	Gain	34,979,298	36,236,729	4	ERBB2, RARA
21	21q22	Gain	39,702,870	41,559,383	4	
1	1p36.3	Loss	1,145,847	7,059,893	3	

(continued)

Table 6 (continued)

Chrom.	Cytogenetic band	CNA	Position start	Position end	No. of cases affected	Genes
1	1q21–1q23	Gain	148,424,834	161,212,962	3	NTRK1, PRCC, TPM3, SDHC, ARNT, FCGR2B, MUC1
3	3p26–3p25	Loss	6,636,703	8,869,825	3	
6	6p21.3	Gain	32,200,774	33,467,588	3	
7	7p22–7p21	Gain	6,234,587	12,919,070	3	PMS2
8	8p23–8p22	Loss	1,706,678	17,782,553	3	
16	16p11.2–16q11.2	Loss	32,345,186	45,067,244	3	
17	17p13	Loss	7,063,252	7,563,870	3	TP53
17	17p11.2	Gain	19,748,613	21,055,083	3	USP6
17	17q21–17q22	Gain	43,043,105	49,753,469	3	COL1A1
17	17q23	Gain	54,872,929	59,626,448	3	CLTC, NACA, BRIP1
17	17q25	Gain	72,271,167	78,774,742	3	MSF
17	17p11.2	Gain	18,114,679	19,211,694	3	

Start and end positions are delineated by the nearby borders of adjacent clones. Genes are the cancer census genes situated in the subsequent MCRs. CAN is copy number alteration previously analyzed using the same aCGH platform (22, 23). This further underscores the notion that the molecular pathogenesis and therefore therapeutic requirements of these cutaneous T-cell lymphomas may be distinct. Combined with gene expression data, histological evaluation, and clinical information, this knowledge can contribute to improved patient management, prognosis, and/or prediction of therapeutic outcomes.

Acknowledgments

We thank MF Benner, MK Borg, R Dijkman, MH Vermeer, R Willemze, W Zoutman (Department of Dermatology, Leiden University Medical Center (LUMC)); EJ Dreef, Dr PM Jansen (Department of Pathology LUMC); Dr JM Boer (Center for Human and Clinical Genetics, LUMC) and J Knijnenburg (Department of Molecular Cell Biology, LUMC) for their excellent assistance.

References

1. Willemze R, Jaffe ES, Burg G et al (2005) Who-eortc classification for cutaneous lymphomas. Blood 105(10):3768–3785. doi:2004-09-3502 (pii) 10.1182/blood-2004-09-3502
2. Mao X, Orchard G, Lillington DM et al (2003) Genetic alterations in primary cutaneous cd30+ anaplastic large cell lymphoma. Genes Chromosomes Cancer 37(2):176–185. doi:10.1002/gcc.10184
3. Prochazkova M, Chevret E, Beylot-Barry M et al (2003) Chromosomal imbalances: a hallmark of tumour relapse in primary cutaneous cd30+ t-cell lymphoma. J Pathol 201(3):421–429. doi:10.1002/path.1469
4. Telenius H, Carter NP, Bebb CE et al (1992) Degenerate oligonucleotide-primed pcr: general amplification of target DNA by a single degenerate primer. Genomics 13(3):718–725. doi:0888-7543(92)90147-K (pii)
5. Fiegler H, Carr P, Douglas EJ et al (2003) DNA microarrays for comparative genomic hybridization based on dop-pcr amplification of bac and pac clones. Genes Chromosomes Cancer 36(4):361–374. doi:10.1002/gcc.10155
6. Knijnenburg J, Szuhai K, Giltay J et al (2005) Insights from genomic microarrays into structural chromosome rearrangements. Am J Med Genet A 132A(1):36–40. doi:10.1002/ajmg.a.30378
7. Knijnenburg J, Van Der Burg M, Tanke HJ et al (2007) Optimized amplification and fluorescent labeling of small cell samples for genomic array-cgh. Cytometry A 71(8):585–591. doi:10.1002/cyto.a.20412
8. Rouveirol C, Stransky N, Hupe P et al (2006) Computation of recurrent minimal genomic alterations from array-cgh data. Bioinformatics 22(7):849–856. doi:btl004 (pii) 10.1093/bioinformatics/btl004
9. Bekkenk MW, Geelen FA, Van Voorst Vader PC et al (2000) Primary and secondary cutaneous cd30(+) lymphoproliferative disorders: a report from the dutch cutaneous lymphoma group on the long-term follow-up data of 219 patients and guidelines for diagnosis and treatment. Blood 95(12):3653–3661
10. Liu HL, Hoppe RT, Kohler S et al (2003) Cd30+ cutaneous lymphoproliferative disorders: the stanford experience in lymphomatoid papulosis and primary cutaneous anaplastic large cell lymphoma. J Am Acad Dermatol 49(6):1049–1058. doi:10.1016/S0190 S0190962203024848 (pii)
11. Yu JB, Blitzblau RC, Decker RH et al (2008) Analysis of primary cd30+ cutaneous lymphoproliferative disease and survival from the surveillance, epidemiology, and end results database. J Clin Oncol 26(9):1483–1488. doi:26/9/1483 (pii) 10.1200/JCO.2007.14.1374
12. Boni R, Xin H, Kamarashev J et al (2000) Allelic deletion at 9p21-22 in primary cutaneous cd30(+) large cell lymphoma. J Invest Dermatol 115(6):1104–1107. doi:jid197 (pii) 10.1046/j.1523-1747.1998.00167.x

13. Fischer TC, Gellrich S, Muche JM et al (2004) Genomic aberrations and survival in cutaneous t cell lymphomas. J Invest Dermatol 122(3):579–586. doi:10.1111/j.0022-202X.2004.22301.x JID22301 (pii)
14. Zettl A, Rudiger T, Konrad MA et al (2004) Genomic profiling of peripheral t-cell lymphoma, unspecified, and anaplastic large t-cell lymphoma delineates novel recurrent chromosomal alterations. Am J Pathol 164(5):1837–1848. doi:S0002-9440(10)63742-X (pii) 10.1016/S0002-9440(10)63742-X
15. Futreal PA, Coin L, Marshall M et al (2004) A census of human cancer genes. Nat Rev Cancer 4(3):177–183. doi:10.1038/nrc1299 nrc1299 (pii)
16. Gogusev J, Telvi L, Nezelof C (2002) Molecular cytogenetic aberrations in cd30+ anaplastic large cell lymphoma cell lines. Cancer Genet Cytogenet 138(2):95–101. doi:S0165460802005897 (pii)
17. Pons E, Uphoff CC, Drexler HG (1998) Expression of hepatocyte growth factor and its receptor c-met in human leukemia-lymphoma cell lines. Leuk Res 22(9):797–804. doi:S014521269800071X (pii)
18. Choi YL, Tsukasaki K, O'neill MC et al (2007) A genomic analysis of adult t-cell leukemia. Oncogene 26(8):1245–1255. doi:1209898 (pii) 10.1038/sj.onc.1209898
19. Kallies A, Hawkins ED, Belz GT et al (2006) Transcriptional repressor blimp-1 is essential for t cell homeostasis and self-tolerance. Nat Immunol 7(5):466–474. doi:ni1321 (pii) 10.1038/ni1321
20. Martins GA, Cimmino L, Shapiro-Shelef M et al (2006) Transcriptional repressor blimp-1 regulates t cell homeostasis and function. Nat Immunol 7(5):457–465. doi:ni1320 (pii) 10.1038/ni1320
21. Salaverria I, Bea S, Lopez-Guillermo A et al (2008) Genomic profiling reveals different genetic aberrations in systemic alk-positive and alk-negative anaplastic large cell lymphomas. Br J Haematol 140(5):516–526. doi:BJH6924 (pii) 10.1111/j.1365-2141.2007.06924.x
22. Vermeer MH, Van Doorn R, Dijkman R et al (2008) Novel and highly recurrent chromosomal alterations in sezary syndrome. Cancer Res 68(8):2689–2698. doi:68/8/2689 (pii) 10.1158/0008-5472.CAN-07-6398
23. Van Doorn R, Van Kester MS, Dijkman R et al (2009) Oncogenomic analysis of mycosis fungoides reveals major differences with sezary syndrome. Blood 113(1):127–136. doi:blood-2008-04-153031 (pii) 10.1182/blood-2008-04-153031

Chapter 14

Genome-Wide DNA Profiling of HIV-Related B-Cell Lymphomas

Andrea Rinaldi, Daniela Capello, Emanuele Zucca, Gianluca Gaidano, and Francesco Bertoni

Abstract

Non-Hodgkin lymphomas represent a frequent complication of human immunodeficiency virus (HIV) infection, occurring at higher frequency than in immunocompetent individuals, and causing morbidity and mortality. Here, we present the method we have followed to analyze the genomic lesions in HIV-related and in other immunodeficiency-related lymphomas, as well in diffuse large B-cell lymphoma (DLBCL) samples derived from immunocompetent hosts. The technology we have used is represented by the GeneChip Human Mapping 250K NspI arrays (Affymetrix, Santa Clara, CA, USA), arrays based on 25mer oligonucleotides initially designed for large-scale genotyping, that is, the detection of thousands of single-nucleotide polymorphisms (SNPs), then shown to be applicable for the detection of cancer alterations. The protocol is shown in all its steps with suggestions and tips. Applications of the technology and obtained results are also briefly summarized.

Key words: Affymetrix, SNP, Microarray, CGH, Lymphoma, HIV, DLBCL, BL, PEL

1. Introduction

Non-Hodgkin lymphomas represent a frequent complication of human immunodeficiency virus (HIV) infection, occurring at higher frequency than in immunocompetent individuals, and causing morbidity and mortality (1, 2). The underlying immunodeficiency, a chronic antigen stimulation, the cytokine dysregulation, and viruses such as Epstein–Barr virus (EBV) or human herpes virus 8 (HHV8) are all factors contributing to a higher incidence of lymphoma in individuals with HIV infection (1–6).

Screening of known genetic lesions has revealed that the molecular pathogenesis of HIV–NHL is characterized by molecular heteroge-

Diponkar Banerjee and Sohrab P. Shah (eds.), *Array Comparative Genomic Hybridization: Protocols and Applications*, Methods in Molecular Biology vol. 973, DOI 10.1007/978-1-62703-281-0_14,

neity, involving different pathogenetic pathways preferentially associated with specific clinicopathologic entities (7–12).

Here, we present the method we have followed to analyze the genomic lesions in HIV-related and in other immunodeficiency-related lymphomas, the posttransplant DLBCL (PT-DLBCL), and in DLBCL samples derived from immunocompetent hosts (IC-DLBCL) (13–16). The technology we have used is represented by the GeneChip Human Mapping 250K NspI arrays (Affymetrix, Santa Clara, CA, USA). These arrays are based on 25mer oligonucleotides initially designed for large-scale genotyping, that is, the detection of thousands of single-nucleotide polymorphisms (SNPs) (17, 18), and later shown to be applicable for the detection of cancer genome alterations (19–21). The resolution of the 250K array is relatively high, with a median physical distance between interrogated SNPs of approximately 5 kb. Importantly, since the arrays are SNP based, they can provide not only information on the copy number, but also on the loss of heterozygosity (LOH), allowing the identification of regions of copy-neutral LOH (or uniparental disomy). All these features make the platform a very good choice to study genomic alterations in human cancers. Briefly, high-molecular weight genomic DNA is digested with a restriction enzyme to reduce the genomic complexity and ligated to adaptors that recognize the cohesive four base-pair overhangs. A generic primer that recognizes the adaptor sequence is then used to amplify adaptor—ligated DNA fragments. The amplified DNA is then fragmented, labeled, and hybridized to a GeneChip Human Mapping 250K array. Finally, the data are analyzed first with Affymetrix proprietary tools, and then tumor genomic profiles can be obtained following different workflows.

2. Materials

2.1. Instruments

- Affymetrix GeneChip Scanner 3000 7 G.
- Affymetrix GeneChip Fluidic Station 450.
- Affymetrix GeneChip Hybridization Oven 640.
- Affymetrix computer workstation.
- Affymetrix GeneChip Operating Software (GCOS) v1.4.
- Affymetrix Genotyping Analysis Software (GTYPE) v4.1.
- Affymetrix GeneChip Chromosome Copy Number Analysis Tool (CNAT) v4.01.
- Microwave.
- Plate Centrifuge Eppendorf 5804 or 5810.
- Tabletop ultracentrifuge capable with rotor for 1.5 mL tubes.

- Ice baskets.
- Spectrophotometer: NanoDrop 1000 or higher.
- Power supply and apparatus for gel electrophoresis.
- Gel imaging UV transilluminator and image capture camera.
- MJR PCT-225 Tetrad Thermal Cycler or similar.
- Microplate shaker: Jitterbug Boekel Scientific, model 130000.
- Manifold: QIAvac multiwell unit (Qiagen).
- Vacuum Regulator for use during PCR clean-up step (Qiagen).
- Pipettes (multichannel and single channel, variable volumes) and pipette tips
- Vortexer for plates and tubes (VWR).

2.2. Reagents and Plastics

- Affymetrix GeneChip Human Mapping 250K NspI arrays (see Note 1).
- Affymetrix GeneChip Human Mapping 250K Nsp Assay Kit.
- 1.5 mL microcentrifuge tubes.
- Multiplate 96-Well Unskirted PCR Plates (Bio-Rad).
- MicroAmp Clear Adhesive Films (Applied Biosystems).
- 12 tubes strip.
- Gloves.
- Solution basin, 55 and 100 mL (Labcor).
- Cooling chamber, double gold block, Diversified Biotech CHAM-1020 Quantity 2.
- Cooling chamber, single gold block, Diversified Biotech CHAM-1000.
- 0.5 M EDTA (Ambion).
- NspI (10,000 U/mL) (New England Biolabs).
- T4 DNA Ligase (New England Biolabs).
- Clontech Titanium DNA Amplification Kit.
- Clontech DNA Amplification Clean-Up Kit.
- Molecular Biology Grade Water.
- Electrophoresis running buffer (e.g., 0.5× TBE).
- Agarose powder.
- Gel loading solution (Sigma).
- DNA Marker: Hi-Lo Bionexus 50–10,000 bp.
- 5 M TMACL (Tetramethyl Ammonium Chloride) (Sigma).
- 10% Tween-20 Pierce P/N 28320 diluted to 3% in molecular biology grade water.

- MES Hydrate sigma Ultra (Sigma).
- MES Sodium Salt (Sigma).
- DMSO (Sigma).
- Denhardt's Solution 50× concentrate (Sigma).
- HSDNA (herring sperm DNA) (Promega).
- Human Cot-1 DNA (Invitrogen).
- 20× SSPE BioWhittaker.
- Tough-Spots Diversified Biotech SPOT-1000.
- SAPE (Streptavidin, R-phycoerythrin conjugate) (Invitrogen), 1 mg/mL.
- Ab (Biotinylated Anti-Streptavidin) (Vector Laboratories), 0.5 mg.
- 5 M NaCl, RNase-free, DNase-free (Ambion).
- Glassware, bottles for preparing agarose gels and electrophoresis running buffer.

3. Methods

3.1. Set Up of the Laboratory

Appropriate laboratory configuration is important to minimize the possibility of carryover contamination. Use of dedicated areas as pre-PCR Clean Room and main lab is highly recommended, since the most likely potential source of contamination performing the assay is previously amplified PCR product.

In the pre-PCR Clean Room, or in a biosafety hood, which might replace pre-PCR Clean Room, the operator should handle genomic DNA and perform digestion and ligation reactions. Gloves should always be worn in this area. The area should have dedicated equipments (see Note 2).

The main lab area has to be considered contaminated with PCR products and template. The operator should never reenter in the pre-PCR area without having first undergone decontamination by showering and changing into freshly laundered clothes.

3.2. Evaluation of DNA Samples

High purity, non-degraded DNA is required to obtain accurate data (see Note 3). The success of this assay requires the amplification of PCR fragments between 200 and 1,100 bp.

3.2.1. DNA Quality Control

Before starting, genomic DNA should be run on a 1% agarose gel: high-quality genomic DNA will run as a major band at approximately 10–20 kb.

3.2.2. DNA Purification

DNA free of PCR inhibitors such as heme, chelating agents, and high concentration of salts. In order to avoid inhibitors or to concentrate the DNA, the following steps for precipitation can be used.

Add 1/10 volume of 5 M NH4OAc to the genomic DNA (10 μL if the DNA was eluted in 100 μL).

Then, add 2.5 volumes of 100% ethanol (275 μL if the DNA was eluted in 100 μL).

Mix the tube well and incubate at −20°C over night.

Centrifuge the tube at maximal speed for 15 min at 4°C.

Carefully remove and discard the supernatant.

Wash the pellet with 500 μL 70% cold ethanol. Centrifuge again and remove and discard the supernatant.

Air-dry the pellet (vacuum centrifugation might be needed).

Resuspend the DNA pellet using appropriate solution and volume.

3.2.3. DNA Quantification

Determine the concentration of genomic DNA solution by measuring its absorbance at 260 nm. Recommended instrument is NanoDrop spectrophotometer for its convenience: no dilutions or cuvette are needed; direct measurement of 1.5 μL of genomic DNA sample is sufficient (see Note 4).

3.3. SNP-Array Protocol

See Notes 5 and 6.

3.3.1. DNA Digestion

Prepare the Digestion Master Mix on ice as follows (see Note 7):

- Water 9.75 μL.
- NE Buffer 2 (10×) 2 μL.
- BSA (10× 1 mg/mL) 2 μL.
- NspI (10 U/μL) 1 μL.

Then, add 5 μL genomic DNA samples (50 ng/μL, total 250 ng of DNA) in individual wells of a 96-well plate.

Finally, add 14.75 μL of the prepared Digestion Master Mix to each well.

Cover the plate with a plate cover, tightly seal, vortex for 2 s, briefly spin at 2,000 rpm (400×*g*), and then put in a PCR thermal cycler and run according to the following digestion program: 37°C, 120 min; 65°C, 20 min.

For the next step, program the thermal cycler in advance so that the lid is heated.

3.3.2. DNA Ligation

Prepare the Ligation Master Mix on ice since ligase buffer contains ATP which has to be thawed/held at 4°C:

- Adaptor NspI (50 μM) 0.75 μL.
- T4 DNA Ligase Buffer 2.5 μL.

– T4 DNA Ligase (400 U/μL) 2 μL.

Aliquot 5.25 μL of the Ligation Master Mix into each digested DNA sample.

Cover the plate with a plate cover, tightly seal, vortex for 2 s, and briefly spin at 2,000 rpm.

Transfer the plate into a thermal cycler and run the following ligation program: 16°C, 180 min; 70°C, 20 min; 4°C, on hold.

Dilute each DNA ligation reaction (total volume 25 μL, each sample) by adding 75 μL of water.

3.3.3. PCR

Amplify each sample in triplicate.

Prepare PCR Master Mix by mixing the following reagents in three 96-well plates (volumes are shown for one PCR reaction, consider three reactions for each sample):

– Water 39.5 μL.
– Clontech Titanium Taq PCR Buffer (10×) 10 μL.
– G-C Melt (5 M) 20 μL.
– dNTP (2.5 mM each) 14 μL.
– PCR primer 002 (100 μM) 4.5 μL.
– Clontech Titanium Taq DNA Polymerase (50×) 2 μL

Transfer 10 μL of each diluted ligated DNA from the 96-well plate into the corresponding three individual wells, one in each of the three PCR plates using a 12-channel pipette.

Add 90 μL of PCR Master Mix to obtain a total volume of 100 μL.

Cover the plate with a plate cover, tightly seal, vortex for 2 s, and briefly spin at 2,000 rpm.

Run the following 250K PCR program on the MJR Research DNA Tetrad Engine: (1) 94°C, 3 min; (2) 94°C, 30 s; (3) 60°C, 30 s; (4) 68°C, 15 s; (5) go to (2)–(4) for 29 times; (6) 68°C, 7 min; (7) 4°C, on hold.

Once, the PCR is done, move from the pre-PCR area to the main lab.

3.3.4. PCR Products Quality Control

Mix 3 μL of each PCR products with 3 μL of 2× gel loading dye on a 2% TBE gel at 120 V for 1 h. The PCR products must have an average size between 200 and 1,100 bp.

3.3.5. PCR Purification and Quantification

Add 8 μL of 0.1 M EDTA to each PCR reaction. Wells which are not going to be used must be covered with adhesive film.

Vortex the plate at medium speed for 2 s, and spin at 2,000 rpm for 1 min.

With the vacuum switched on (600 mbar), transfer the three PCR reactions from each individual DNA sample into a single well of the Clean-Up Plate.

Once the wells are completely dry, wash the PCR products by adding 50 μL of water and dry the wells completely (it takes approximately 20 min). Repeat this step twice for a total of three washes with water.

Add 45 μL of RB buffer to each well. Cover the plate with a plate cover, tightly seal, and shake it moderately on a plate shaker for 10 min at room temperature.

Recover the purified PCR products in a fresh 96-well plate by pipetting the eluate out of each well and transferring it to the corresponding well in the fresh 96-well plate.

Quantify the purified PCR products. The expected quantity is 2 μg of DNA/μL (see Note 8).

3.3.6. Fragmentation

Preheat thermal cycler to 37°C, and perform all the steps as quickly as possible. Due to the sensitive nature of the Fragmentation Reagent (DNase I), all the steps must be performed on ice.

Prepare the Fragmentation Mix using 45 μL PCR product and 5 μL of 10× Fragmentation Buffer.

Prepare the Fragmentation Reagent. Since the concentration of stock Fragmentation Reagent (U/μL) may vary from lot to lot, it is mandatory to check the concentration on the tube before conducting the dilution. Dilute the stock of Fragmentation Reagent to 0.05 U/μL by adding water and 12 μL of 10× Fragmentation Buffer having a final volume of 120 μL. Add 5 μL of the Fragmentation Mix to the diluted Fragmentation Reagent.

Run the following 250K Fragmentation program on the MJR Research DNA Tetrad Engine: 37°C, 35 min; 95°C, 15 min; 4°C, Hold.

3.3.7. Fragmentation Quality Control

Mix 4 μL of fragmented PCR product with 4 μL gel loading dye and run on 4% TBE gel at 120 V for 30–60 min. Average fragment size should be less than 180 bp.

3.3.8. Labeling

Prepare the Labeling Mix on ice by adding 14 μL of 5× TdT Buffer, 2 μL of DNA labeling Reagent, and 3.5 μL of TdT.

Add 19.5 μL of Labeling Master Mix into the fragmentation plate containing 50.5 μL of each fragmented DNA samples in individual wells. Run the following 250K label program on the thermo cycler: 37°C, 4 h; 95°C, 15 min; 4°C, hold.

3.3.9. Hybridization

Prepare the Hybridization Cocktail Master Mix with:

- MES (12×; 1.22 M), 12 μL.
- DMSO (100%), 13 μL.
- Denhardt's Solution (50×), 13 μL.
- EDTA (0.5 M), 3 μL.
- HSDNA (10 mg/mL), 3 μL.

- Oligo Control Reagent OCR, 0100, 2 μL.
- Human Cot-1 DNA (1 mg/mL), 3 μL.
- Tween 20 (3%), 1 μL.
- TMACL (5 M), 140 μL.

The total volume has to be of 190 μL per sample.

Then, transfer each of the labeled samples from the plate to individual 1.5 mL Eppendorf tubes.

Add the 190 μL of the Hybridization Cocktail Master Mix into the 70 μL of labeled DNA samples.

Heat the 260 μL of Hybridization Mix and labeled DNA at 99°C in a heat block for exactly 10 min to denature, cool on ice for 10 s, and then at 49°C for 1 min.

Inject 200 μL of denatured hybridization cocktail into the array cartridge.

Store the remaining Hybridization Mix at −20°C for future use.

Place the array into the Affymetrix hybridization oven with the following settings: temperature at 49°C for 16–18 h at 60 rpm.

3.3.10. Washing and Staining

Prepare the following solutions:

- Wash A: Non-Stringent Wash Buffer (6× SSPE, 0.01% Tween 20).
- Wash B: Stringent Wash Buffer (0.6× SSPE, 0.01% Tween 20).
- 2× MES Stock Buffer (1.25 M MES, 0.89 M [Na+]).
- 1× Array Holding Buffer (Final 1× concentration: 100 mM MES, 1 M [Na+], 0.01% Tween 20).

Filter the first three solutions using 0.2 μM filters, and store them at room temperature. Store the fourth one at 2°C to 8°C, and protect it from light.

After 16–18 h of hybridization, remove the hybridization cocktail from the array, and set it aside in a microcentrifuge vial. You can store it on ice during the procedure or at −80°C for long-term storage. Fill the array completely with 270 ìL of Array Holding Buffer.

Before proceeding, create a new experiment using GCOS. From the File menu, click New Experiment and fill all the fields in the new experiment window (sample name, project, user, experiment name, probe array type, barcode). The experiment name must be unique. Appropriate library files must be installed for a probe array to appear in the drop-down menu.

On GCOS, select "Fluidics" from the menu bar and run a Prime Protocol to fill all the fluidics station lines with the appropriate buffers. The fluidics station is now ready to run fluidics station protocols.

Prepare the following reagents for each array.

Stain Buffer.

- Water 800.04 μL.
- SSPE (20×) 360 μL.
- Tween 20 (3%) 3.96 μL.
- Denhardt's (50×) 24 μL.

Split the total volume (1,188 μL) in two different 1.5 mL vials with 594 μL each. Add 6 μL of 1 mg/mL Streptavidin Phycoerythrin (SAPE) in one vial to have SAPE Solution Mix. Add 6 μL of 0.5 mg/mL biotinylated antibody in the second vial to have Antibody Solution Mix.

Prepare 100 mL of Array Holding Buffer adding:

- MES Stock Buffer (12×) 8.3 mL.
- 5 M NaCl 18.5 mL.
- Tween 20 (10%) 0.1 mL.
- Water 73.1 mL.

Add 820 ìL of Array Holding Buffer to each microcentrifuge tube. One tube is needed per each module which is being used.

For each module, place a vial containing SAPE Stain Solution in sample holder 1, a vial containing Antibody Stain Solution in sample holder 2, and a vial containing Array Holding Buffer in sample holder 3.

In the Fluidics Station dialog box on the workstation, select the correct experiment name from the drop-down Experiment list. The Probe Array Type will automatically appear.

In the Protocol drop-down list, select Mapping500Kv1_450 to control the washing and staining of the probe array.

Select Run in the Fluidics Station dialog box to begin the washing and staining.

Follow the instructions presented on the computer screen.

When the staining phase is over, remove the microcentrifuge vials containing the stain. Replace with three empty microcentrifuge vials as prompted.

Remove the arrays from fluidics station modules by pressing down the cartridge lever to the eject position.

Check the array windows for the presence of bubbles. If bubbles, fill with Array Holding Buffer (see Note 9).

3.3.11. Scanning

Once the wash protocols are complete, the array can be scanned (see Note 10).

Be sure that laser is warmed up prior to scanning by turning the GeneChip Scanner 3000 7 G laser on at least 10 min before use. Using GCOS, select Scanner from the menu bar.

Before scanning the array, clean the glass surface of the probe array with a nonabrasive towel or tissue. Tough-Spots can be applied to each of the two septa on the array cartridge to prevent leaking of fluids from the cartridge during scanning.

Once the warm-up is finish, open the sample door of the scanner and insert the array into the holder. The door of the GeneChip Scanner 3000 7 G will automatically close. Click the Start Scan icon in the tool bar. Select the experiment name that corresponds to the probe array being scanned. Let the scanner acquire the images and crate the .cel files (see Note 11).

4. Notes

4.1. Potential Pitfalls and Troubleshooting

1. The 250K arrays are now to be considered as "old" arrays. However, they still represent a good choice due to the quality of data obtained, cost, and easier data interpretation when compared with profiles obtained with the newer Affymetrix SNP6 arrays due to the relatively lower resolution.
2. It is very important to have different instruments for pre- and post-PCR, especially pipettes.
3. Genomic DNA extraction methods such as phenol-chloroform or QIAmp DNA Mini kit yield DNA samples of optimal quality and amount. Preamplified genomic DNA (e.g., using QIAGEN Repli-g Kit) has been reported to give results comparable to the standard DNA preparation methods (22, 23).
4. Considering that the NanoDrop and other UV spectrophotometers use absorbance which cannot distinguish between DNA, RNA, degraded nucleic acids, free nucleotides, and other contaminants (24), for a correct concentration, measurement has been more recently indicated the Qubit Quantitation Platform which uses fluorescent dyes.
5. The described protocol largely follows the original Affymetrix Genome-Wide Human Mapping 500K Assay Manual User Guide (obtained at http://www.affymetrix.com/support/downloads/manuals/500k_assay_manual.pdf), which has to be downloaded by the operator and that can contain updated suggested by the manufacturer.
6. Here, we present the protocol version to be performed on 96-well plates. It is also possible to run the assay with the use of single tubes; however, we strongly recommend the use of plates and multichannel pipettes to decrease the risk of mistakes.
7. In the presented protocol, all volumes and amounts of reagents shown are meant for one sample.
8. The PCR purification part of the protocol is the section that can often undergo improvements. Thus, the latest versions of the protocol have to be periodically reviewed.

9. If the operator is unfamiliar in handling GeneChip arrays and Fluidics Station modules, it is strongly advised to refer to the Affymetrix user's guide.
10. Scan must be preformed within 24 h: if the array cannot be scanned promptly, it must be kept at 4°C and in the darkness until ready for scanning. Before scanning, the array must be allowed to warm to room temperature.
11. The image file generated by scanner is a .dat file. A .cel file is automatically generated from the .dat file. These are image files: the .dat files represent the raw image, while the .cel files contain intensity values for the probes on the array. Both .dat and .cel files can be useful for troubleshooting and for quality controls.

5. Data Analysis and Interpretation

The .cel files obtained with GTYPE can be processed following different strategies. The Affymetrix GeneChip Chromosome Copy Number Analysis Tool (CNAT) can provide the genomic profile of the individual samples as well the LOH estimate. These data might already be satisfactory for the analysis of small number of samples or for diagnostic purposes. However, for large studies, different algorithms would be better used to estimate the copy number and LOH (25–28) and to identify the regions recurrently targeted by DNA gains, losses, or LOH, which more likely contains genes relevant for the tumor pathogenesis (29–31). The use of the 250K arrays have contributed to the discovery of new tumor lesions, such as the inactivation of *PAX5* in acute lymphoblastic leukemias (32) or of *TNFAIP3* in lymphomas (33–36).

Regarding HIV-lymphomas, using the 250K arrays, we have analyzed 28 HIV-DLBCL, 11 HIV-BL, four HIV-primary central nervous system DLBCL (PCNSL), eight cell lines derived from HIV-related primary effusion lymphomas (HIV-PEL), and six derived from HIV-BL (13). Genomic complexity varied across HIV-lymphoma subtypes. HIV-BL showed a significantly lower number of lesions than HIV-DLBCL, whereas the median number of copy-number changes was significantly higher in EBV-negative HIV-DLBCL compared to EBV-positive cases. Compared to IC-DLBCL, HIV-DLBCL displayed a distinct genomic profile with no gains of 18q and specific genetic lesions. In particular, fragile sites-associated genes, including *FHIT* (FRA3B), WWOX (FRA16D), *DCC* (FRA18B), and *PARK2* (FRA6E) were frequently inactivated in HIV-NHL by interstitial deletions, and a significantly higher prevalence of *FHIT* alterations was observed in HIV-DLBCL compared to IC-DLBCL. The same genes involved by fragile site deletions were also frequently affected by aberrant methylation of regulative regions and by downregulation of the transcripts. Similar results regarding *FHIT* and *WWOX* have also been later confirmed by others (37, 38).

In conclusion, the use of the Affymetrix GeneChip Human Mapping 250K NspI arrays, as well of other similar and newer alternatives, represent impressive tools to obtain genome-wide information of unbalanced genomic lesions which can help in identifying new pathogenetic mechanisms or lesions bearing prognostic relevance.

Acknowledgments

Work supported by: Oncosuisse grant OCS-1939-8-2006 (Bern, Switzerland); Cantone Ticino "Computational life science/Ticino in rete" program (Bellinzona, Switzerland); Fondazione per la Ricerca e la Cura sui Linfomi (Bellinzona, Switzerland); Nelia et Amadeo Barletta Foundation (Lausanne, Switzerland); Ricerca Sanitaria Finalizzata, Regione Piemonte (Torino, Italy); VI Programma Nazionale di Ricerca sull'AIDS, ISS (Rome, Italy); PRIN-MIUR2006; Novara-AIL Onlus (Novara, Italy); Fondazione CRT (Torino, Italy).

References

1. Dunleavy K, Wilson WH (2010) HIV-associated lymphoid neoplasms. In: Magrath IT (ed) The lymphoid neoplasms, 3rd edn. Hodder Arnold, London, pp 1421–1439
2. Levine AM, Said JW (2010) Management of acquired immunodeficiency syndrome-related lymphoma. In: Armitage JO, Coiffier B, Dalla Favera R, Harris NL, Mauch PM (eds) Non-Hodgkin lymphomas, 2nd edn. Lippincott Williams & Wilkins, Philadelphia, PA, pp 507–526
3. Carbone A, Gloghini A (2005) AIDS-related lymphomas: from pathogenesis to pathology. Br J Haematol 130(5):662–670
4. Spano JP, Costagliola D, Katlama C, Mounier N, Oksenhendler E, Khayat D (2008) AIDS-related malignancies: state of the art and therapeutic challenges. J Clin Oncol 26(29):4834–4842
5. Grogg KL, Miller RF, Dogan A (2007) HIV infection and lymphoma. J Clin Pathol 60(12): 1365–1372
6. Cheung MC, Pantanowitz L, Dezube BJ (2005) AIDS-related malignancies: emerging challenges in the era of highly active antiretroviral therapy. Oncologist 10(6):412–426
7. Carbone A, Gloghini A, Capello D, Gaidano G (2001) Genetic pathways and histogenetic models of AIDS-related lymphomas. Eur J Cancer 37(10):1270–1275
8. Carbone A, Cesarman E, Spina M, Gloghini A, Schulz TF (2009) HIV-associated lymphomas and gamma-herpesviruses. Blood 113(6): 1213–1224
9. Nair P, Pan H, Stallings RL, Gao SJ (2006) Recurrent genomic imbalances in primary effusion lymphomas. Cancer Genet Cytogenet 171(2):119–121
10. Klein U, Gloghini A, Gaidano G, Chadburn A, Cesarman E, Dalla-Favera R, Carbone A (2003) Gene expression profile analysis of AIDS-related primary effusion lymphoma (PEL) suggests a plasmablastic derivation and identifies PEL-specific transcripts. Blood 101(10):4115–4121
11. Jenner RG, Maillard K, Cattini N, Weiss RA, Boshoff C, Wooster R, Kellam P (2003) Kaposi's sarcoma-associated herpesvirus-infected primary effusion lymphoma has a plasma cell gene expression profile. Proc Natl Acad Sci USA 100(18):10399–10404

12. Capello D, Gloghini A, Martini M, Spina M, Tirelli U, Bertoni F, Rinaldi A, Morra E, Rambaldi A, Sinigaglia F, Larocca LM, Carbone A (2011) Mutations of CD79A, CD79B and EZH2 genes in immunodeficiency-related non-Hodgkin lymphomas. Br J Haematol 152(6):777–780
13. Capello D, Scandurra M, Poretti G, Rancoita PM, Mian M, Gloghini A, Deambrogi C, Martini M, Rossi D, Greiner TC, Chan WC, Ponzoni M, Moreno SM, Piris MA, Canzonieri V, Spina M, Tirelli U, Inghirami G, Rinaldi A, Zucca E, Favera RD, Cavalli F, Larocca LM, Kwee I, Carbone A, Gaidano G, Bertoni F (2010) Genome wide DNA-profiling of HIV-related B-cell lymphomas. Br J Haematol 148(2):245–255
14. Rinaldi A, Capello D, Scandurra M, Greiner TC, Chan WC, Bhagat G, Rossi D, Morra E, Paulli M, Rambaldi A, Rancoita PMV, Inghirami G, Ponzoni M, Moreno SM, Piris MA, Mian M, Chigrinova E, Zucca E, Dalla Favera R, Gaidano G, Kwee I, Bertoni F (2010) SNP-arrays provide new insights in the pathogenesis of post-transplant diffuse large B-cell lymphoma. Br J Haematol 149(7):569–577
15. Scandurra M, Mian M, Greiner TC, Rancoita PMV, de Campos CP, Chan WC, Vose JM, Chigrinova E, Inghirami G, Chiappella A, Baldini L, Ponzoni M, Ferreri AJM, Franceschetti S, Gaidano G, Moreno SM, Piris MA, Facchetti F, Tucci A, Nomdedeu JF, Lazure T, Lambotte O, Uccella S, Pinotti G, Pruneri G, Martinelli G, Young KH, Tibiletti MG, Rinaldi A, Zucca E, Kwee I, Bertoni F (2010) Genomic lesions associated with a different clinical outcome in diffuse large B-cell lymphoma treated with R-CHOP. Br J Haematol 151(3):221–231
16. Kwee I, Capello D, Rinaldi A, Rancoita PM, Bhagat G, Greiner TC, Spina M, Gloghini A, Chan WC, Paulli M, Zucca E, Tirelli U, Carbone A, Gaidano G, Bertoni F (2012) Genomic aberrations affecting the outcome of immunodeficiency-related diffuse large B-cell lymphoma. Leuk Lymphoma 53(1):71–76
17. Matsuzaki H, Loi H, Dong S, Tsai YY, Fang J, Law J, Di X, Liu WM, Yang G, Liu G, Huang J, Kennedy GC, Ryder TB, Marcus GA, Walsh PS, Shriver MD, Puck JM, Jones KW, Mei R (2004) Parallel genotyping of over 10,000 SNPs using a one-primer assay on a high-density oligonucleotide array. Genome Res 14(3):414–425
18. Kennedy GC, Matsuzaki H, Dong S, Liu WM, Huang J, Liu G, Su X, Cao M, Chen W, Zhang J, Liu W, Yang G, Di X, Ryder T, He Z, Surti U, Phillips MS, Boyce-Jacino MT, Fodor SP, Jones KW (2003) Large-scale genotyping of complex DNA. Nat Biotechnol 21(10): 1233–1237
19. Huang J, Wei W, Zhang J, Liu G, Bignell GR, Stratton MR, Futreal PA, Wooster R, Jones KW, Shapero MH (2004) Whole genome DNA copy number changes identified by high density oligonucleotide arrays. Hum Genomics 1(4):287–299
20. Zhao X, Li C, Paez JG, Chin K, Janne PA, Chen TH, Girard L, Minna J, Christiani D, Leo C, Gray JW, Sellers WR, Meyerson M (2004) An integrated view of copy number and allelic alterations in the cancer genome using single nucleotide polymorphism arrays. Cancer Res 64(9):3060–3071
21. Bignell GR, Huang J, Greshock J, Watt S, Butler A, West S, Grigorova M, Jones KW, Wei W, Stratton MR, Futreal PA, Weber B, Shapero MH, Wooster R (2004) High-resolution analysis of DNA copy number using oligonucleotide microarrays. Genome Res 14(2):287–295
22. Treff NR, Su J, Tao X, Northrop LE, Scott RT Jr (2011) Single-cell whole-genome amplification technique impacts the accuracy of SNP microarray-based genotyping and copy number analyses. Mol Hum Reprod 17(6): 335–343
23. Dean FB, Hosono S, Fang L, Wu X, Faruqi AF, Bray-Ward P, Sun Z, Zong Q, Du Y, Du J, Driscoll M, Song W, Kingsmore SF, Egholm M, Lasken RS (2002) Comprehensive human genome amplification using multiple displacement amplification. Proc Natl Acad Sci USA 99(8):5261–5266
24. Glasel JA (1995) Validity of nucleic acid purities monitored by 260nm/280nm absorbance ratios. Biotechniques 18(1):62–63
25. Olshen AB, Venkatraman ES, Lucito R, Wigler M (2004) Circular binary segmentation for the analysis of array-based DNA copy number data. Biostatistics 5(4):557–572
26. Rancoita PMV, Hutter M, Bertoni F, Kwee I (2009) Bayesian DNA copy number analysis. BMC Bioinformatics 10(1):10
27. Lin M, Wei LJ, Sellers WR, Lieberfarb M, Wong WH, Li C (2004) dChipSNP: significance curve and clustering of SNP-array-based loss-of-heterozygosity data. Bioinformatics 20(8): 1233–1240
28. Rancoita PM, Hutter M, Bertoni F, Kwee I (2010) An integrated Bayesian analysis of LOH and copy number data. BMC Bioinformatics 11:321
29. Lenz G, Wright GW, Emre NC, Kohlhammer H, Dave SS, Davis RE, Carty S, Lam LT, Shaffer AL, Xiao W, Powell J, Rosenwald A,

Ott G, Muller-Hermelink HK, Gascoyne RD, Connors JM, Campo E, Jaffe ES, Delabie J, Smeland EB, Rimsza LM, Fisher RI, Weisenburger DD, Chan WC, Staudt LM (2008) Molecular subtypes of diffuse large B-cell lymphoma arise by distinct genetic pathways. Proc Natl Acad Sci USA 105(36): 13520–13525

30. Beroukhim R, Getz G, Nghiemphu L, Barretina J, Hsueh T, Linhart D, Vivanco I, Lee JC, Huang JH, Alexander S, Du J, Kau T, Thomas RK, Shah K, Soto H, Perner S, Prensner J, Debiasi RM, Demichelis F, Hatton C, Rubin MA, Garraway LA, Nelson SF, Liau L, Mischel PS, Cloughesy TF, Meyerson M, Golub TA, Lander ES, Mellinghoff IK, Sellers WR (2007) Assessing the significance of chromosomal aberrations in cancer: methodology and application to glioma. Proc Natl Acad Sci USA 104(50):20007–20012

31. Diskin SJ, Eck T, Greshock J, Mosse YP, Naylor T, Stoeckert CJ Jr, Weber BL, Maris JM, Grant GR (2006) STAC: a method for testing the significance of DNA copy number aberrations across multiple array-CGH experiments. Genome Res 16(9):1149–1158

32. Mullighan CG, Zhang J, Kasper LH, Lerach S, Payne-Turner D, Phillips LA, Heatley SL, Holmfeldt L, Collins-Underwood JR, Ma J, Buetow KH, Pui CH, Baker SD, Brindle PK, Downing JR (2011) CREBBP mutations in relapsed acute lymphoblastic leukaemia. Nature 471(7337):235–239

33. Novak U, Rinaldi A, Kwee I, Nandula SV, Rancoita PMV, Compagno M, Cerri M, Rossi D, Murty VV, Zucca E, Gaidano G, Dalla Favera R, Pasqualucci L, Bhagat G, Bertoni F (2009) The NF-KB negative regulator TNFAIP3 (A20) is commonly inactivated by somatic mutations and genomic deletions in marginal zone B-cell lymphomas. Blood 113(20):4918–4921

34. Compagno M, Lim WK, Grunn A, Nandula SV, Brahmachary M, Shen Q, Bertoni F, Ponzoni M, Scandurra M, Califano A, Bhagat G, Chadburn A, Dalla-Favera R, Pasqualucci L (2009) Mutations of multiple genes cause deregulation of NF-kappaB in diffuse large B-cell lymphoma. Nature 459(7247): 717–721

35. Kato M, Sanada M, Kato I, Sato Y, Takita J, Takeuchi K, Niwa A, Chen Y, Nakazaki K, Nomoto J, Asakura Y, Muto S, Tamura A, Iio M, Akatsuka Y, Hayashi Y, Mori H, Igarashi T, Kurokawa M, Chiba S, Mori S, Ishikawa Y, Okamoto K, Tobinai K, Nakagama H, Nakahata T, Yoshino T, Kobayashi Y, Ogawa S (2009) Frequent inactivation of A20 in B-cell lymphomas. Nature 459(7247):712–716

36. Rinaldi A, Mian M, Chigrinova E, Arcaini L, Bhagat G, Novak U, Rancoita PM, De Campos CP, Forconi F, Gascoyne RD, Facchetti F, Ponzoni M, Govi S, Ferreri AJ, Mollejo M, Piris MA, Baldini L, Soulier J, Thieblemont C, Canzonieri V, Gattei V, Marasca R, Franceschetti S, Gaidano G, Tucci A, Uccella S, Tibiletti MG, Dirnhofer S, Tripodo C, Doglioni C, Dalla Favera R, Cavalli F, Zucca E, Kwee I, Bertoni F (2011) Genome-wide DNA profiling of marginal zone lymphomas identifies subtype-specific lesions with an impact on the clinical outcome. Blood 117(5):1595–1604

37. Deffenbacher KE, Iqbal J, Liu Z, Fu K, Chan WC (2010) Recurrent chromosomal alterations in molecularly classified AIDS-related lymphomas: an integrated analysis of DNA copy number and gene expression. J Acquir Immune Defic Syndr 54(1):18–26

38. Roy D, Sin SH, Damania B, Dittmer DP (2011) Tumor suppressor genes FHIT and WWOX are deleted in primary effusion lymphoma (PEL) cell lines. Blood 118(7): e32–e39

Chapter 15

Array Comparative Genomic Hybridization in Osteosarcoma

Bekim Sadikovic, Paul C. Park, Shamini Selvarajah, and Maria Zielenska

Abstract

Osteosarcoma, the most frequent primary bone tumor, is a malignant mesenchymal sarcoma with a peak incidence in young children and adolescents. Left untreated, it progresses relentlessly to local and systemic disease, ultimately leading to death within months. Genomically, osteosarcomas are aneuploid with chaotic karyotypes, lacking the pathognomonic genetic rearrangements characteristic of most sarcomas. The familial genetics of osteosarcoma helped in elucidating some of the etiological molecular disruptions, such as the tumor suppressor genes *RB1* in retinoblastoma and *TP53* in Li–Fraumeni, and *RECQL4* involved in DNA repair/replication in Rothmund–Thomson syndrome. Genomic profiling approaches such as array comparative genomic hybridization (aCGH) have provided additional insights concerning the mechanisms responsible for generating complex osteosarcoma genomes. This chapter provides a brief introduction to the clinical features of conventional osteosarcoma, the predominant subtypes, and a general overview of materials and analytical methods of osteosarcoma aCGH, followed by a more detailed literature overview of aCGH studies and a discussion of emerging genes, molecular mechanisms, and their clinical implications, as well as more recent application of integrative genomics in osteosarcoma. aCHG is helping elucidate genomic events leading to tumor development and evolution as well as identification of prognostic markers and therapeutic targets in osteosarcoma.

Key words: Osteosarcoma, aCGH, Array comparative genomic hybridization, Copy number, RUNX2, Genomic instability, Genomic integration, Integrative analysis

1. Introduction

Array comparative genomic hybridization (aCGH) has emerged as a reliable and reproducible high-resolution molecular approach for detecting genomic imbalances in cancer. However, like all new technologies, there are inevitable challenges which must be overcome or minimized for aCGH to provide an accurate genomic signature of a tumor sample. In order to discuss the utility of aCGH

Diponkar Banerjee and Sohrab P. Shah (eds.), *Array Comparative Genomic Hybridization: Protocols and Applications*, Methods in Molecular Biology vol. 973, DOI 10.1007/978-1-62703-281-0_15, © Springer Science+Business Media, LLC 2013

in the study of osteosarcoma (OS), it is important to review some of the clinical features of OS, as well as to address some of the challenges it poses, such as its paucity, high level of genomic instability, and heterogeneity. Therefore, this chapter will provide a brief clinical overview of OS, followed by a generalized description of the materials, methods, and data analysis for aCGH. In addition, there will also be a focus on the description of some of the studies performed using this technology, followed by a discussion of the body of knowledge that has emerged from these studies, concerning the molecular and genetic pathways associated with the genomic alterations in OS.

OS, the most frequent of primary bone tumors, is a malignant mesenchymal sarcoma characterized by the formation of bone or osteoid by malignant tumor cells. The peak incidence of OS is in the second decade of life, corresponding to the peak period of skeletal growth, with a gradual decline in incidence thereafter [1]. OS in patients over 40 years of age is mostly secondary since it occurs after exposure to radiation, chemical agents, and viruses or arises in areas of preexisting Paget's disease of bone [2, 3]. It should therefore be considered as a disease different from OS in young patients with regard to histology, though its clinical course is comparable [4]. Although most OS are sporadic, several risk factors have been associated with their incidence, including prior radiation therapy and prior chemotherapy, particularly with alkylating agents or anthracyclines [1]. In addition, OS may be associated with specific clinical syndromes, including Rothmund–Thomson syndrome (a rare syndrome of erythematous and maculopapular skin lesions, hyperpigmentation, cataracts, photosensitivity, hypogonadism, as well as psychomotor and skeletal abnormalities) [5], Li–Fraumeni syndrome (a rare familial cancer syndrome associated with mutation of the tumor suppressor gene *TP53*), as well as retinoblastoma, particularly the familial type [6].

Classification systems for OS have evolved based on the recognition of the heterogeneous nature of the clinical, radiological, and histological features of the disease. It may be classified by predominant matrix constituents, histological grade, location within bone (intramedullary/surface), distribution/number (unifocal or multifocal), and primary or secondary to malignant transformation of a preexistent osseous lesion. Each of these factors, along with distinct clinical–pathologic features, serves to categorize and subtype the different forms of OS (Table 1).

Conventional OS, also termed central or high-grade intramedullary OS, represents the most common variant of OS, accounting for approximately 75% of all lesions [3]. It is most commonly seen in the metaphyseal portions of long tubular bones. Histologically, conventional OS is subclassified into osteoblastic, chondroblastic, and fibroblastic subtypes depending on the predominant type of extracellular matrix: osteoblastic if the majority of the tumor cells

Table 1
Clinical classification of osteosarcoma

Osteosarcoma subtype	Occurrence	Age of incidence	Comments
Conventional	75% of OS	80% in second decade of life (childhood OS)	Mostly high grade and aneuploid
Small cell	Rare	70% in first and second decade of life	90% diploid
High-grade surface	Rare	Childhood and adult	Mostly high grade
Telangiectactic	Rare	Childhood and adult	Mostly high grade
Multifocal	Rare	Childhood and adult	Mostly high grade
Intracortical	Rare	10–30 years	Mostly high grade
Periosteal	<2% of OS	~21 years	Good prognosis
Parosteal	Up to 5% of OS	More common in adults	Good prognosis
Well-differentiated intramedullary/low grade	1–2% of OS	Adult	Good prognosis

OS subtypes and the corresponding occurrence frequency, age of incidence, and comments related to clinical features. Summarized from [6, 7]

are producing osteoid (50%), chondroblastic if chondroid matrix predominates (25%), and fibroblastic when the spindle cell proliferation is producing scant osteoid and is similar to that seen with high-grade fibrosarcomas (25%) [6]. The histological subtype is predictive in survival, with chondroblastic OS having a better prognosis than fibroblastic OS [8]. The tumor cells exhibit varying extent of pleomorphism and differentiation, ranging from epitheloid, plasmacytoid, fusiform, ovoid, small round cell, clear cell, and giant cell (both mono- and multinucleated) types to frank anaplasia [4].

Besides the conventional high-grade OS, other subtypes are known. The telangiectatic OS is characterized by the occurrence of large spaces filled with blood with or without septa. Survival indications are similar to conventional high-grade OS. Small cell OS have a slightly worse prognosis compared to conventional OS and are composed of small cells with variable degrees of osteoid production [4].

When untreated, OS aggressively progresses to local and systemic disease, ultimately leading to death within a matter of months. Aside from the histologic subtype, the prognosis depends on the tumor location with axial sites faring worse, alkaline phosphatase levels, and tumor size [9]. In addition, approximately 20% of OS patients have detectable metastases at the time of diagnosis, and these tend to have a worse prognosis. For the remaining proportion of patients, 80% harbor micrometastases undetectable by conventional imaging modalities. The route of metastasis is primarily hematogenous and involves lung and bone, although rare instances of soft tissue, visceral, and lymphatic spread have been described, usually in late-stage disease [3, 6].

Currently, the treatment typically involves a multimodal approach, consisting of neoadjuvant systemic polychemotherapy, followed by surgical excision, and then adjuvant chemotherapy. This introduction of chemoregimens to the surgical treatment has significantly improved the long-term disease-free survival rate to 60–70% of patients. In fact, histological response to chemotherapy, characterized by the degree of necrosis detected by histological examination of multiple sections in the tumor after treatment, is now the single most important prognostic indicator for the outcome of disease in patients without metastases at the initial diagnosis [10]. However, current chemotherapeutic regimens have reached a plateau in terms of improving survival, because of resistance to treatment in a subset of patients [4]. Recent developments in the treatment of cancers involve drugs directed at specific molecules implicated in the pathogenesis of tumors [11]. For OS, such targeted drugs have not yet been identified, but some promising results have been obtained with in vitro studies on human OS cell lines. Imatinib mesylate (Gleevec) is a small molecule inhibitor for c-kit and PDGFR and was shown to be able to induce apoptosis in OS cells [12]. Another potential receptor tyrosine kinase in OS is Ezrin, which was identified by microarray analysis of an OS mouse model and shown to be associated with the metastatic capacity of the tumor [13] and expressed in 44% of high-grade OS. Currently, no small molecule inhibitors are available for Ezrin [4]. Recent data have provided evidence for a role for decitabine, an inhibitor of DNA methylation, in inducing apoptosis in OS [14].

High level of tissue heterogeneity and genomic instability, as outlined above, requires some special consideration when performing and analyzing the aCGH data of this tumor. The following suggestions in regards to the materials and analytical methodologies should serve as a basic guideline for designing and conducting aCGH analysis in OS.

2. Materials

In general, materials required for aCGH analysis of OS involve standard reagents and equipment and follow standard protocols supplied with most major commercial aCGH platforms. One special consideration involves the procurement, selection, and processing of the tumor samples used in the study. The clinical, radiological, and histological features of OS play a central role in providing the context in which the molecular alterations are interpreted. Therefore, in cohort studies of OS samples from multiple subjects, detailed clinical and careful histopathological assessment are key to ensuring a relative homogeneity of the study samples. This is particularly emphasized in consideration of the (as discussed in more detail later) high level of genomic instability characteristic of OS [15] and high level of interindividual variability that may be associated with different histopathological and clinical subtypes. The second major consideration involves the intra-tumor variability with respect to the levels of normal (non-tumor) cell types, as well as the presence of multiple clonal populations with distinct histopathological features. While aCGH is high-resolution and high-throughput, one of its major limitations is that it indiscriminately quantitates genomic content across the various distinct cell subtypes within the sample being analyzed, which is important when one considers the level of heterogeneity present in most tumors including the presence of non-tumor tissue [16]. Molecular analyses could be compromised when the entire sample is analyzed, without tumor enrichment methods, thereby diluting and underestimating the genetic properties of the tumor. In other words, "purity" of the tumor specimen will be directly correlated to the sensitivity and, in some cases, specificity of the aCGH analysis and requires special analytical and bioinformatics considerations (discussed in more detail in data analysis section). In general, a tissue specimen that reaches 90% and higher tumor content, of unifocal origin in a histopathological sense, will result in a decreased intra-tumor genomic variability and increase analytical sensitivity and specificity. The third major consideration involves the tumor tissue preservation methods. Snap freezing of the fresh biopsy specimens is recommended and is preferred methodology over the formalin fixation and paraffin embedding (FFPE) methods. Genomic DNA content in the freshly frozen tissues is generally preserved with minimal degradation for years and possibly decades and will generally yield high-quality DNA for highest-resolution aCGH protocols. FFPE tissue preservation protocols will result in increased levels of DNA degradation and may result in increased technical artifacts in high-resolution aCGH analyses. Selection of clinically and histopathologically uniform samples across the patient cohort, minimizing the intra-tumor variability, and use of well-preserved DNA, as outlined above, are key for generation of robust data for bioinformatics analysis and will maximize the opportunity for identification of clinically relevant genomic copy number alterations using aCGH.

3. Methods

There are currently multiple commercially available platforms for aCGH analysis with detailed specifications for sample preparation and methodology. The following section provides a general overview of the common steps. DNA extraction from tumor tissue specimens of OS cell lines can be performed using the standard phenol chloroform extraction or any of the commercially available DNA purification kits. In general, it is recommended that DNA is treated with RNase to remove any contaminating RNA. Two to three micrograms of DNA are generally required, and it is preferred to avoid use of the whole-genome amplification (WGA) protocols. However, when the tissue sample is limited, WGA can be used to amplify as little as tens of nanograms of DNA sample to micrograms range. While there are a number of commercial kits for the WGA, some of which claim minimal amplification bias, all of the current techniques are inferior to using the non-WGA DNA [17]. Regions with high CpG content as well as repetitive genomic sequences have different chemical properties resulting in different amplification rates, which can create the characteristic "wave" effect across the normalized aCGH dataset—which can be mistaken for copy number gains and losses across what actually is the copy number normal DNA—thereby complicating the data analysis and resulting in decreased sensitivity and specificity (personal observation, unpublished data). This effect of WGA can be further exacerbated in partially degraded (often in FFPE) genomic DNA [18].

The next step involves the fragmentation of the genomic DNA using restriction enzymes to a mean range of approximately 2–300 nucleotides, commonly *Alu*I and *Rsa*I. Theoretically, any set of restriction enzymes producing the desired range of fragmentation can be used. Next, the fragmented DNA is labeled using DNA dyes, commonly Cy5-dUT or Cy3-dUTP, for the test and control sample specifically. At this stage, it is important to decide the type of control to be used in the aCGH reaction, specifically, single sex-matched control or pooled DNA from multiple sex-matched controls. While different protocols, research groups, as well as diagnostic laboratories prescribe to using either single or pooled normal controls, it is important to note that the either approach should yield no analytical differences across majority of the genome that is generally included in most commercial aCGH platforms. However, if the platform design includes probes in the copy number variable (CNV) regions, including segmental duplications, there will be some differences between the two approaches, namely, the copy number for a particular CNV will be determined relative to the individual control or alternatively to the average copy number across the pooled control cohort. While the CNVs can show a "normal" rage across the population, some CNVs have been associated with disease, and while this is a rapidly evolving field of human genomics research, it is important to consider these caveats when deciding on the type of the control sample in your aCGH analysis. After labeling, equimolar amounts of sample and control DNA are co-hybridized to the array platform (generally in presence of blocking agent such as human CotI or salmon sperm DNA to prevent nonspecific hybridization). After washing and drying steps, specified by the manufacturer, the arrays are scanned using the specified optical microarray scanners. The high-resolution images are analyzed using specialized software, generally supplied by the array manufacturer, and ration of the signal intensity for test versus control sample is generated for individual array probes, which are commonly "spotted" in multiple copies on individual array chips to account for technical variation. Such generated data files are used for subsequent data analysis using the aCHG manufacturer's software or among many of the commercial software suites designed for genomic data analysis.

4. Data Analysis and Interpretation

While there are currently many commercial aCGH platforms available, each with possible variations in sample preparation and technical protocols, there are even more bioinformatic, statistical, and analytical variations involved in data processing and interpretation. Additional sections in this book volume deal with specifics of aCGH data analysis, but it is important to highlight some analytical and interpretative caveats associated with analysis of OS tissue samples. The general steps involved in aCGH data processing include data import and in some cases, normalization across the tumor cohort, probe annotation to the reference genome, detection of genomic regions with significant copy number alterations, and finally, annotation of the significant regions to genomic features of interest (generally genes). Two major considerations in bioinformatics analysis of individual OS tumor tissues involve intratumor heterogeneity and technical variability resulting in data "noise." As previously mentioned, it is important to obtain tumor with very low levels of contaminating normal tissues and avoid multifocal tumors. Normal tissue contamination will result in "normalization" of the copy number signal in the copy number disrupted regions in the tumor tissue, which can decrease analytical sensitivity and increase the number of false negatives. Furthermore, OS is a highly genomically unstable which can result in multifocal tumors with differing genomic copy number profiles, as well as genomically distinct cell lineages evolving from the single primary. This can also result in normalization of the copy number level for the copy number discordant regions in these cell lineages. This is a significant technical challenge to overcome, but analytically and in regards to clinical diagnostics of OS, it provides an interesting theoretical issue. Specifically, tumors with highly unstable genomes display high level of copy number alterations and cell lineage heterogeneity. However, copy number alterations providing selective advantages involved in tumor progression will be preserved and selected for across tumor cell lineages. Therefore, decreased analytical sensitivity in the copy number variable regions across evolving tumor cell lineages will result in what could be referred as biological filtering of the data against random genomic variations and for clinically significant copy number alterations. This same analytical concept can be expanded in the aCGH analysis and delineation of "driver" relative to "passenger" copy number alterations in patient cohorts. While OS also displays a high level of interindividual variation in genomic content, probe-level analysis of copy number in a patient cohort relative to the normal control cohort can be used as an alternative approach to identification of recurrent and presumably clinically relevant copy number alterations driving this tumor.

5. Research and Clinical Applications

Most members of the sarcoma family have well-defined chromosomal aberrations, usually consisting of recurrent chromosomal translocations resulting in fusion genes with oncogenic fusion protein products. However, no recurrent translocations have been identified so far in OS. Cytogenetic studies reveal that the majority of OS are characterized by heterogeneous complex chromosomal abnormalities that often vary widely from cell-to-cell [19]. They exhibit a high degree of aneuploidy, gene amplification, and multiple unbalanced chromosomal structural rearrangements that have precluded definitive identification of any recurrent cytogenetic aberrations [20].

Despite the complex pattern of genetic changes in OS, certain chromosomal regions appear to be affected more often than others and may highlight locations of genes involved in the development and progression of these tumors. Conventional cytogenetic techniques have shown that the most common chromosomal abnormalities were gain of +1 and loss of −9, −10, −13, and −17. Partial or complete loss of the long arm of chromosome 6 was also observed [21]. Common chromosomal rearrangements were found in 1p11-13, 1q10-12, 1q11, 1q21-22, 4q27-33, 6p23-25,

7p13-22, 7q11-36, 11p10-5, 11p14-15, 12p13, 15p11-13, 17p12-13, 19q13, and 22q11-13 [21, 22]. Metaphase comparative genomic hybridization (CGH) has been systematically applied and has identified numerous regions of copy number change, including high-level gain or amplification at chromosomal regions 1p, 1q21-q31, 3q26, 6p, 6p12.1, 6p12-21, 8p, 8q12.21.3, 8q22-q23, 8q24.2, 12q12-13, 12q13-q14, 14q24-qter, 17p11-12, Xp11.2-21, and Xq12 [22–29]. The improved mapping resolution of array CGH has shown that 1p36, 6p21, 8q24, 16p13, 17p11, and 19p13 are recurrently gained/amplified [30–34]. In addition, DNA sequence loss has also been observed at 2q, 6p, 8p, 10p, and 17p13 [26, 32]. Spectral karyotyping demonstrated that OS has a complex pattern of clonal and non-clonal rearrangements and helped confirm the structural basis for the imbalances detected by CGH. SKY analysis of primary OS and cell lines identified at least 531 rearrangements and 300 breakpoints with frequent gain of 8q23-24 and 1p11-13 and rearrangement of 20q. In addition, centromeric rearrangements were often observed involving chromosomes 1, 6, 13, 14, 17, and 10 [35]. These data suggest that genomic instability is an important feature of OS and implicates pathways that normally maintain genomic integrity.

The genomic instability model modifies the classical multi-hit hypothesis, described as an ordered sequence of mutations in different oncogenes and tumor suppressor genes, each associated with a defined step in tumorigenesis [36]. In the genomic instability model, one of the hits affects a gene responsible for maintenance of genomic stability [37]. A consequence of such an event is a higher-than-average rate of mutations, leading to cell-to-cell heterogeneity that would predispose the tissue towards an accelerated accumulation of genetic aberrations and tumorigenesis.

To account for the karyotypic and genetic heterogeneity observed among and within tumors, a genomic instability model for OS, via dysfunction of genomic fidelity systems, has gained acceptance through numerous cytogenetic studies. This instability is believed to provide the biological basis for the tumor's capacity to adapt to aggressive treatment regimens. Aberrant double-strand break repair processes can lead to repetitive chromosome end-to-end fusions and precipitate the formation of structural changes indicative of the breakage–fusion–bridge (BFB) cycle. Intermediary structures supportive of this mechanism had been observed in OS cell lines and tumors, which would enhance phenotypic diversity for tumor evolution and adaptation to selective pressures in the microenvironment [38–41]. Repeated breaks and fusions via the BFB cycles can lead to varying levels of genomic imbalance, where a greater number of cycles are expected to lead to high copy number for some regions undergoing repeated cycles. Utilization of whole-genome analysis of OS-derived cell lines as well as OS tumors has enabled the identification of signatures consistent with BFB events [15, 39–42]. Low-level gains situated next to peaks of high-level amplifications in such regions are consistent with the idea of amplification of selective regions via BFB cycles, repeated DNA double-stranded breaks, and subsequent loss of genomic material at or near the point of chromosome breakage. In keeping with this, high copy gains and/or amplification of chromosome arms 6p, 8q, and 17p have been one of the most frequently reported observations in OS tumors [23, 25–28, 32, 35, 43, 44] and are believed to confer a more aggressive disease course [24].

In addition, an in silico analysis of the distribution of segmental duplications within these chromosomal regions of interest highlights the role that genomic architectural elements may play in facilitating the rearrangements leading to concomitant genomic imbalance [15]. Interestingly, data from a cohort of ten tumors exhibited a specific trend which suggests a correlation between the extent of genomic aberration and the clinical, phenotypic behavior. The tumors that exhibited more BFB-associated gains showed very poor response to preoperative chemotherapy, whereas those with relatively low numbers of these events had an optimal response [15]. This suggests the presence of genes within these regions whose increased dosage may contribute to tumor progression and confer a resistance to chemotherapeutic agents by yet unknown mechanisms. This observation needs extensive validation on a larger cohort to identify a predictive gene set that may have clinical utility for outcome prediction in OS.

Genome profiling provides a readily accessible platform for the dissection of consistent genomic patterns that can provide clues about underlying molecular mechanisms of cancer development. This technology could play an important role in medical decisions regarding rational therapy stratification. It is particularly promising for CIN-characterized tumors such as OS, where instability leads to complex karyotypic and genetic changes. With no definitive molecular markers of either diagnostic or prognostic value, most OS patients are initially provided with the same treatment regimen. The observation that OS tumors exhibit varying levels of genomic imbalance, and may respond differently to treatment, highlights the need to stratify patients for more rational treatment regimens. For instance, tumors showing genomic signatures associated with metastasis or recurrence could receive more aggressive treatments. While those patients with more benign genomic profiles could be spared aggressive treatments that can cause complications or serious side effects. One must keep in mind, though, that frequent use of genomic arrays for diagnostic use might be exorbitantly expensive. Results from studies on larger cohorts may make it possible to derive a heterogeneity index for particular chromosomal regions that can differentiate between poor and good responders to treatment. The review of literature of OS focusing on the studies using aCGH is summarized in Table 2.

Table 2
Summary of the aCGH studies in osteosarcoma

Reference	Sample: type/number	Cytogenetic loci	Relevant genes	Follow-up studies
Angstadt [45]	123 canine OS Tumors; aCGH	6p21.1 (gain); 8q21-q24 (gain); 13q13-q13 (loss)	*RUNX2, TUSC3 PTEN*	qRT-PCR; FISH
Atiye [30]	22 OS tumors; cDNA aCGH	12q-q15, 8q24, 6p12-p13, and 17p11-p13 (gain); 3p and 13q (loss)	*MDM2, SAS, CDK4, CYP27B1, COX10, PMP22, COPS3*	
Dujardin [46]	18 OS; CGH	12q13-q15 (gain)	*CDK4, MDM2*	Immunohistochemistry, FISH, qRT-PCR
Hattinger [47]	U-2OS and Saos-2 OS cell lines; CGH	5q12-q15, 11q14-qter, 8q22-qter (gain)	*MLL, DHFR, MYC*	RT-PCR, FISH
Hattinger [48]	6-doxorubicin and 8-methotrexate resistant variants of OS cell lines U-2OS and Saos-2; aCGH	7p11.2-q21, 7q11.2-q22, 19p, X (gain); 8q23-qter (loss)	*MDR1, DHFR, MLL, MYC*	FISH
Kresse [49]	36 OS tumors and 20 cell lines; aCGH	1q21.2-q22, 8q22.1 (gain); 10q22.1-q222.2, 6q16.1; 13q14.2, 13q21.1 (loss)	*LSAMP*	RT-PCR, methylation-specific PCR
Lim [40]	MG-63 OS cell line; aCGH	6p21, 6p22-pter, 8q23-q24 (gain)		mBAND, SKY
Lim [39]	MG-63, Saos-2, U-2 OS and HOS OS cell line, and one OS tumor	6p21, 8q24, 9p21-p22 (gain)		mBAND, FISH
Lockwood [50]	22 OS tumors; whole-genome tiling array	6p21.1, 8q21.2, 8q24.2, 17p11.2, 19q12 (gain)	*CDC5L, SLC29A1, MRPL14, HSP90AB1, SLC35B2, MYC, DDEF1, ATPAF2, RAI1*	FISH; immunohistochemistry
Lu [51]	OS tumors and cell lines	6p12-p21, 12q13-q15, 17p11.2 (gain)	*CDC5L, MAPK14, MAPK13, CDKN1A, PIM1, MDGA1, BTB9, DNAH8, CCND3, PTK7, CDC5L, and RUNX2*	qRT-PCR, Western blotting, immunohistochemistry
Ma [52]	Murine and human OS tumors	11q22 amplicon	*MMP13, cIAP1, cIAP2*	Functional assay (viral transduction)

(continued)

Table 2 (continued)

Reference	Sample: type/number	Cytogenetic loci	Relevant genes	Follow-up studies
Man [31]	48 OS tumors; aCGh	1p36, 6p21, 8q24, 12q14.3, 16p13 and 17p11.2 (gain)	*PRDM16, CDC5L, HSPCB, NFKBIE, IFNG, MGRN 1, PMP22, MYCD, SOX1, ELAC27*	FISH
Pasic [53]	OS tumors	3q13.31 (loss)	*LOC285194 or BC040587, LSAMP*	FISH, qPCR
Sadikovic [54]	Promoter tiling arrays, aCGH, gene expression arrays in OS cell lines HOS, MG-63 and U-2 OS	8q24.3	*GADD45A, WT1, LHX9, C-MYC*	Integrative aCGH whole-genome DNA methylation, gene expression, gene network analysis; RT-PCR, FISH
Sadikovic [55]	Promoter tiling arrays, aCGH, gene expression arrays in OS tumors	1q21.1-q21.3, 6p12.3-p21.1 (gain); 8p21.2-p21.3 (loss)	*DOCK5, TNFRSF10A/D, RUNX2*	Integrative aCGH whole-genome DNA methylation, gene expression, gene network analysis; RT-PCR, FISH
Selvarajah [42]	OS cell lines HOS, MG-63, U-2 OS, Saos-2; aCGH	5p15.3, 8q24 (gain), 9p21-p23 (loss)	*TERT, C-MYC, RAD21*	mBAND, FISH
Selvarajah [15]	OS tumors; aCGH	6p21-p22, 17p11.2-p12 (gain); 8q24 (loss)	*DEK, E2F3, RUNX2, C-MYC, RECQL4, TOP3A, MAPK7, COPS3*	FISH, SKY
Squire [32]	OS tumors; cDNA aCGH	6p12.1, 8q12-q21.3, 8q22-q23, 8q24, 17p11.2-p12 (gain); 17p13 (loss)	*MYC, TP53, MAP2K4, PMP22, COX10, COPS3, TOP3A*	
Thomas [56]	Canine OS tumors; canine BAC array	8q24 (gain); 13q14, 17p13 (loss)	*MYC, KIT, WT1, HRAS, PTEN, CDKN2A*	FISH
Yang [57]	Human OS tumors; aCGH	1p35-p35, 1q23.1-q21.1, 6p22.1-p21.31, 19p13.11-p13.2 (gain); 5q12.3-q13.2, 5q14.3-q22.2, and 13p14.3-q13.2 (loss)	*WWOX*	IHC on FFPE OS tumors
Yasuda [58]	OS cell line UTOS-1	5q13, 5q13, 12p13, 12q14.3-q15, 17p11.2-p12, Xq25 (gain); 6q13, 6q16.3-q21, 18ptel, 20q13.2-q13.3	*DAB2, CCND2, MDM2, FLI, TOP3A, HTR1B, STK6*	IHC, RT-PCR

aCGH studies of OS predominantly human, but also some relevant studies in canine and rodent OS models, are summarized. Reference column indicates the specific study; sample column indicates the type and number of samples used in the study; cytogenetic loci column summarizes the most significant loci identified and indicates gain or loss; relevant genes indicates the genes proposed to play a role in the region by the authors of the referred manuscripts; and follow-up studies column indicates whether and which additional/follow-up studies are presented in the manuscript

Literature survey of genomic copy number aberrations associated with OS primary tumors and cell lines, in human and other species, was performed. Inconsistency in the orthologous regions of gains and losses highlights the marked instability characteristic for this tumor type, while recurrent, albeit non-consensus aberrations of 12q13-14, 13q13-14, 17p11, 6p21, and 8q21-24 suggest a central mechanistic role for genes such as *MYC* and *RUNX2* in the pathogenesis. The challenges in translating the emerging data into clinical application lie in discerning the core molecular mechanism driving the tumorigenesis from those specific to clinical and histopathological subtypes and spurious noise attributable to genomic instability.

6. Discussion

Understanding the genetic basis of cancer is fundamental to the study of tumor development and pathogenesis, as well as to the development of effective treatment strategies. The body of knowledge concerning molecular findings in OS has grown significantly in the past couple of decades, despite the difficulty of obtaining tissues suitable for molecular studies following preoperative chemotherapy and the necessity for decalcification of specimens for tissue processing. The etiology of sporadic OS is unclear. These tumors may have arisen from osteoblasts or cells driven towards the osteoblastic lineage, compliant with their capacity to produce osteoid, alkaline phosphatase, osteocalcin, osterix, and bone sialoprotein [4]. The development of OS is likely a multistep process that arises from the accumulation of critical genetic changes. The molecular changes leading to the development of OS have been shown to occur in several of the cell cycle regulatory genes [59–62]. An association between retinoblastoma (RB) and OS has long been recognized. Children with hereditary RB are at increased risk of developing OS and exhibit up to 1,000 times the incidence of OS as the general population. This association was confirmed by the identification of the retinoblastoma gene (*RB1*) on human chromosome 13 which permitted several groups to demonstrate that mutations in the RB1 gene occurred in a high percentage of OS [63]. LOH at the *RB1* gene has been suggested as a poor prognostic indicator [64]. The *RB1* gene functions as a tumor suppressor gene as is located on the q14 band of chromosome 13. The gene product participates in a cell cycle regulatory pathway that functions to inhibit entry of cells into the S phase. Other proteins in the cell cycle regulatory pathway in which the RB protein functions include p16 protein, a cyclin-dependent kinase inhibitor, and cyclin-dependent kinase 4. Loss of p16 expression and gene deletions have been reported in some OS lacking *RB* mutations, which may allow phosphorylation and inactivation of the *RB* protein [65]. The second gene significantly associated with OS is the *TP53* gene, which is commonly mutated in the sporadic tumors [63]. p53 is central to many of the cell's anticancer mechanisms. It can induce growth arrest, apoptosis, and cell senescence. It is believed to function as a cell cycle checkpoint after DNA damage following irradiation with cells appearing to enter a sustained arrest in the G2 phase of the cell cycle. Thus, expression of mutant forms of p53 may alter cellular resistance to the DNA damage [66]. As with *RB1*, mutations in genes that regulate p53 have been identified in OS. The *MDM2* gene, located on chromosome 12q13 along with *CDK4*, encodes a protein that binds p53 and blocks the activity of the p53 by directing it to the ubiquitination pathway. Overexpression of *MDM2* in OS provides an alternative means to disrupt the normal p53 pathway [63]. Ladanyi et al. reported that amplification of *MDM2* is more frequent in metastatic or recurrent tumors and suggest that amplification in metastatic or recurrent tumors may be associated with tumor progression [67]. Another protein involved in this pathway is the p14 product of the *INK4A* gene, which is transcribed from the same gene producing the p16 protein involved in the *RB1* pathway [68]. The p14 protein exerts a protective effect on p53 by binding to the *MDM2* gene product. Alterations consistent with inactivation of p14 have been found in OS tumors and cell lines [3, 63]. Genes other than p53 and RB1 have also been associated with OS. High frequencies of allelic loss have been detected at 3q, suggesting that other tumor suppressor genes important in OS may exist [69]. *HER2/*neu (c-erbB-2) overexpression has been observed in several cases and has been associated with early pulmonary metastases and decreased survival [70]. However, results for HER2 expression in OS remain controversial, with some studies reporting up to 61% of positive cases and others reporting only negative results. Further, expression of HER2 has been reported to be a favorable prognostic indicator by some groups and unfavorable by others. For example, a study by our group analyzing HER2 expression and gene copy number in 18 patients found expression in only 2 of the samples and gene amplification in none [71]. Bone morphogenetic proteins (BMPs) are important in the induction of cartilage and bone formation and patterning of skeletal elements. Expression of BMP type II receptor was found to correlate with metastasis in OS [72]. High copy number gain of the *MYC-C* oncogene at 8q24 is also frequently demonstrated in OS and observed to be associated with poor outcome [25]. Significantly, this same cytoband is also the location of two genes: *RecQL4* and *EXT1*. RecQL4 helicase protein-inactivating mutations lead to the Rothmund–Thomson syndrome [73, 74] and *EXT1* to multiple exostoses [75]; both syndromes have strong predisposition to OS. Another oncogene implicated in OS is *FOS*. Functionally, c-fos dimerizes with Jun family members forming the AP-1 transcription complex, and it is involved in regulation of cell proliferation, differentiation, and transformation. Overexpression of c-fos was found in over half of the cases in one study, particularly from patients who developed metastatic disease [3, 76]. Studies have shown a number of growth factors to be elevated in OS. High levels of TGF beta-1 in high-grade OS and low or absent mRNA in low-grade tumors suggest a role for TGF beta-1 in tumor progression [77]. In a study analyzing 62 cases of OS for P-glycoprotein expression, 27 of the cases exhibited overexpression. The patients with elevated expression had a significantly higher recurrence rate than those without increased expression [78]. Another study showed that increased P-glycoprotein expression at the time of diagnosis is associated with a higher rate of relapse and that such patients may benefit from additional chemotherapeutic modalities [79]. A study of gene expression in metastatic OS identified an association between the differential gene expression pattern of three genes with a phenotypic role in OS metastasis and invasion. THBS3 was expressed at significantly high levels in patients with metastasis at diagnosis, which is a predictor of worse overall survival, event-free survival, and relapse-free survival at diagnosis. High SPARC expression was found in 96.3% of OS samples and correlated with the worst event-free survival and relapse-free survival. Overexpression of SPP1 was found in 89% of the samples and correlated with better overall survival, event-free survival, and relapse-free survival at diagnosis. Interestingly, all encode for proteins involved in extracellular remodeling, suggesting potential roles in OS progression [80]. In another recent study, OS samples from surgery showed correlation between the expression of a major vault protein (MVP), metastatic disease at diagnosis, and event-free survival. MVP gene expression was found to correlate with metastatic disease at diagnosis after neoadjuvant chemotherapy and was also associated with worse event-free survival. These findings suggest that MVP expression is involved in one of the mechanisms of drug resistance in OS and is induced by chemotherapy [81].

Application of aCGH technology has had a huge impact in delineation of genomic copy number changes in cancer in general and OS in particular. Due to its high levels of genomic instability, intra- and inter-tumor variability, and the shier burden of karyotype abnormalities, delineation of the copy number abnormalities responsible for driving the tumorigenesis and evolution has been a challenge. One way to approach this challenge was to define copy number changes with high level of recurrence, across patient cohorts, which would be expected to be selected for during the process of tumor cell evolution and selection. One of the limitations of this approach is that even the minimal recurrent copy number alterations are often many megabases in length and often contain dozens to hundreds of genes, large fraction of which are likely "passenger" mutations and do not contribute to the phenotype. Secondly, overlaid on top of the genomic information are the epigenetic changes that in combination with genetic changes can drive the tumorigenic process and drive genomic instability. Epigenetic changes in DNA methylation associated with carcinogenesis involve hypermethylation and inactivation of tumor suppressor genes, hypomethylation and activation of oncogenes, and genome-wide hypomethylation which is thought to play a key role in instability of the cancer genomes through destabilization and euchromatinization of repetitive elements including LINE, SINE, and large fraction of segmental duplications that are rich in CpG dinucleotides [82]. Thirdly, it is important to recognize that although recurrent copy number alterations are generally associated with promotion of tumorigenesis, they do so by promotion of selective phenotype, generally through the foundation upon which the phenotype is built—the gene expression. Therefore, it follows logically that in order to define the key genetic drivers of tumorigenesis in regions of the recurrent copy number alterations, integration of multiple genomic datasets may provide to be useful. Currently, much of the efforts in cancer genomics research is focused on analysis and integration of genetic, epigenetic, and gene expression studies, including the recently launched efforts of the International Cancer Genome Consortium (ICGC) whose function is summarized as "to obtain a comprehensive description of genomic, transcriptomic, and epigenomic changes in 50 different tumor types and/or subtypes which are of clinical and societal importance across the globe." Two of the first studies in cancer research in general, and first and only ones thus far in OS that successfully integrated genome-wide epigenetic, copy number, and gene expression analyses in OS cell lines [54], and tumor tissues [55], provided an important proof of concept and defined some key genetic events in OS tumorigenesis. One, these studies, both in cell lines and tumor tissues, showed that the strongest correlation in the genome-wide copy number, DNA methylation, and gene expression was in the form of copy number gain, hypomethylation, and gene overexpression, while demonstrating evidence of genomic hypomethylation in the segmental duplications surrounding the genomic copy number alteration events. Secondly, it was demonstrated that only a small fraction of the genes in regions of copy number alterations in cell lines and OS tumors showed corresponding changes in the levels of gene expression relative to the normal osteoblasts. Finally, using integrative analyses and gene network analysis, these studies defined multiple genes in key gene networks involved in OS oncogenesis. For example, in the MG-63 cell line, we demonstrated that *MYC-N* oncogene which was found to be gained and overexpressed was the central node in the gene network which exhibited deregulation of gene expression in multiple genes concurrent with copy number alterations and disruption of DNA methylation. In the study involving patient tumor tissues, we identified significant changes including hypomethylation, gain, and overexpression of histone cluster 2 genes at chromosome 1q21.1-q21.3; loss of chromosome 8p21.2-p21.3 and underexpression of *DOCK5* and TNFRSF10A/D genes; and amplification-related overexpression of RUNX2 at chromosome 6p12.3-p21.1. Gene network analysis pointed to the *RUNX2* gene as the key node in the most significantly deregulated gene network, providing a model in which amplification and overexpression of RUNX2 could disrupt G2/M cell cycle checkpoints and downstream OS-specific changes, such as failure of bone differentiation and genomic polyploidization, while failure of DOCK5-signaling, together with p53 and TNFRSF10A/D-related cell cycle and death pathways, may play a critical role in abrogating apoptosis. We concluded that the RUNX2 interactome is constitutively activated in OS and that the downstream intracellular pathways are strongly associated with the regulation of osteoblast differentiation and control of cell cycle and apoptosis in OS. In a subsequent study of gene expression of 16 genes that were previously reported to be involved in OS tumorigenesis in a larger group of OS patient tumor samples, we showed that RECQL4, SPP1, RUNX2, and IBSP were significantly overexpressed, and DOCK5, CDKN1A, RB1, P53, and LSAMP showed significant loss of expression relative to normal osteoblasts [83]. Importantly, RUNX2 was the only gene of the 16 to show significant overexpression in tumors that had a poor response to chemotherapy relative to good responders.

In conclusion, aCGH has played and will undoubtedly continue to play an important role in delineating the molecular and cytogenetic events associated with OS tumorigenesis. While some key recurrent cytogenetic features have now been firmly established, including 12q13-14, 13q13-14, 17p11, 6p21, and 8q21-24, the genes driving these events are just beginning to be deciphered. High level of cytogenomic variability and genomic instability in this tumor further complicates the picture and underlies the necessity for functional studies and integrative genomics approaches to augment and fully utilize the aCGH data in both diagnostic and clinical setting.

7. Notes

Special considerations regarding the high-level interindividual and intra-tumor tissue heterogeneity as well as genomic instability require special considerations when it comes to both histopathological assessment for DNA isolation as well as data analysis. This was elaborated in detail in the introduction, methods, and analysis sections.

Acknowledgments

B.S. was a past recipient of the National Cancer Institute of Canada (currently Canadian Cancer Society) and the Terry Fox Foundation Fellowship in osteosarcoma research. S.S. was a past recipient of the Canadian Institute of Health Research Fellowship for Graduate Training in Molecular Medicine. M.Z is funded by the Canadian Cancer Society grant CCRI-020247.

References

1. Skubitz KM, D'adamo DR (2007) Sarcoma. Mayo Clin Proc 82(11):1409–1432
2. Horvai A, Unni KK (2006) Premalignant conditions of bone. J Orthop Sci 11(4):412–423. doi:10.1007/s00776-006-1037-6
3. Kansara M, Thomas DM (2007) Molecular pathogenesis of osteosarcoma. DNA Cell Biol 26(1):1–18. doi:10.1089/dna.2006.0505
4. Rozeman LB, Cleton-Jansen AM, Hogendoorn PC (2006) Pathology of primary malignant bone and cartilage tumours. Int Orthop 30(6): 437–444. doi:10.1007/s00264-006-0212-x
5. Dick DC, Morley WN, Watson JT (1982) Rothmund-thomson syndrome and osteogenic sarcoma. Clin Exp Dermatol 7(1):119–123
6. White LM, Kandel R (2000) Osteoid-producing tumors of bone. Semin Musculoskelet Radiol 4(1):25–43. doi:smr00105 [pii]
7. Klein MJ, Siegal GP (2006) Osteosarcoma: anatomic and histologic variants. Am J Clin Pathol 125(4):555–581. doi:UC6KQHLD9LV 2KENN [pii] 10.1309/UC6K-QHLD-9LV2-KENN
8. Hauben EI, Weeden S, Pringle J et al (2002) Does the histological subtype of high-grade central osteosarcoma influence the response to treatment with chemotherapy and does it affect overall survival? A study on 570 patients of two consecutive trials of the european osteosarcoma intergroup. Eur J Cancer 38(9):1218–1225. doi:S0959804902000370 [pii]
9. Bielack SS, Kempf-Bielack B, Delling G et al (2002) Prognostic factors in high-grade osteosarcoma of the extremities or trunk: an analysis of 1,702 patients treated on neoadjuvant cooperative osteosarcoma study group protocols. J Clin Oncol 20(3):776–790
10. Posthumadeboer J, Witlox MA, Kaspers GJ et al (2011) Molecular alterations as target for therapy in metastatic osteosarcoma: a review of literature. Clin Exp Metastasis 28(5):493–503. doi:10.1007/s10585-011-9384-x
11. Fabbro D, Ruetz S, Buchdunger E et al (2002) Protein kinases as targets for anticancer agents: from inhibitors to useful drugs. Pharmacol Ther 93(2–3):79–98. doi:S0163725802001791 [pii]
12. Mcgary EC, Weber K, Mills L et al (2002) Inhibition of platelet-derived growth factor-mediated proliferation of osteosarcoma cells by the novel tyrosine kinase inhibitor sti571. Clin Cancer Res 8(11):3584–3591
13. Khanna C, Wan X, Bose S et al (2004) The membrane-cytoskeleton linker ezrin is necessary for osteosarcoma metastasis. Nat Med 10(2): 182–186. doi:10.1038/nm982 nm982 [pii]
14. Al-Romaih K, Somers GR, Bayani J et al (2007) Modulation by decitabine of gene expression and growth of osteosarcoma u2os cells in vitro and in xenografts: Identification of apoptotic genes as targets for demethylation. Cancer cell Int 7:14
15. Selvarajah S, Yoshimoto M, Ludkovski O et al (2008) Genomic signatures of chromosomal instability and osteosarcoma progression detected by high resolution array cgh and interphase fish. Cytogenet Genome Res 122(1):5–15. doi:000151310 [pii] 10.1159/000151310
16. Nowak NJ, Miecznikowski J, Moore SR et al (2007) Challenges in array comparative genomic hybridization for the analysis of cancer samples. Genet Med 9(9):585–595. doi:10.1097/GIM.0b013e3181461c4a 00125817-200709000-00005 [pii]
17. Pugh TJ, Delaney AD, Farnoud N et al (2008) Impact of whole genome amplification on analysis of copy number variants. Nucleic Acids Res 36(13):e80. doi:gkn378 [pii] 10.1093/nar/gkn378
18. Picard F, Lebarbier E, Hoebeke M et al (2011) Joint segmentation, calling, and normalization of multiple cgh profiles. Biostatistics (Oxford, England) 12(3):413–428. doi:kxq076 [pii] 10.1093/biostatistics/kxq076

19. Ragland BD, Bell WC, Lopez RR et al (2002) Cytogenetics and molecular biology of osteosarcoma. Lab Invest 82(4):365–373
20. Sandberg AA, Bridge JA (2003) Updates on the cytogenetics and molecular genetics of bone and soft tissue tumors: Osteosarcoma and related tumors. Cancer Genet Cytogenet 145(1):1–30
21. Bridge JA, Nelson M, Mccomb E et al (1997) Cytogenetic findings in 73 osteosarcoma specimens and a review of the literature. Cancer Genet Cytogenet 95(1):74–87. doi:S0165460896003068 [pii]
22. Batanian JR, Cavalli LR, Aldosari NM et al (2002) Evaluation of paediatric osteosarcomas by classic cytogenetic and cgh analyses. Mol Pathol 55(6):389–393
23. Forus A, Weghuis DO, Smeets D et al (1995) Comparative genomic hybridization analysis of human sarcomas: II. Identification of novel amplicons at 6p and 17p in osteosarcomas. Genes Chromosomes Cancer 14(1):15–21
24. Ozaki T, Schaefer KL, Wai D et al (2002) Genetic imbalances revealed by comparative genomic hybridization in osteosarcomas. Int J Cancer 102(4):355–365. doi:10.1002/ijc.10709
25. Stock C, Kager L, Fink FM et al (2000) Chromosomal regions involved in the pathogenesis of osteosarcomas. Genes Chromosomes Cancer 28(3):329–336
26. Tarkkanen M, Karhu R, Kallioniemi A et al (1995) Gains and losses of DNA sequences in osteosarcomas by comparative genomic hybridization. Cancer Res 55(6):1334–1338
27. Tarkkanen M, Kiuru-Kuhlefelt S, Blomqvist C et al (1999) Clinical correlations of genetic changes by comparative genomic hybridization in ewing sarcoma and related tumors. Cancer Genet Cytogenet 114(1):35–41. doi:S016546089900031X [pii]
28. Zielenska M, Bayani J, Pandita A et al (2001) Comparative genomic hybridization analysis identifies gains of 1p35 approximately p36 and chromosome 19 in osteosarcoma. Cancer Genet Cytogenet 130(1):14–21
29. Hulsebos TJ, Bijleveld EH, Oskam NT et al (1997) Malignant astrocytoma-derived region of common amplification in chromosomal band 17p12 is frequently amplified in high-grade osteosarcomas. Genes Chromosomes Cancer 18(4):279–285. doi:10.1002/(SICI)1098-2264(199704)18:4<279::AID-GCC5>3.0.CO;2-Y [pii]
30. Atiye J, Wolf M, Kaur S et al (2005) Gene amplifications in osteosarcoma-cgh microarray analysis. Genes Chromosomes Cancer 42(2):158–163. doi:10.1002/gcc.20120
31. Man TK, Lu XY, Jaeweon K et al (2004) Genome-wide array comparative genomic hybridization analysis reveals distinct amplifications in osteosarcoma. BMC Cancer 4:45. doi:10.1186/1471-2407-4-45 1471-2407-4-45 [pii]
32. Squire JA, Pei J, Marrano P et al (2003) High-resolution mapping of amplifications and deletions in pediatric osteosarcoma by use of cgh analysis of cdna microarrays. Genes Chromosomes Cancer 38(3):215–225
33. Zielenska M, Marrano P, Thorner P et al (2004) High-resolution cdna microarray cgh mapping of genomic imbalances in osteosarcoma using formalin-fixed paraffin-embedded tissue. Cytogenet Genome Res 107(1–2):77–82
34. Lau CC, Harris CP, Lu XY et al (2004) Frequent amplification and rearrangement of chromosomal bands 6p12-p21 and 17p11.2 in osteosarcoma. Genes Chromosomes Cancer 39(1):11–21. doi:10.1002/gcc.10291
35. Bayani J, Zielenska M, Pandita A et al (2003) Spectral karyotyping identifies recurrent complex rearrangements of chromosomes 8, 17, and 20 in osteosarcomas. Genes Chromosomes Cancer 36(1):7–16
36. Lengauer C, Kinzler KW, Vogelstein B (1997) Genetic instability in colorectal cancers. Nature 386(6625):623–627
37. Loeb LA (2001) A mutator phenotype in cancer. Cancer Res 61(8):3230–3239
38. Gisselsson D, Bjork J, Hoglund M et al (2001) Abnormal nuclear shape in solid tumors reflects mitotic instability. Am J Pathol 158(1):199–206. doi:S0002-9440(10)63958-2 [pii] 10.1016/S0002-9440(10)63958-2
39. Lim G, Karaskova J, Beheshti B et al (2005) An integrated mband and submegabase resolution tiling set (smrt) cgh array analysis of focal amplification, microdeletions, and ladder structures consistent with breakage-fusion-bridge cycle events in osteosarcoma. Genes Chromosomes Cancer 42(4):392–403
40. Lim G, Karaskova J, Vukovic B et al (2004) Combined spectral karyotyping, multicolor banding, and microarray comparative genomic hybridization analysis provides a detailed characterization of complex structural chromosomal rearrangements associated with gene amplification in the osteosarcoma cell line mg-63. Cancer Genet Cytogenet 153(2):158–164
41. Selvarajah S, Yoshimoto M, Park PC et al (2006) The breakage-fusion-bridge (bfb) cycle as a mechanism for generating genetic heterogeneity in osteosarcoma. Chromosoma 115(6):459–467. doi:10.1007/s00412-006-0074-4

42. Selvarajah S, Yoshimoto M, Maire G et al (2007) Identification of cryptic microaberrations in osteosarcoma by high-definition oligonucleotide array comparative genomic hybridization. Cancer Genet Cytogenet 179(1):52–61. doi:S0165-4608(07)00492-X [pii] 10.1016/j.cancergencyto.2007.08.003
43. Tarkkanen M, Bohling T, Gamberi G et al (1998) Comparative genomic hybridization of low-grade central osteosarcoma. Mod Pathol 11(5):421–426
44. Tarkkanen M, Elomaa I, Blomqvist C et al (1999) DNA sequence copy number increase at 8q: a potential new prognostic marker in high-grade osteosarcoma. Int J Cancer 84(2): 114–121
45. Angstadt AY, Motsinger-Reif A, Thomas R et al (2011) Characterization of canine osteosarcoma by array comparative genomic hybridization and rt-qpcr: signatures of genomic imbalance in canine osteosarcoma parallel the human counterpart. Genes Chromosomes Cancer 50(11):859–874. doi:10.1002/gcc.20908
46. Dujardin F, Binh MB, Bouvier C et al (2011) Mdm2 and cdk4 immunohistochemistry is a valuable tool in the differential diagnosis of low-grade osteosarcomas and other primary fibro-osseous lesions of the bone. Mod Pathol 24(5):624–637. doi:modpathol2010229 [pii] 10.1038/modpathol.2010.229
47. Hattinger CM, Reverter-Branchat G, Remondini D et al (2003) Genomic imbalances associated with methotrexate resistance in human osteosarcoma cell lines detected by comparative genomic hybridization-based techniques. Eur J Cell Biol 82(9):483–493
48. Hattinger CM, Stoico G, Michelacci F et al (2009) Mechanisms of gene amplification and evidence of coamplification in drug-resistant human osteosarcoma cell lines. Genes Chromosomes Cancer 48(4):289–309. doi:10.1002/gcc.20640
49. Kresse SH, Ohnstad HO, Paulsen EB et al (2009) Lsamp, a novel candidate tumor suppressor gene in human osteosarcomas, identified by array comparative genomic hybridization. Genes Chromosomes Cancer 48(8):679–693. doi:10.1002/gcc.20675
50. Lockwood WW, Stack D, Morris T et al (2011) Cyclin e1 is amplified and overexpressed in osteosarcoma. J Mol Diagn 13(3):289–296. doi:S1525-1578(11)00026-2 [pii] 10.1016/j.jmoldx.2010.11.020
51. Lu XY, Lu Y, Zhao YJ et al (2008) Cell cycle regulator gene cdc5l, a potential target for 6p12-p21 amplicon in osteosarcoma. Mol Cancer Res 6(6):937–946. doi:6/6/937 [pii] 10.1158/1541-7786.MCR-07-2115
52. Ma O, Cai WW, Zender L et al (2009) Mmp13, birc2 (ciap1), and birc3 (ciap2), amplified on chromosome 9, collaborate with p53 deficiency in mouse osteosarcoma progression. Cancer Res 69(6):2559–2567. doi:0008-5472.CAN-08-2929 [pii] 10.1158/0008-5472.CAN-08-2929
53. Pasic I, Shlien A, Durbin AD et al (2010) Recurrent focal copy-number changes and loss of heterozygosity implicate two noncoding rnas and one tumor suppressor gene at chromosome 3q13.31 in osteosarcoma. Cancer Res 70(1):160–171. doi:0008-5472.CAN-09-1902 [pii] 10.1158/0008-5472.CAN-09-1902
54. Sadikovic B, Yoshimoto M, Al-Romaih K et al (2008) In vitro analysis of integrated global high-resolution DNA methylation profiling with genomic imbalance and gene expression in osteosarcoma. PLoS One 3(7):e2834. doi:10.1371/journal.pone.0002834
55. Sadikovic B, Yoshimoto M, Chilton-Macneill S et al (2009) Identification of interactive networks of gene expression associated with osteosarcoma oncogenesis by integrated molecular profiling. Hum Mol Genet 18(11):1962–1975. doi:ddp117 [pii] 10.1093/hmg/ddp117
56. Thomas R, Wang HJ, Tsai PC et al (2009) Influence of genetic background on tumor karyotypes: Evidence for breed-associated cytogenetic aberrations in canine appendicular osteosarcoma. Chromosome Res 17(3):365–377. doi:10.1007/s10577-009-9028-z
57. Yang J, Cogdell D, Yang D et al (2010) Deletion of the wwox gene and frequent loss of its protein expression in human osteosarcoma. Cancer Lett 291(1):31–38. doi:S0304-3835(09)00617-X [pii] 10.1016/j.canlet.2009.09.018
58. Yasuda T, Kanamori M, Nogami S et al (2009) Establishment of a new human osteosarcoma cell line, utos-1: cytogenetic characterization by array comparative genomic hybridization. J Exp Clin Cancer Res 28:26. doi:1756-9966-28-26 [pii] 10.1186/1756-9966-28-26
59. Benassi MS, Molendini L, Gamberi G et al (2001) Involvement of ink4a gene products in the pathogenesis and development of human osteosarcoma. Cancer 92(12):3062–3067
60. Radig K, Schneider-Stock R, Oda Y et al (1996) Mutation spectrum of p53 gene in highly malignant human osteosarcomas. Gen Diagn Pathol 142(1):25–32
61. Molendini L, Benassi MS, Magagnoli G et al (1998) Prognostic significance of cyclin expression in human osteosarcoma. Int J Oncol 12(5):1007–1011
62. Radig K, Schneider-Stock R, Mittler U et al (1998) Genetic instability in osteoblastic

tumors of the skeletal system. Pathol Res Pract 194(10):669–677
63. Hansen MF (2002) Genetic and molecular aspects of osteosarcoma. J Musculoskelet Neuronal Interact 2(6):554–560
64. Feugeas O, Guriec N, Babin-Boilletot A et al (1996) Loss of heterozygosity of the rb gene is a poor prognostic factor in patients with osteosarcoma. J Clin Oncol 14(2):467–472
65. Nielsen GP, Burns KL, Rosenberg AE et al (1998) Cdkn2a gene deletions and loss of p16 expression occur in osteosarcomas that lack rb alterations. Am J Pathol 153(1):159–163. doi:S0002-9440(10)65556-3 [pii] 10.1016/S0002-9440(10)65556-3
66. Levine AJ (1997) P53, the cellular gatekeeper for growth and division. Cell 88(3):323–331. doi:S0092-8674(00)81871-1 [pii]
67. Ladanyi M, Cha C, Lewis R et al (1993) Mdm2 gene amplification in metastatic osteosarcoma. Cancer Res 53(1):16–18
68. Maitra A, Roberts H, Weinberg AG et al (2001) Loss of p16(ink4a) expression correlates with decreased survival in pediatric osteosarcomas. Int J Cancer 95(1):34–38. doi:10.1002/1097-0215(20010120)95:1<34::AID-IJC1006>3.0.CO;2-V [pii]
69. Kruzelock RP, Murphy EC, Strong LC et al (1997) Localization of a novel tumor suppressor locus on human chromosome 3q important in osteosarcoma tumorigenesis. Cancer Res 57(1):106–109
70. Gorlick R, Huvos AG, Heller G et al (1999) Expression of her2/erbb-2 correlates with survival in osteosarcoma. J Clin Oncol 17(9):2781–2788
71. Onda M, Matsuda S, Higaki S et al (1996) Erbb-2 expression is correlated with poor prognosis for patients with osteosarcoma. Cancer 77(1):71–78. doi:10.1002/(SICI)1097-0142(19960101)77:1<71::AID-CNCR13>3.0.CO;2-5 [pii] 10.1002/(SICI)1097-0142(19960101)77:1<71::AID-CNCR13>3.0.CO;2-5
72. Guo W, Gorlick R, Ladanyi M et al (1999) Expression of bone morphogenetic proteins and receptors in sarcomas. Clin Orthop Relat Res 365:175–183
73. Mohaghegh P, Hickson ID (2001) DNA helicase deficiencies associated with cancer predisposition and premature ageing disorders. Hum Mol Genet 10(7):741–746
74. Wang LL, Gannavarapu A, Kozinetz CA et al (2003) Association between osteosarcoma and deleterious mutations in the recql4 gene in rothmund-thomson syndrome. J Natl Cancer Inst 95(9):669–674
75. Tiet TD, Alman BA (2003) Developmental pathways in musculoskeletal neoplasia: Involvement of the indian hedgehog-parathyroid hormone-related protein pathway. Pediatr Res 53(4):539–543. doi:10.1203/01.PDR.0000054688.93486.18 01.PDR.0000054688.93486.18 [pii]
76. Wu JX, Carpenter PM, Gresens C et al (1990) The proto-oncogene c-fos is over-expressed in the majority of human osteosarcomas. Oncogene 5(7):989–1000
77. Franchi A, Arganini L, Baroni G et al (1998) Expression of transforming growth factor beta isoforms in osteosarcoma variants: Association of tgf beta 1 with high-grade osteosarcomas. J Pathol 185(3):284–289. doi:10.1002/(SICI)1096-9896(199807)185:3<284::AID-PATH94>3.0.CO;2-Z [pii] 10.1002/(SICI)1096-9896(199807)185:3<284::AID-PATH94>3.0.CO;2-Z
78. Chan HS, Grogan TM, Haddad G et al (1997) P-glycoprotein expression: Critical determinant in the response to osteosarcoma chemotherapy. J Natl Cancer Inst 89(22):1706–1715
79. Baldini N, Scotlandi K, Serra M et al (1999) P-glycoprotein expression in osteosarcoma: a basis for risk-adapted adjuvant chemotherapy. J Orthop Res 17(5):629–632. doi:10.1002/jor.1100170502
80. Dalla-Torre CA, Yoshimoto M, Lee CH et al (2006) Effects of thbs3, sparc and spp 1 expression on biological behavior and survival in patients with osteosarcoma. BMC Cancer 6:237. doi:1471-2407-6-237 [pii] 10.1186/1471-2407-6-237
81. Dalla-Torre CA, De Toledo SR, Yoshimoto M et al (2007) Expression of major vault protein gene in osteosarcoma patients. J Orthop Res 25(7):958–963. doi:10.1002/jor.20371
82. Sadikovic B, Al-Romaih K, Squire JA et al (2008) Cause and consequences of genetic and epigenetic alterations in human cancer. Curr Genomics 9(6):394–408. doi:10.2174/138920208785699580
83. Sadikovic B, Thorner P, Chilton-Macneill S et al (2010) Expression analysis of genes associated with human osteosarcoma tumors shows correlation of runx2 overexpression with poor response to chemotherapy. BMC Cancer 10:202. doi:1471-2407-10-202 [pii] 10.1186/1471-2407-10-202

Chapter 16

Comparative Genomic Hybridization of Wilms' tumor

Shahrad Rod Rassekh and Evica Rajcan-Separovic

Abstract

Cytogenetic analysis of solid tumors including Wilms' tumor is challenging due to poor chromosome morphology, complexity of abnormalities, and to the possibility of stromal cell overgrowth in tissue culture. Molecular cytogenetic techniques such as chromosomal comparative genomic hybridization (CGH) have improved the diagnosis of chromosomal aberrations in Wilms' tumor since they can provide results based on the analysis of DNA from nondividing cells. However, chromosomal CGH provides only a limited resolution across the whole genome, which is not different than routine cytogenetic analysis (gains or losses of less than one chromosome band or 10 Mb are not detectable by routine cytogenetics or chromosomal CGH). More recently, the development of genomic arrays opened the possibility of assessing the whole genome at a much higher resolution at a sub-microscopic or sub-band level. Based on the principle of chromosomal CGH, this approach, frequently termed array-CGH, opens the possibility to find invisible changes at the whole genome level not only in abnormal but also in normal tumor karyotypes. Here, we discuss the main technical features, benefits, and limitations of the above three techniques as applied to Wilms' tumor and summarize the main advances in our knowledge about the genetic changes of Wilms' tumor and their clinical relevance.

Key words: Wilms' tumor, Nephroblastoma, Kidney, Pediatric, Cancer, Comparative genomic hybridization (CGH), Array CGH, Cytogenetics

1. Introduction

1.1. Pathology and Clinical Findings in Wilms' Tumor

Tumors of the kidney are the fifth most common cancer in children. Wilms' tumor is the most frequently occurring renal tumor in children with an incidence of 1 in 10,000 (1, 2) and represents approximately 5.9% of all childhood malignancies (3). Wilms' tumor, also referred to as nephroblastoma, was first described in 1814 by Rance (4) and subsequently further described by Max Wilms who identified nephroblastoma as a mixture of three tissues and after whom this tumor is named (5). The three tissues that characterize Wilms' tumor include blastemal, epithelial, and stromal components. Classical nephroblastoma includes all three components although

Diponkar Banerjee and Sohrab P. Shah (eds.), *Array Comparative Genomic Hybridization: Protocols and Applications*, Methods in Molecular Biology vol. 973, DOI 10.1007/978-1-62703-281-0_16, © Springer Science+Business Media, LLC 2013

sometimes one or two components are predominant (6). If the tumor cells contain multipolar mitotic figures, nuclei that are three times larger than adjacent cells and hyperchromasia with enlarged cells, then the tumor is referred to as an Anaplastic Wilms' tumor (6). This anaplasia can be either diffuse or focal. The definition of anaplastic changes in Wilms' Tumor was revised in 1996 to state that in focal anaplasia the changes are confined to circumscribed regions within the primary tumor with surrounding non-anaplastic tumor. Diffuse anaplasia is when it is found in an extrarenal site, the random biopsy specimen reveals unequivocal anaplasia, the tumor is coupled with extreme nuclear unrest, and there is nuclear atypia elsewhere in the tumor (7).

The creation of cooperative groups in Europe (International Society of Pediatric Oncology; SIOP) and North America (National Wilms' Tumor Study Group; NWTS, now merged into Children's Oncology Group; COG) to perform clinical trials in the therapy of Wilms' tumor and the use of multimodal treatment plans have resulted in an improvement of survival rates from 30% to its current rate of over 90% for stage 1–3 disease, and survival of over 70% in stage 4 metastatic disease (8).

Wilms' tumor is largely a sporadic tumor although rare familial cases have been described. Congenital abnormalities are commonly found in individuals diagnosed with Wilms' tumor at rates of 9–17% (9, 10). In addition, Wilms' tumor is associated with a few specific congenital syndromes including Denys-Drash syndrome and Wilms' tumor/Aniridia/GU abnormalities/Mental Retardation (WAGR) syndrome (11–13). It has also been associated with overgrowth syndromes such as Beckwith–Wiedemann syndrome, Perlman's syndrome, Simpson–Golabi–Behmel syndrome, and Hemihypertrophy (14–18). Scott et al. published an excellent review of syndromes associated with Wilms' tumor (19). Approximately 5–10% of Wilms' tumor cases have bilateral disease which usually presents at an early age (20, 21).

2. Techniques for Analyzing the Wilms' Tumor Genome and Their Applications

2.1. Conventional Cytogenetic Analysis

Conventional cytogenetic analysis of the Wilms' tumor genome involves metaphase chromosome analysis in order to detect larger scale chromosomal gains and losses (i.e., >10 Mb or one chromosomal band). Previous cytogenetic investigations resulted in identification of a number of recurrent chromosomal gains and losses such as gains of 1q, 6, 7q, 8, 12, 13, and 18 and losses of chromosome 22, 1p,11p, 11q, and 16q (22–24) (Table 1). Gain of 1q is frequently a result of i(1)(q10) or an unbalanced product of a t(1;16)(q10;p10), resulting in a gain of 1q and loss of 16q (i.e., der(1;16)(q10;p10)). Although routine cytogenetics analysis is not

Table 1
Recurrent chromosomal gains and losses in Wilms' tumor

Method	Gains	Losses	Clinical associations	References
Conventional cytogenetics	1q, 6, 7q, 8, 12, 13, and 18	22, 1p,11p/11q, and 16q	No association with histologic subtype, stage, or clinical outcome	(22–24)
Chromosomal CGH	1q, 7q, 8, 12 (3p, 3q)	1p, 4p/4q, 7p, 11p/11q, 16q, 18q, 21q, 22q (5p, 15q)	No correlation between chromosome abnormalities and tumor stage	(30, 31)
	1q		1q gain associated with relapse of favorable histology Wilms	(32)
Array CGH-whole genome	1q, 13q31	12q24, 18q21, 1q32.1, 2q36.3-2q37.1	Associated with relapse	(35)
	Increased number of copies for 15q26.3 containing IGFIR detected in 13% Wilms' tumors	NR	Increase in IGFIR copy number associated with adverse outcome	(38)
	2p23-p24	NR	Uncertain clinical relevance	(40)
	Gain, MYCN amplification			
	15q	17p, 11p13	Associated with relapse	(41)
	Gain of the entire chromosome, 9, gain of 9q34, or a more complex combination of gains/losses.	loss of 9p	None of these tumor groups showed any statistically significant correlation with clinicopathological variables	(49)
	1q, 7q, 18q21.1	7p, 11q, 14q, 16q, 2q37.1, 7q31-q32, 11q22.3	NR	(25)
	MYCN	17p, 4q31.3 and 14q	17p increased risk of relapse	(39)
			Loss of 17p,4q, 14q, 7p, and 20p are most specific for anaplasia	

(continued)

Table 1 (continued)

Method	Gains	Losses	Clinical associations	References
			MYCN gain seen in anaplastic and favorable histology	
	na	7p, deletions narrowed down to 2.1 Mb	No clinical differences in those with or without 7p deletions, but clinical status differs depending on IGF2 copy number status	(50)
	1q, 7q, 8, 12, and 18	1p, 4q, 7p, 9q, 11p, 11q,14q, 16q, and 21q	Stable karyotypes in stage 1 More complex in stage 3–5 tumors	(51)
Region specific array CGH				
Array: 57 genes commonly amplified in cancer	16p13.1 gain of MRP1 gene	NR	MRP1 gene amplification in a metachronous tumor	(42)
Array: entire chromosome 22	Partial gain of 22q; 7/8 tumors delimiting common amplified regions	Monosomy 22 (3/28 tumors); overlapping regions of partial loss of 22q	NR	(37)
Array: entire chromosome 1	Focal areas of gain for 1q associated with relapse (1q22-23.1)	NR	Poor outcome associated with gain of 1q	(35)
	Four breakpoints within a ~1 Mb region identified, leading to concurrent loss of 1p and gain of 1q		1p loss poor outcome	(36)

NR clinical association not reported

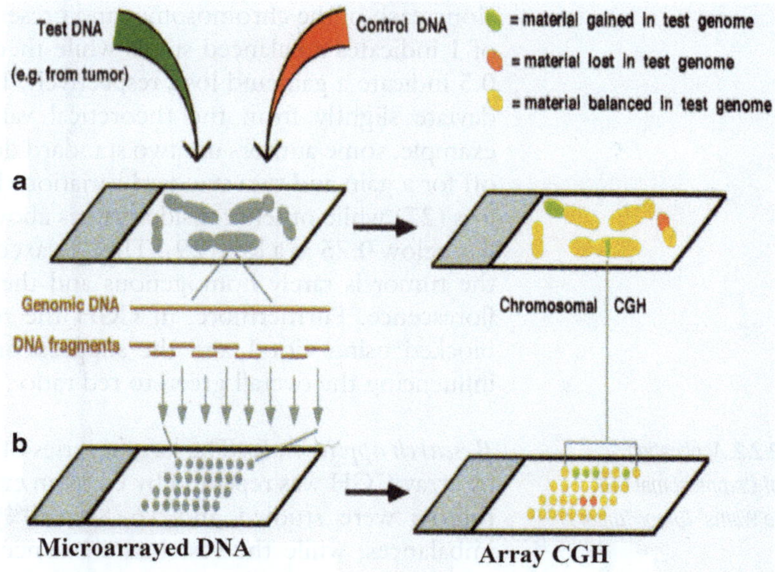

Fig. 1. (**a**) Chromosomal CGH: differentially labeled test and reference DNA are applied to denataured chromosomes. (**b**) Array CGH: differentially labeled test and reference DNA are hybridized to arrays of DNA segments (Modified from Lichter et al. 2000, with permission).

required to diagnose Wilms' tumor tumor, it may have a prognostic value as specific chromosomal changes, particularly losses have been associated with poor outcome (23). Tissue culture failure and artifacts hamper traditional cytogenetic analysis of Wilms' tumor tumors, as demonstrated by a normal cytogenetic finding in over 30–50% of Wilms' tumor tumors analyzed (24, 25).

2.2. Chromosomal Comparative Hybridization (Chromosomal CGH)

2.2.1. Methods

The technique of chromosomal CGH overcomes the problems of tissue culture failure or artifacts as it is based on using tumor DNA extracted directly from either fresh or archival tumor tissue (26). DNAs from the tumor and a normal reference tissue sample are differentially labeled using a green (FITC) and red (Texas red) fluorescent dye and denatured. Their simultaneous hybridization to denatured normal metaphase spreads results in "green" and "red" DNA competing to hybridize to homologous sites on the metaphase chromosomes (Fig. 1a). If there is more tumor DNA for a specific chromosomal region due to its gain, the "green" (tumor) DNA hybridizes more abundantly to that region on the normal metaphase spread and the region gained in cancer appears green on the normal metaphase spread. If there is a loss of a chromosomal region in the tumor there is more reference ("red") DNA than tumor DNA ("green") for that chromosomal region and the red DNA hybridizes to most of its homologous sites on the metaphase spread, making them to appear red (27). The copy number changes for chromosomal segments bigger than 10 Mb (28) are established by calculating the ratio of the green to red fluorescent intensity along each of the chromosomes and presenting it as a profile. Ratio of 1 indicates a balanced status while theoretical ratios of 1.5 and 0.5 indicate a gain and loss, respectively. In practice the ratios that deviate slightly from the theoretical values are acceptable. For example, some authors use two standard deviations above 1 as a cut off for a gain and two standard variations below 1 as a cut off for a loss (27) while others consider values above 1.25 as a gain and values below 0.75 as a loss (29). This "relaxed" cut-off is used because the tumor is rarely homogenous and there is always background florescence. Furthermore, in CGH the repetitive DNA is always blocked using Cot-1 and the suppression is never absolute thus influencing the overall green to red ratio (28).

2.2.2. Application of Chromosomal CGH to Wilms' Tumor Tumor

Research applications: The largest series of Wilms' tumors analyzed by array CGH was reported by Getman et al. (30). Eighty sporadic tumors were studied and 16/80 (20%) showed chromosomal imbalances, while the rest had a balanced genome. The authors compared their CGH findings with the CGH findings in 46 Wilms' tumor tumors reported by Steenman et al. 1997(31). They concluded that the most common chromosomal changes detected by CGH are compatible with the conventional cytogenetic findings with 1q, 7q, 8, and 12 being the most common gains while 1p,7p,11p,11q, and 16q representing the most frequent losses. Gains and losses of novel regions (e.g., loss of 15q and 5p and gains of 3p and 3q) were also detected.

Clinical applications: Getman et al. (30) showed no correlation of CGH findings and tumor stage, while Hing et al. (32) associated gain of 1q (with critical region defined to 1q21-25) with poor outcome of Wilms' tumor with favorable histology.

2.2.3. Limitations of Chromosomal CGH

Although chromosomal CGH increases the potential for identifying new chromosomal abnormalities, it has the following limitations:

- It is very time consuming as it requires identification of DAPI banded chromosome, and this requires significant cytogenetic expertise.
- It does not improve the resolution considerably in comparison to routine G band chromosome analysis.
- CGH does not allow the detection of balanced chromosomal and ploidy changes (although such changes are not common in Wilms' tumor tumor, based on routine cytogenetic analysis).
- CGH does not allow accurate detection of clonal heterogeneity (clones representing at least 35% cells are reliably detected) (28).
- Copy number changes of centromeric and telomeric regions cannot be assessed, as these regions are highly polymorphic and are variably suppressed by Cot-1 DNA.

- Some chromosome regions are prone to artifactual shifts (e.g., telomeres and centromeres of all chromosomes, 1p, 16p, chromosomes 22 and 19, see Note 1) (27).

The development of the array based Comparative Genomic Hybridization (array CGH) helped overcome most of the above issues and resulted in identification of new and much smaller DNA copy number changes as well as improved sensitivity to detecting abnormal clones in Wilms' tumor.

2.3. Array CGH

2.3.1. Methods

The principle of array CGH is the same as for the chromosomal CGH, the difference being that differentially labeled tumor and reference DNA are co-hybridized to an array of small segments of DNA sampled from a normal human genome and arrayed on a glass slide (Fig. 1b). The arrayed DNA segments could be larger (150 kb) human DNA segments inserted into a bacterial artificial chromosome (BAC clones) or smaller DNA fragments (20–60 oligonucleotides in size). The number, size, and distribution of the DNA segments on the glass slide determines the array resolution, but usually the higher the number of DNA fragments the higher the resolution. Gains and losses of chromosomal DNA >1 kb detected by array CGH are called copy number variants (CNVs).

2.3.2. Applications of Array CGH to Wilms' Tumor

Research applications of array CGH: The advantages of improved resolution of array CGH in comparison to chromosomal CGH are demonstrated in Figs. 2 and 3, for two Wilms' tumor tumors studied by cytogenetics, chromosomal CGH, and array CGH using BAC and oligo array (1 and 0.02 Mb resolution). For example, in Fig. 2, the translocation between 6q and 15q appeared balanced by cytogenetic (2a), chromosomal CGH (2b) and 1 Mb BAC array analysis (not shown), but a 3 Mb deletion at the chromosome 15q26 breakpoint was detected by oligo array. Furthermore, the improved resolution of the BAC and oligo array allowed the refinement of the chromosomal breakpoints of the common chromosomal abnormality in Wilms' tumor der(16)(1;16)(q10;p10). As presented in Fig. 3, the breakpoints on chromosome 1 were comparable in two tumors while the breakpoints on chromosome 16 varied-in one tumor they occurred on 16p, while in the other on 16q. Such refined mapping of breakpoints was not possible by chromosomal CGH or cytogenetics.

Array CGH also has an increased sensitivity for detecting clonal changes in Wilms' tumor in comparison to chromosomal CGH (25). For example, in 2/10 tumors with normal karyotypes and normal chromosomal CGH, the presence of a clonal change was detected using BAC array CGH. The clonal chromosomal abnormality was confirmed to be present in only ~30% cells as determined by FISH analysis. The increased sensitivity of array CGH for detecting cell

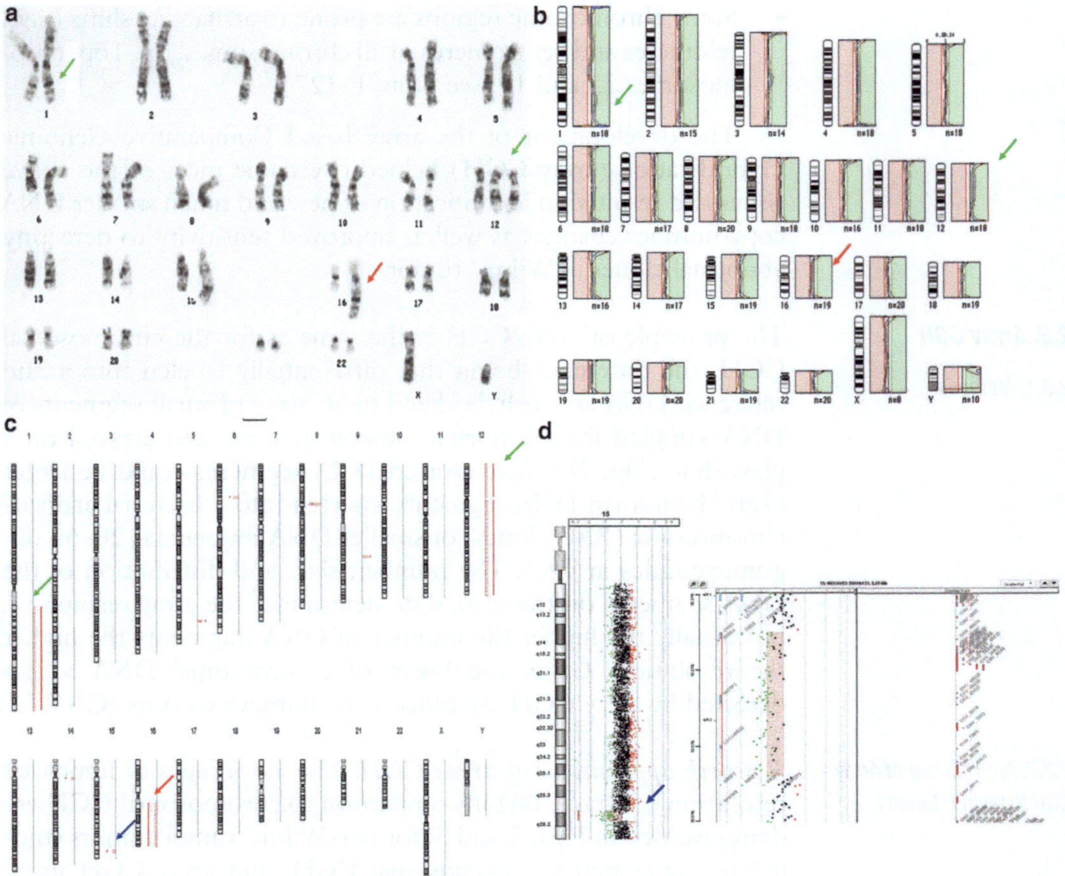

Fig. 2. Case G25. Different methods of establishing genomic content of Wilms' tumor: G-banded karyotype (**a**); chromosomal CGH profile (**b**); Agilent oligo array CGH (**c**). Chromosomal gains seen by all three methods are indicated with *green arrows* (gain of chromosome 12 and 1q). Chromosomal losses seen by all methods are indicated with red arrows (loss of 16q). Note the uneven gain of chromosome 12 by chromosomal CGH (**b**), which is more continuous by array CGH (**c**). Chromosome 16 appears to be completely lost, as per chromosomal CGH, due to artifactual loss which is known to affect 16p. The loss of 16q only was seen by array CGH (**c**). In (**d**) an enlarged area of chromosome loss at 15q26 occurring at the breakpoint of the t(6;15)(q21;q26) is shown. This loss could not be detected by either conventional cytogenetic or chromosomal CGH. *Middle panel* indicates genes in the 15q26 region, while the right-hand panel indicates DNA copy number variants detected in general population.

lines with chromosomal abnormality in peripheral blood was also noted by Ballif et al. (33) and Cheung et al. (34), as chromosomal abnormalities present in only 5–7% of cells could be detected.

In the past couple of years, there was a steady increase in the number of studies using array CGH for determining genomic changes in Wilms' tumor tumor. The application of whole genome arrays as well as targeted arrays representing densely sampled DNA segments from a particular chromosome (e.g., chromosome 1 or

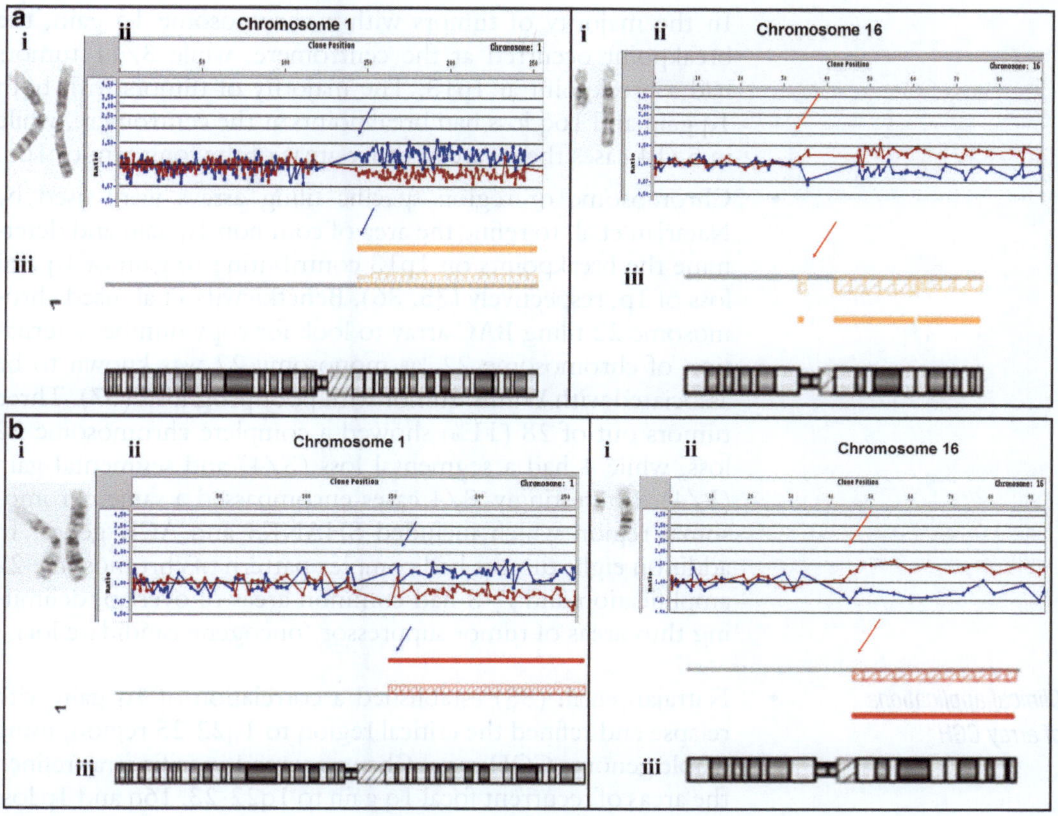

Fig. 3. Breakpoint refinement of the der(16)t(1;16) in two Wilms' tumors (**a** and **b**) as determined by conventional cytogenetics (*i*), 1 Mb BAC array CGH (*ii*), and 0.02 Mb oligo array CGH (*iii*). Chromosome 1 breakpoints are in the pericentromeric region of the long arm of both cases (*blue arrows*). Chromosome 16 breakpoint occurred in the pericentromeric region of the short arm of 16q in one case (**a**) and in the pericentomeric region of the long arm of the other (**b**).

chromosome 22) or genes known to undergo amplification in Wilms' tumor is summarized in brief in Table 1 and some examples are provided below.

- Natrajan et al. used a 1 Mb resolution whole genome BAC array to study 76 Wilms' tumor tumor samples with favorable histology, 37 relapsed and 39 non-relapsed tumors (35). Overall, the most common copy number changes involved chromosomes previously identified to show copy number alternations i.e., gain of 1q, 12, 8, and loss of 16q, 11 and 1p. In addition, focal changes of chromosomal regions known to be affected in Wilms' tumor (8p21.1-12, 12p13.33-12p13.31, 11p13) as well as a number of novel focal changes such as high level gains of 15q11.2-12 and 15q14-21.1 were identified.

In the majority of tumors with a chromosome 1q gain, the breakpoint occurred at the centromere, while 3/11 tumors had a breakpoint at 1p13. The majority of tumors with both 1q gain and 16q loss had breakpoints at the centromere, while in 5/14 cases they were at other areas of the long arm of 16q.

- Chromosome or region specific tiling arrays were used by Natarjan et al. to refine the area of common 1q gain and determine the breakpoints on 1p13 contributing to gain of 1q and loss of 1p, respectively (35, 36). Benetkiewitz et al. used chromosome 22 tiling BAC array to look for copy number alternation of chromosome 22, as monosomy 22 was known to be associated with Wilms' tumor with poor prognosis (37). Three tumors out of 28 (11%) showed a complete chromosome 22 loss, while 4 had a segmental loss (3/4) and segmental gain (1/4). Interestingly, 3/4 cases encompassed a same chromosomal region which included SHANK3 and ACR genes. In addition eight tumors had complex pattern of chromosome 22 amplification and 7/8 had common areas of overlap, delimiting thus areas of tumor suppressor/oncogene candidate loci.

Clinical applications of array CGH

- Natrajan et al. (35) established a correlation of 1q gain with relapse and refined the critical region to 1q22-25 region, using whole genome CGH array. Chromosome 1 specific array refined the areas of recurrent focal 1q gain to 1q22-23. 16q and 1p loss were associated with 1q gain, however 1q gain showed stronger association with shorter survival than 1p and 16q loss. Additional loci associated with relapse were also reported by the authors and included 12q24.13-24.31, 21q22.11-21q22, 14q32.12, and a region on chromosome 2q36.3-37.1 and 18q21.1. Rassekh et al. (25) also reported a copy number change of the last two regions providing further evidence that they likely carry tumor associated genes. Natrajan et al., (38) identified gain of 15q26 containing the IGF2 gene in 9/68 cases of Wilms' tumor (13%) using 1 Mb array CGH (38). This chromosomal change was found to correlate with shorter relapse free survival time in favorable histology Wilms' tumor tumor.
- Using whole genome array CGH Williams et al. identified anaplasia specific changes of chromosome 17p, 4q and 14q regions (39). In addition, this study revealed frequent gain of MYCN in both anaplastic and favorable Wilms' tumor tumor. MYCN amplification in Wilms' tumor was also noted by Schaub et al. (40).
- The difference between primary and relapsed tumor pairs was studied by Natrajan et al. using a whole genome BAC arrays (41). Increase in the number of alternations in the relapsed tumors was evident, amounting to a 21.2% of the genome

showing changes in primary tumors and 33% in the relapsed tumors. However, no consistent specific alternations could be associated with tumor progression. Abnormalities of p53 and 15q26 including the IGF gene were noted in more than one paired primary-relapse samples.

- Goldstein et al. using a targeted array with genes commonly amplified in cancer showed amplification of multi-drug associated protein 1 (MDAP1) in 1/3 tumors (42). Lower level amplification was present at diagnosis but increased dramatically (100× in comparison to normal tumor) after chemotherapy, suggesting the role of this gene in therapy resistance.

2.3.3. Limitation of Array CGH in Studying Wilms' Tumor Tumor

Limitations of array CGH in studying tumors includes: (a) balanced translocations and ploidy changes cannot be detected (but are not a common finding in Wilms' tumor as per conventional cytogenetics, (b) distinction between benign and pathogenic as well as germ line and acquired gains and losses, particularly if they are small, can be challenging (see Note 2). Benetkiewitz et al. (37) studied DNA extracted from tumor and blood in patients with Wilms' tumor using a chromosome 22 specific high resolution array and showed that in one case the partial gain of 22q (660 kb) is acquired and therefore most likely pathogenic, while copy number variation of 2 smaller, single clone changes in the tumor were also detected in the blood, so their contribution to development of the tumor was less likely. However, constitutional deletions affecting regions with tumor suppressor genes regardless of the size should be carefully evaluated and tested to rule out a mutation of the second copy of the gene in the tumor.

3. Materials

- BAC microarrays at 1-Mb resolution (SpectralChip 2600; PerkinElmer–Spectral Genomics Waltham, MA).
- Agilent array kit G4425B-031750 (Unrestricted HD-CGH Microarray ISCA v2, ×105 K, at no G4425B-031750).
- Genomic DNA Enzymatic Labeling Kit (Agilent cat no 5190-0449).

4. Methods

- Chromosomal CGH was described on several instances in this series (43, 44).
- Array CGH: Method as per the commercial array supplier Agilent is used (protocol Version 6.2.1, February 2010 Agilent

Oligonucleotide Array-Based CGH for Genomic DNA Analysis Enzymatic Labeling for Blood, Cells or Tissues).

The arrays are scanned using an Axon B scanner and the images of hybridized slides extracted Agilent feature extraction software. The selection of CNV is based on application of a filter setting as shown in Table 2, after the data were processed using Agilent's CGH Analytics Software (see Note 3).

5. Notes: Data Analysis and Interpretation

Note 1: Chromosomal CGH: When performing chromosomal CGH it is necessary to consider the artifactual shifts, which involve AT rich chromosomal regions of 1p, 16p, 19, and 22. Shifts of the above chromosomes are most likely artifactual if they simultaneously affect more than one chromosome (27). An artifactual shift of 16p is shown in Fig. 2. It could not be confirmed in the tumor by array CGH.

Note 2: Array CGH: Interpretation: The widespread use of array CGH revealed that a large proportion of the human genome (~15%) contains regions of copy number variability (45) and the distinction between pathogenic and benign gains and losses can be challenging. CNVs detected in the general population are catalogued in the Database of Genomic Variants (DGV http://projects.tcag.ca/cgi-bin/variation/) and this database should be consulted in order to determine if the copy number variation seen in tumor has been reported in controls, in which case it is less likely to be pathogenic. Guidelines for distinguishing pathogenic and benign changes used for constitutional CNVs can also be used for tumor CNVs (46–48). For example, complete overlap of the tumor CNV with CNVs in DGV reported in at least two studies indicates that the CNV is most likely benign. In addition, CNVs detected in tumors should be compared to CNVs detected in non-cancer tissues (e.g., blood from the patient) in order to distinguish tumor specific versus constitutional (germ line) CNVs. The availability of constitutional DNA from blood is therefore desired. The size of the CNV may also indicate pathogenicity as CNVs bigger than 1.5 Mb occur at a very low frequency in controls (0.6%) (45). Cataloguing of CNVs that are recurrent and specific for an array platform is desired, to be able to identify artifacts. Table 2 presents all CNVs detected in a Wilms' tumor for which the profile was shown in Fig. 2. Copy number changes for chromosome 12 and chromosomal regions of 1q, 15q, and 16q are indicated with an arrow. Copy number changes for one region on chromosome 9 is an artifact, as it occurs in many cases analyzed using the platform. The role of the intronic CNV on chromosome 5 involving gene

Table 2
Agilent array CGH parameters for calling CNVs and list of CNVs obtained using the Agilent 105 K array in Wilms' tumor (profile for this tumor is presented in Fig. 2)

Aberration *N*	Chr	Cytoband	Start	Stop	Size (bp)	#Probes	Amplification	Deletion
Aberration Algorithm: ADM-2								
Threshold: 6.0								
GC Correction: OFF								
Centralization: ON								
Bin Size: 10								
Centralization Threshold: 6.0								
Fuzzy Zero: ON								
Nesting Level: OFF								
Combine Replicates (Intra Array): OFF								
Combine Replicates (Inter Array): OFF								
Genome: hg18								
Aberration Filters: minProbes = 3 AND minAvgAbsLogRatio = 0.25								
AND maxAberrations = 10,000 AND percentPenetrance = 0								
Feature Level Filters: NONE								
Design Level Filters: NONE								
Array Level Filters: NONE								
Genomic Boundaries: Not Applied								

(continued)

Table 2 (continued)

Aberration *N*	Chr	Cytoband	Start	Stop	Size (bp)	#Probes	Amplification	Deletion
1[a]	1	q21.1–q44	143,697,723	247,179,291	103,481,568	3,758	0.491,954	0
2[b]	5	q33.3	157,192,125	157,192,184	59	3	1.182,005	0
3	6	p21.32	32,519,935	32,633,740	113,805	5	0	−1.213,007
4	8	p11.23	39,356,595	39,499,811	143,216	7	0	−3.906,098
5[c]	9	q22.2	92,604,082	92,604,139	57	3	1.383,444	0
6[a]	12	p13.33–p11.1	33,854	33,816,331	33,782,477	1,277	0.484,021	0
7[a]	12	q12–q24.33	37,003,348	132,231,339	95,227,991	3,417	0.517,418	0
8	14	q11.2	19,273,289	19,484,072	210,783	12	0	−0.805,493
9[a]	15	q26.1–q26.2	91,229,339	94,466,005	3,236,666	83	0	−0.738,109
10[a]	16	q11.2–q24.3	45,063,118	88,156,510	43,093,392	1,526	0	−0.518,109

[a]Acquired oncogenic change
[b]Affects intron, gene has no known role in cancer
[c]Recurrent in noncancerous and cancerous tissues from many individuals, array specific artifact
Unlabelled CNVs are common polymorphisms

CLINT1 is likely not relevant for the tumor development, while the remaining CNVs on chromosome 6, 8, 14, and 15 represent common CNVs, frequently detected in the general population.

Note 3. Table 2 shows the parameters used for making CNV calls using the commercial Agilent's CGH Analytics Software. The minimal number of probes (3 in this case) and the human genome version represent (e.g., hg 18) the essential information for determining the array resolution and genomic position of the CNV. The selection of filters to be used can be determined based on successful detection of known CNVs.

Acknowledgments

The authors wish to thank the BC Children's Hospital Telethon grant for funding the Wilms' tumor chromosomal and array CGH study, Dr Dagmar Kalousek's research laboratory for developing the chromosomal CGH method, and Dr Suzanne Chan, Chansonette Harvard, and Sally Martell for Wilms' tumor cytogenetic, chromosomal, and array CGH analysis.

References

1. Stiller CA, Parkin DM (1990) International variations in the incidence of childhood renal tumors. Br J Cancer 62(6):1026–1030. doi:10.1038/bjc.1990.432
2. Jemal A, Siegel R, Ward E et al (2006) Cancer statistics, 2006. CA Cancer J Clin 56(2): 106–130
3. Jg G, Ml B (2006) Epidemiology of childhood cancer. Principles and practice of pediatric oncology, 5th edn. Lippincott Williams & Wilkins, Philadelphia
4. Rance T (1814) Case of fungus haematodes of the kidneys. Med Phys J 32:19–25
5. Wilms M (1899) *Die Mischgeschwülste der Niere*. Leipzig: Verlag von Arthur Georgi; monograph, pp. 1–90.
6. Varan A (2008) Wilms' Tumor in children: an overview. Nephron Clin Pract 108(2):C83–C90. doi:10.1159/000113012
7. Faria P, Beckwith JB, Mishra K et al (1996) Focal versus diffuse anaplasia in wilms tumor: new definitions with prognostic significance—a report from the National Wilms Tumor Study Group. Am J Surg Pathol 20(8):909–920. doi:10.1097/00000478-199608000-00001
8. Pritchard-Jones K (2002) Controversies and advances in the management of wilms' tumour. Arch Dis Child 87(3):241–244. doi:10.1136/adc.87.3.241
9. Narod SA, Hawkins MM, Robertson CM et al (1997) Congenital anomalies and childhood cancer in great britain. Am J Hum Genet 60(3):474–485
10. Merks JHM, Caron HN, Hennekam RCM (2005) High incidence of malformation syndromes in a series of 1,073 children with cancer. Am J Med Genet A 134A(2):132–143. doi:10.1002/ajmg.a.30603
11. Denys P, Malvaux P, Vandenbe H et al (1967) Association dun syndrome anatomopathologique de pseudohermaphrodisme masculin dune tumeur de wilms dune nephropathie parenchymateuse et dun mosaicisme xx/xy. Arch Fr Pediatr 24(7):729–739
12. Drash A, Sherman F, Hartmann WH et al (1970) A syndrome of pseudohermaphroditism, wilms tumor, hypertension, and degenerative renal disease. J Pediatr 76(4):585–593. doi:10.1016/s0022-3476(70)80409-7
13. Riccardi VM, Sujansky E, Smith AC et al (1978) Chromosomal imbalance in aniridia-wilms tumor association—11p interstitial deletion. Pediatrics 61(4):604–610
14. Wiedeman H (1964) Familial malformation complex with umbilical hernia and macroglos-

sia—a "new syndrome"? J Genet Hum 13: 223–232
15. Beckwith J (1969) Macroglossia, omphalocele, adrenal cytomegally, gigantism, and hyperplastic visceromegaly. Birth Def 5:188–196
16. Grundy RG, Pritchard J, Baraitser M et al (1992) Perlman and Wiedemann–Beckwith syndromes—2 distinct conditions associated with wilms-tumor. Eur J Pediatr 151(12):895–898. doi:10.1007/bf01954125
17. Hughesbenzie RM, Pilia G, Xuan JY et al (1996) Simpson-golabi-behmel syndrome: genotype/phenotype analysis of 18 affected males from 7 unrelated families. Am J Med Genet 66(2):227–234. doi:10.1002/(sici)1096-8628(19961211)66:2<227::aid-ajmg20>3.0.co;2-u
18. Hoyme HE, Seaver LH, Jones KL et al (1998) Isolated hemihyperplasia (hemihypertrophy): report of a prospective multicenter study of the incidence of neoplasia and review. Am J Med Genet 79(4):274–278. doi:10.1002/(sici)1096-8628(19981002)79:4<274::aid-ajmg8>3.0.co;2-m
19. Scott RH, Stiller CA, Walker L et al (2006) Syndromes and constitutional chromosomal abnormalities associated with wilms tumour. J Med Genet 43(9):705–715. doi:10.1136/jmg.2006.041723
20. Coppes MJ, Dekraker J, Vandijken PJ et al (1989) Bilateral wilms tumor—long-term survival and some epidemiological features. J Clin Oncol 7(3):310–315
21. Ritchey ML, Shamberger RC, Hamilton T et al (2005) Fate of bilateral renal lesions missed on preoperative imaging: a report from the National Wilms Tumor Study Group. J Urol 174(4):1519–1521. doi:10.1097/01.ju.0000179536.97629.c5
22. Hoglund M, Gisselsson D, Hansen GB et al (2004) Wilms tumors develop through two distinct karyotypic pathways. Cancer Genet Cytogenet 150(1):9–15. doi:10.1016/j.cancergencyto.2003.08.017
23. Md Zin R, Murch A, Charles A (2011) Pathology, genetics and cytogenetics of wilms' tumour. Pathology 43(4):302–312. doi:10.1097/PAT.0b013e3283463575
24. Gow KW, Murphy JJ (2002) Cytogenetic and histologic findings in wilms' tumor. J Pediatr Surg 37(6):823–827. doi:10.1053/jpsu.2002.32880
25. Rassekh S, Chan S, Harvard C et al (2008) Screening for submicroscopic chromosomal rearrangements in wilms tumor using whole-genome microarrays. Cancer Genet Cytogenet 182(2):84–94. doi:10.1016/j.cancergencyto.2007.12.015
26. Kallioniemi A (2008) Cgh microarrays and cancer. Curr Opin Biotechnol 19(1):36–40. doi:10.1016/j.copbio.2007.11.004
27. Lestou VS, Lomax BL, Barrett IJ et al (1999) Screening of human placentas for chromosomal mosaicism using comparative genomic hybridization. Teratology 59(5):325–330. doi:10.1002/(sici)1096-9926(199905)59:5<325::aid-tera3>3.0.co;2-n
28. Lichter P, Joos S, Bentz M et al (2000) Comparative genomic hybridization: uses and limitations. Semin Hematol 37(4):348–357. doi:10.1016/s0037-1963(00)90015-5
29. Wessendorf S, Lichter P, Schwanen C et al (2001) Potential of chromosomal and matrix-based comparative genomic hybridization for molecular diagnostics in lymphomas. Ann Hematol 80:B35–B37
30. Getman ME, Houseal TW, Miller GA et al (1998) Comparative genomic hybridization and its application to wilms' tumorigenesis. Cytogenet Cell Genet 82(3–4):284–290. doi:10.1159/000015120
31. Steenman M, Redeker B, Demeulemeester M et al (1997) Comparative genomic hybridization analysis of wilms tumors. Cytogenet Cell Genet 77(3–4):296–303. doi:10.1159/000134602
32. Hing S, Lu YJ, Summersgill B et al (2001) Gain of 1q is associated with adverse outcome in favorable histology wilms' tumors. Am J Pathol 158(2):393–398. doi:10.1016/s0002-9440(10)63982-x
33. Balliff BC, Rorem EA, Sundin K et al (2006) Detection of low-level mosaicism by array CGH in routine diagnostic specimens. Am J Med Genet A 140A(24):2757–2767. doi:10.1002/ajmg.a.31539
34. Cheung SW, Shaw CA, Scott DA et al (2007) Microarray-based CGH detects chromosomal mosaicism not revealed by conventional cytogenetics. Am J Med Genet A 143A(15):1679–1686. doi:10.1002/ajmg.a.31740
35. Natrajan R, Williams RD, Hing SN et al (2006) Array CGH profiling of favourable histology wilms tumours reveals novel gains and losses associated with relapse. J Pathol 210(1):49–58. doi:10.1002/path.2021
36. Natrajan R, Williams RD, Grigoriadis A et al (2007) Delineation of a 1 Mb breakpoint region at 1p13 in wilms tumors by fine-tiling oligonucleotide array CGH. Genes Chromosomes Cancer 46(6):607–615. doi:10.1002/gcc.20446
37. Benetkiewicz M, De Stahl TD, Gordor A et al (2006) Identification of limited regions of genetic aberrations in patients affected with wilms' tumor using a tiling-path chromosome

22 array. Int J Cancer 119(3):571–578. doi:10.1002/ijc.21868

38. Natrajan R, Reis-Filho JS, Little SE et al (2006) Blastemal expression of type I insulin-like growth factor receptor in wilms' tumors is driven by increased copy number and correlates with relapse. Cancer Res 66(23):11148–11155. doi:10.1158/0008-5472.can-06-1931
39. Williams R, Al-Saadi R, Natrajan R et al (2011) Molecular profiling reveals frequent gain of MYCN and anaplasia-specific loss of 4q and 14q in wilms tumor. Genes Chromosomes Cancer. doi:10.1002/gcc.20907
40. Schaub R, Burger A, Bausch D et al (2007) Array comparative genomic hybridization reveals unbalanced gain of the MYCN region in wilms tumors. Cancer Genet Cytogenet 172(1):61–65. doi:10.1016/j.cancergencyto.2006.08.010
41. Natrajan R, Little SE, Sodha N et al (2007) Analysis by array CGH of genomic changes associated with the progression or relapse of wilms' tumour. J Pathol 211(1):52–59. doi:10.1002/path.2087
42. Goldstein M, Rennert H, Bar-Shira A et al (2003) Combined cytogenetic and array-based comparative genomic hybridization analyses of wilms tumors: Amplification and overexpression of the multidrug resistance associated protein 1 gene (MRP1) in a metachronous tumor. Cancer Genet Cytogenet 141(2):120–127. doi:10.1016/s0165-4608(02)00667-2
43. Rao P (2005) Comparative genomic hybridization for analysis of changes in DNA copy number in multiple myeloma. Methods Mol Med 113:71–83
44. Carless M (2009) Analysis of genomic aberrations using comparative genomic hybridization of metaphase chromosomes. Methods Mol Biol 523:177–202
45. Cooper GM, Coe BP, Girirajan S et al (2011) A copy number variation morbidity map of developmental delay. Nat Genet 43(9):838–846. doi:10.1038/ng.909
46. Koolen DA, Pfundt R, De Leeuw N et al (2009) Genomic microarrays in mental retardation: a practical workflow for diagnostic applications. Hum Mutat 30(3):283–292. doi:10.1002/humu.20883
47. Qiao Y, Harvard C, Tyson C et al (2010) Outcome of array cgh analysis for 255 subjects with intellectual disability and search for candidate genes using bioinformatics. Hum Genet 128(2):179–194. doi:10.1007/s00439-010-0837-0
48. Kearney HM, Thorland EC, Brown KK et al (2011) American college of medical genetics standards and guidelines for interpretation and reporting of postnatal constitutional copy number variants. Genet Med 13(7):680–685. doi:10.1097/GIM.0b013e3182217a3a
49. Natrajan R, Warren W, Messahel B et al (2008) Complex patterns of chromosome 9 alterations including the p16(INK4a) locus in wilms tumours. J Clin Pathol 61(1):95–102. doi:10.1136/jcp. 2007.047159
50. Ohshima J, Haruta M, Arai Y et al (2009) Two candidate tumor suppressor genes, MEOX2 and SOSTDC1, identified in a 7p21 homozygous deletion region in a wilms tumor. Genes Chromosomes Cancer 48(12):1037–1050. doi:10.1002/gcc.20705
51. Hawthorn L, Cowell JK (2011) Analysis of wilms tumors using SNP mapping array-based comparative genomic hybridization. PLoS One 6(4):doi:e18941 10.1371/journal.pone.0018941

Chapter 17

Array-CGH in Childhood MDS

Marcel Tauscher, Inka Praulich, and Doris Steinemann

Abstract

To study genomic imbalances potentially involved in disease development and/or progression of childhood MDS, array-based comparative genomic hybridization (aCGH) is a helpful tool. Copy number alterations (CNA) of subtle chromosomal regions containing potential candidate genes, e.g., *TP53* or *RUNX1* can be detected. However, characterizing small and/or heterogeneous tumor subpopulations by high-resolution aCGH within a majority of normal cells is a challenge in MDS and requires validation by independent methods like FISH or quantitative PCR. For the identification of tumor-relevant CNA, the analysis of DNA isolated from purified granulocytes or myeloid populations instead of DNA from whole bone marrow (BM) cells is helpful to overcome some of these limitations.

Key words: Copy number alterations, Copy number variation, Childhood MDS, Array-CGH, Monosomy 7, TP53

1. Introduction

Myelodysplastic syndromes (MDS) are a heterogeneous group of myeloid disorders with cytopenias affecting one or more hematopoietic lineages. Patients are at an increased risk of developing acute myeloid leukemia (AML). In the elderly, MDS constitutes one of the most frequent hematologic malignancies, but it is a rare disease in childhood. Cytogenetic abnormalities are found in approximately 50% of patients, with monosomy 7 (-7) being the most frequent alteration in childhood MDS (1, 2). However, gross karyotypic changes may represent clonal evolution in a population with karyotypically undetectable alterations. Small copy number changes can be detected by aCGH in a genome-wide manner (3–6). One major advantage of applying aCGH to the analysis of MDS, a disease characterized by enhanced levels of apoptosis and often by a hypocellular marrow (7–9), is that no dividing cells are needed. However, the percentage of malignant cells must remain above the

Diponkar Banerjee and Sohrab P. Shah (eds.), *Array Comparative Genomic Hybridization: Protocols and Applications*, Methods in Molecular Biology vol. 973, DOI 10.1007/978-1-62703-281-0_17,

aCGH detection limit of about 25–35% (10–12). This may be problematic in some clinical situations, particularly in refractory cytopenias (RC) and other low-grade MDS.

2. Materials

2.1. Step 1: Separation of Peripheral Blood Mononuclear Cells (PBMCs) and Granulocytes

Leucosep™ tubes (Greiner Bio-One).
Centrifuge (Beckman Coulter Allegra 21 r).
Centrifugation tube (Falcon 15 mL tube).
PBS buffer (Biochrom L 1820).
Ficoll-Paque™ PLUS (GE Healthcare 17-1440-03).
Overhead shaker (Heidolph Reax 2).

2.2. Step 2: DNA Extraction

QIAamp® DNA Blood Midi Kit (QIAGEN 51183).
Pipettes (Gilson p10, p100, p1000).
Centrifuge (Eppendorf 5415 D).
Water baths (GFL type 1012).
15 mL tube (Falcon).
Centrifugal filters (Amicon Ultra 30 K Ultracel).

2.3. Step 3: Assessment of DNA Quality and Quantity

UV–vis spectrophotometer (Implen NanoPhotometer).
TE buffer.
Microwave (Samsung MW81W).
Agarose (Fermentas TopVision R0492).
TBE buffer.
Power supply (Consort E835).
Electrophoresis buffer chamber (Owl Separation Systems B1A).
Gel documentation (BioRad Universal Hood Item ID 18168).
Pipettes (Gilson Pipetman® p10, p100).
Centrifuge (Eppendorf 5415 D).

2.4. Step 4: Fluorescent Labeling of DNA

Genomic DNA enzymatic labeling kit (Agilent 5190-0449).
Water baths (GFL type 1012).
Heat block (QBT digital block heater).
Thermomixer (Eppendorf 5355).
Pipettes (Gilson Pipetman® p10, p100, p1000).
Ice pan (Kleinfeld 5425744).
Ice (Manitowoc imq 09).
Aluminum block (neoLab 0.5 mL).
Centrifuge (Eppendorf 5415 D).

2.5. Step 5: Cleanup of Labeled Genomic DNA and Determination of Yield, Degree of Labeling, and Specific Activity

Centrifugal filters (Amicon Ultra 30 K Ultracel).
TE buffer.
Centrifuge (Eppendorf 5415 D).
Pipettes (Gilson Pipetman® p10, p100, p1000).
UV–vis spectrophotometer (Implen NanoPhotometer).
SpeedVac (Christ RVC 2-18).

Ice pan (Kleinfeld 5425744).
Aluminum block (neoLab 0.5 mL).

2.6. Step 6: Preparation of Labeled Genomic DNA for Hybridization

Hybridization kit (Agilent aCGH 5188-5220).
Water baths (GFL type 1012).
Heat block (QBT digital block heater).
Centrifuge (Eppendorf 5415 D).
Pipettes (Gilson Pipetman® p10, p100, p1000).
Aluminum block (neoLab 0.5 mL).
Ice pan (Kleinfeld 5425744).

2.7. Step 7: Microarray Hybridization

Microarray slides (Agilent g4411b, eArray).
Gasket slides (Agilent G2534-60003).
Hybridization chamber (Agilent G2534A).
Hybridization oven (Agilent G2545A).
Pipettes (Gilson Pipetman® p100, p1000).
Centrifuge (Eppendorf 5415 D).

2.8. Step 8: Microarray Washing

Wash buffer (Agilent 5188-5226).
Slide staining dishes (MHH, Zentrale Forschungswerkstätten).
Magnetic stir plate (Heidolph MR 2000).
Slide rack (OMNILAB).
Magnetic stir bar (OMNILAB).
Water baths (GFL type 1012).

2.9. Step 9: Microarray Scanning

Microarray scanner (Agilent G2565A).
Slide holder (Agilent G2505-98009).
Ozone-barrier slide cover (Agilent G2505-60550).

2.10. Step 10: Data Analysis

Data extraction (Agilent Feature Extraction 10.10.1.1).
Copy number analysis (Agilent Genomic Workbench 6.5.0.58).
PC (Dell Precision T5400 Intel® Xeon® CPU E5430 2.66 GHz 16 GB RAM).
Microsoft Windows XP Professional x64 Edition.

3. Methods

3.1. Step 1: Cell Separation

Mononuclear cells and/or granulocytes from whole blood or bone marrow are purified by gradient centrifugation according to (13):

- Fill the Leucosep tube with 3 mL Ficoll equilibrated to room temperature (RT).
- Close the tube and centrifuge for 1 min at 1000 × *g* at RT (see Note 1).
- Carefully overlay Ficoll with the sample (up to 6–7 mL).
- Centrifuge 20 min at 400 × *g* in a swinging-bucket rotor. Switch off brakes.

- Harvest cell fraction by pouring the supernatant into a new 15 mL Falcon tube.
- Centrifuge the supernatant for 10 min at 400 × *g*.
- Discard the supernatant.
- Wash the cell pellet with 10 mL PBS.
- Centrifuge for 10 min at 400 × *g*.
- Discard the supernatant.
- Resuspend cell pellet in 2 mL PBS (see Note 2).

3.2. Step 2: Isolation of DNA

- Pipet 200 μL QIAGEN protease into the bottom of a 15 mL centrifuge tube.
- Add 2 mL cell sample and mix briefly.
- Add 2.4 mL buffer AL, and mix thoroughly by inverting the tube 15 times, followed by additional vigorous shaking for at least 1 min.
- Incubate at 70°C for 10 min.
- Add 2 mL ethanol (96–100%).
- Mix by inverting the tube ten times, followed by additional vigorous shaking.
- Carefully transfer one half onto the QIAamp column placed in a 15 mL centrifuge tube.
- Centrifuge at 1,850 × *g* for 3 min and discard the filtrate.
- Place the QIAamp column back in the 15 mL centrifuge tube.
- Load the remaining solution onto the column.
- Centrifuge again at 1,850 × *g* for 3 min and discard the filtrate.
- Add 2 mL buffer AW1 to the column.
- Centrifuge at 4,500 × *g* for 1 min.
- Add 2 mL buffer AW2 to the column.
- Centrifuge at 4,500 × *g* for 15 min.
- Place the QIAamp column in a clean 15 mL centrifuge tube.
- Add 300 μL distilled water directly onto the membrane of the column.
- Incubate at RT for 5 min.
- Centrifuge at 4,500 × *g* for 2 min.
- Add 200 μL distilled water again directly onto the membrane of the column.
- Incubate at RT for 5 min.
- Centrifuge at 4,500 × *g* for 2 min.
- Transfer the 500 μL flow into an Amicon Ultra 30 K column.
- Centrifuge at 14 000 × *g* for 10 min.

- Place the filter upside down in a new microcentrifuge tube.
- Centrifuge at 1,000 × *g* for 2 min.
- Measure DNA quality and quantity.

3.3. Step 3: Assessment of DNA Quality and Quantity

DNA concentration and purity are measured using a UV–vis spectrophotometer.
The *A*260/*A*280 ratio of high quality DNA samples should be in the range of 1.8–2.0 (see Note 3).
The *A*260/*A*230 values are commonly in the range of 2.0–2.2 (see Note 4).
DNA integrity is controlled by running 100 ng DNA in a 1% agarose gel.

4. Array-CGH

Genome CGH 244A microarrays (Agilent) containing 236,000 probes across coding and noncoding human regions (hg19 assembly) represent the human genome at an average of 8.9 kb probe spacing (7.4 kb in Refseq genes). The following steps are performed according to the protocol provided by Agilent (Agilent G4410 90010_CGH_Enzymatic_Protocol).

4.1. Step 1: Fluorescence Labeling of Genomic DNA

1.5 µg of DNA is labeled by random priming according to Agilent Genomic DNA Enzymatic Labeling Kit (see Note 5).

- Adjust each sample and each reference DNA to 1,500 ng in a volume of 26 µL (see Note 6).
- Add 5 µL of random primers to each reaction tube.
- Mix well by gently pipetting up and down.
- Spin for a short run in a microcentrifuge.
- Incubate sample tubes at 95°C for 10 min.
- Incubate sample tubes on ice for 5 min.
- Per reaction add:

5× buffer	10.0 µL
10× dNTPs	5.0 µL
Exo-Klenow fragment	1.0 µL
Cy3- or cy5-UTP	3.0 µL
Final volume	19.0 µL

- Add 19.0 µL of this labeling master mix to each reaction tube.
- Mix well by gently pipetting up and down.

- Spin for a short run in a microcentrifuge.
- Incubate at 37°C (water bath) for 2 h.
- Incubate at 65°C for 10 min to inactivate the enzyme (see Note 7).

4.2. Step 2: Cleanup of Labeled Genomic DNA

Labeled DNA is purified using Amicon Ultra 30 K filters. To determine the yield of dye incorporation and specific activity of dyes, DNA concentration and dye concentrations are measured putting 3.5 μL (of the final 41 μL) into the NanoPhotometer with settings 1-label-guard applications_1-nucleic acids_5-dsdna-dye_dye-type (cy3/cy5). The expected yield of labeled DNA is 6 μg, the expected dye concentration is 3–6 pmol/μL, and the calculated dye activity ((dye)/(DNA)) should be 20–50 pmol/μg.

4.3. Step 3: Preparation of Labeled Genomic DNA for Hybridization

- Equilibrate water bath and heat block to 37° and 95°C, respectively.
- Add the following components in the indicated order.

Labeled reference and sample	79.0 μL
Cot-1 DNA (1 μg/μL)	25.0 μL
Agilent 10× blocking agent	26.0 μL
Agilent 2× hybridization buffer	130.0 μL

- Mix sample by pipetting up and down.
- Spin for a short run in a microcentrifuge.
- Incubate at 95°C for 3 min (heat block).
- Immediately transfer sample tubes to water bath at 37°C for 30 min.
- Spin down for 1 min at 13,000 × *g*.

4.4. Step 4: Microarray Hybridization

- Equilibrate the hybridization oven to 65°C.
- Load a clean gasket slide into the Agilent hybridization chamber base (see Note 8).
- Dispense 245 μL of hybridization sample mixture slowly onto the gasket slide.
- Place a microarray slide – active side down – onto the gasket slide.
- Place the chamber cover onto the sandwiched slides and hand-tighten the clamp onto the chamber.
- Rotate the assembly vertically and assess the mobility of the bubbles (see Note 9).
- Place assembled slide chamber in the rotator rack in a hybridization oven set to 65°C.
- Incubate at 65°C with a rotation of 20 rpm for 40 h.

4.5. Step 5: Microarray Washing

Slide staining dishes:

#1	Disassembly	Wash buffer 1	RT	
#2	1st wash	Wash buffer 1	RT	5 min
#3	2nd wash	Wash buffer 2	37°C	1 min

- Heat the Agilent wash buffer 2 under magnetic stirring in a slide staining dish to 37°C in a water bath.
- Fill slide staining dish #1 completely with wash buffer 1.
- Put a slide rack into slide staining dish #2, add a magnetic stir bar, and put it on a magnetic stir plate (no more than five slides per round).
- Remove one hybridization chamber from incubator.
- Remove the array gasket sandwich from the chamber and transfer it quickly to slide staining dish #1.
- Open the submerged array gasket sandwich from the barcode end by slipping one of the blunt ends of the forceps between the slides and twisting the forceps to separate the slides.
- Let the gasket slide drop to the bottom of the staining dish.
- Quickly put the microarrays into dish #2.
- Stir for 5 min.
- Stop stirring and place the dish beside the magnetic stir bar.
- Put the pre-warmed dish #3 on the magnetic stir bar.
- Transfer the slide rack from dish #2 to dish #3 and stir for 1 min.
- Slowly remove the slide rack from dish #3 (see Note 10).
- Put the slides in an Agilent slide holder.
- Scan slides immediately.

4.6. Step 6: Microarray Scanning and Image Analysis (See Note 11)

Microarray slides are scanned immediately using an Agilent microarray scanner with 5 μm resolution. For image analysis, default CGH settings of Feature Extraction (FE) Software 10.7.31 are applied. Output files from FE are subsequently imported into Agilent's CGH data analysis software, Genomic workbench 5.0.14 for DNA copy number analysis. As a first step, set the ADM-2 (aberration detection module) algorithm threshold to 5.0 and the aberration filter to ten probes with mean $\log_2$ ratio of 0.3. To reduce background noise, it is often necessary to raise the ADM-2 threshold to 6.0 or even higher (see Note 12).

5. Data Analysis and Interpretation

Molecular karyotyping by means of high-resolution array-CGH offers the potential to identify small CNA in MDS below the detection limit of around 20 Mb of standard karyotyping. One chal-

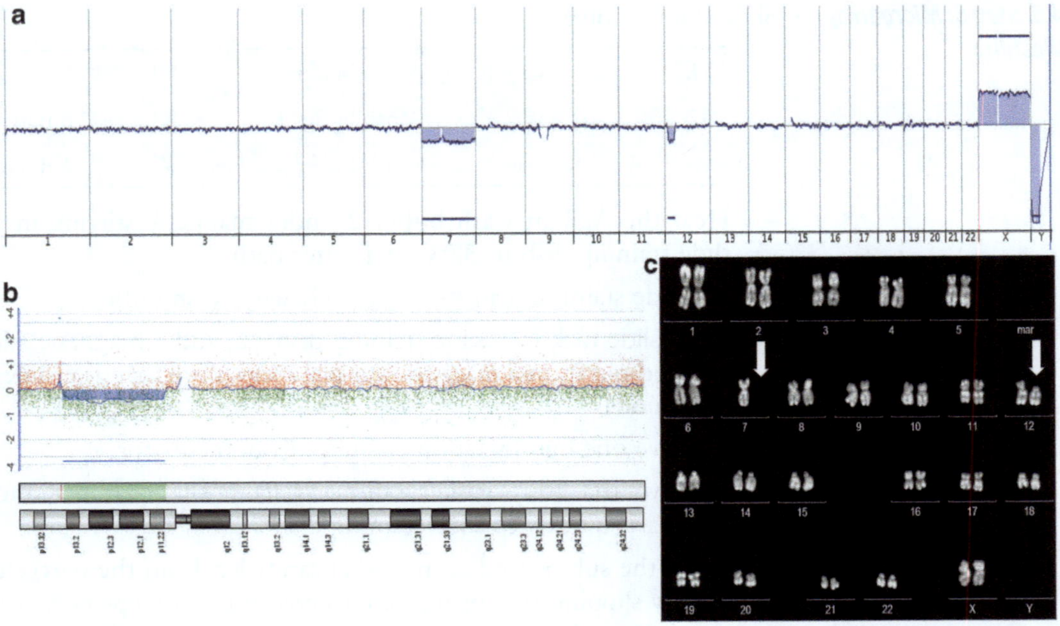

Fig. 1. Genomic profile along chromosomes 1–22 and X, Y generated from a female patient with childhood MDS by means of aCGH. (**a**) Hybridization was performed against a male control DNA pool resulting in +X and –Y. Monosomy 7 and del(12p) are indicated by *blue lines*. (**b**) Genomic profile of chromosome 12 with deletion 12p13.31–p11.21, covered by 1825 probes and spanning 21,563 Mb. (**c**) Fluorescence R-banding: Monosomy 7 and del(12)(p11p13) are marked by *arrows*.

lenge inherent to the analysis of MDS samples is the heterogeneity of cell populations in bone marrow or blood mononuclear cells (MNCs). This is often further complicated by the low percentage of aberrant cells (comparable to low germ-line mosaics). Looking for monosomy 7 in patients with myeloid malignancies by fluorescence in situ hybridization (FISH), a median percentage of 24–76% aberrant cells was reported (14). In accordance with standard cytogenetic techniques, the identification of alterations affecting whole chromosomes or chromosome arms like monosomy 7 and 7q- or 5q- can be certainly identified by aCGH in childhood MDS (Fig. 1). Moreover, smaller aberrations of around a few Mb in size can be detected but should be confirmed by an independent method like FISH, using labeled DNA probes mapping to that region (BAC/PAC clones from BlueGnome, Cambridge, UK). A small hidden loss of 5.7 Mb in the chromosomal region of 6q23.2-q24.1, containing the *MYB* gene, is shown in Fig. 2. This alteration, confirmed by FISH, is not detectable by standard karyotyping, showing the potential of aCGH for identifying important CNA which may be involved in MDS progression.

We were undoubtedly able to detect an alteration, although with 47 kb very small in size, on the 244 K array and to confirm this by a higher resolution eArray. When present in each cell, like the null genotype of *GSTT1*, even these small CNA should be

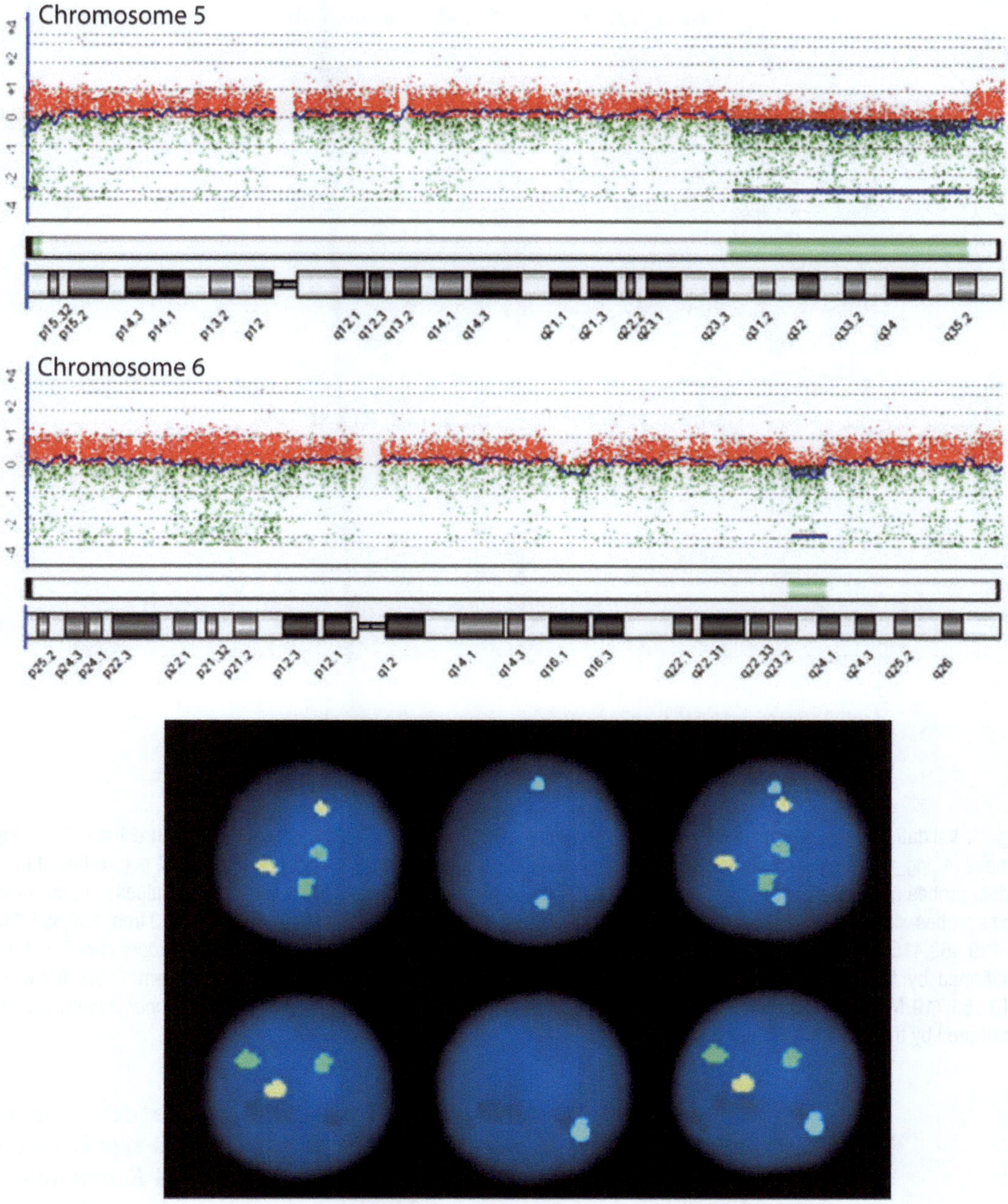

Fig. 2. Genomic profiles of chromosome 5 and chromosome 6 (*upper part*) by means of 244 K aCGH and validation of del(5)(q31.1–q35.2) and del(6)(q23.2–q24.1) by FISH (*lower part*). Log_2 ratios for individual features are plotted against their chromosomal localization along the chromosomes 5 and 6. Chromosomal losses are indicated in *green bars*, as detected with Genomic Workbench under filter settings 50 log 0.3 and threshold 6.0: Deletion of chromosome 5q31.1–q35.2 with average log_2 ratio of −0.372 and deletion around the *MYB* gene locus with an average log_2 ratio of −0.342. Validation of the 5q deletion (*left*) using FISH with a locus-specific probe for *EGR1* (5q31 in *orange*) in combination with a control probe D5S23 (5p15.2 in *green*) and the 6q deletion (*center*) using a locus-specific probe for *MYB* (6q23 in *aqua blue*). The simultaneous hybridization revealed that both deletions occurred in the same malignant clone (the merged image, *right-hand side*). Upper cells: control cells; lower cells: tumor cells.

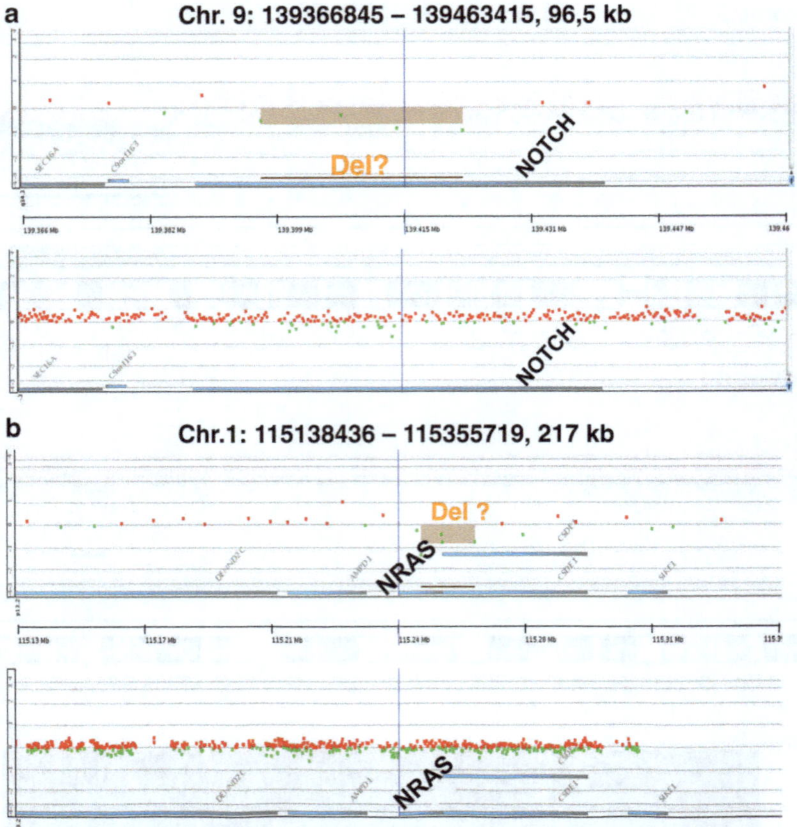

Fig. 3. Validation by a custom-designed high-resolution eArray. Applying filter settings: ADM-2 threshold 5.0, min_probes_4_log_0.3, small hidden deletions were indicated, among them targeting *NOTCH1* and *NRAS*. Log_2 ratios of individual probes are plotted against their chromosomal localization. *Red* spots represent probes with values > 0 and *green* spots probes with values < 0. Bars in *blue* or *brown* indicate deletions. (**a**) The chromosomal region 9q34.3 from 139,355,845 to 139,463,415 Mb containing *NOTCH1* is given in detail. A deletion is indicated by the 244 K array (*upper panel*), but not confirmed by the custom-designed eArray (*lower panel*). (**b**) The chromosomal region 1p13.2 from 115,138,436 to 115,355,719 Mb containing *NRAS* is given in detail. A deletion is indicated by the 244 K array (*upper panel*), but not confirmed by the custom-designed eArray (*lower panel*).

clearly visible (see Note 13). However, sensitive detection of putative disease-relevant alterations of such a small size is hardly possible in MDS. Putative CNA detected on the 244 K array under low filter stringencies, e.g., *NRAS*, emerged as false positives on the eArray (Fig. 3).

6. Notes

1. If using tubes that are prefilled and stored at 4°C, warm up the tubes to RT.
2. The cells can also be pelleted and stored at −80°C in RPMI 1640 containing 8% DMSO and 10% FBS until analyzed.

3. If the *A*260/*A*280 ratio is below 1.8, DNA has to be purified by proteinase K treatment.
4. According to the Agilent protocol, the ideal *A*260/*A*230 nm ratio is >1.0; if the *A*260/*A*230 ratio is <1.5, DNA has to be purified by centrifugal filter units (Amicon Ultra 30 K, Schwalbach, Germany).
5. The reaction can be stored overnight (or longer) at −20°C.
6. We use a pool of five control DNAs from healthy individuals of the same gender as reference sample.
7. The reaction can be stored overnight (or longer) at −20°C.
8. Load all gasket wells before placing the microarray active side down onto the gasket slide.
9. If necessary, tap assembly on a hard surface to move stationary bubbles.
10. If there are droplets on the slides, submerge the slide again and try to minimize the droplets.
11. All exposures of the sensitive cy-dyes to light and ozone must be minimized.
12. The DLRS (derivative log ratio spread) value is a marker for the array quality. As defined by Agilent, the quality is excellent when the DLRS is <0.2, good between 0.2 and 0.3, and poor if >0.3. The DLRS values of MDS samples are normally >0.3. They varied between 0.1497 (good) and 0.4938 (poor) in our study. Therefore, it is important to alter the filter settings for breakpoint calling in case of high noise by (1) increasing the number of consecutive probes showing a defined mean $\log_2$ ratio: ***min_probes_50 with mean log2 ratio of 0.3*** instead of ***min_probes_6 with mean log2 ratio of 0.3*** and/or (2) stepwise increasing the ADM-2 threshold from 6.0 to 10.0 or even higher.
13. eArray: To increase probe density in questionable genomic regions, a customized high-density array (up to bp resolution) can be designed using the Agilent web-based eArray database (http://www.earray.chem.agilent.com/earray). This allows accurate breakpoint identification by sequencing breakpoint-spanning PCR products.

Acknowledgments

The authors are grateful to Prof. Brigitte Schlegelberger and Gillian Teicke for critically reading the manuscript and to Dr. Gudrun Göhring for providing the karyogram.

References

1. Niemeyer CM, Arico M, Basso G, Biondi A, Cantu RA, Creutzig U, Haas O, Harbott J, Hasle H, Kerndrup G, Locatelli F, Mann G, Stollmann-Gibbels B, Veer-Korthof ET, van Wering E, Zimmermann M (1997) Chronic myelomonocytic leukemia in childhood: a retrospective analysis of 110 cases. European Working Group on Myelodysplastic Syndromes in Childhood (EWOG-MDS). Blood 89(10):3534–3543
2. Hasle H (2007) Myelodysplastic and myeloproliferative disorders in children. Curr Opin Pediatr 19(1):1–8
3. Fiegler H, Carr P, Douglas EJ, Burford DC, Hunt S, Scott CE, Smith J, Vetrie D, Gorman P, Tomlinson IP, Carter NP (2003) DNA microarrays for comparative genomic hybridization based on DOP-PCR amplification of BAC and PAC clones. Genes Chromosomes Cancer 36(4):361–374
4. Albertson DG, Pinkel D (2003) Genomic microarrays in human genetic disease and cancer. Hum Mol Genet 12(Spec No 2):R145–R152
5. Wessendorf S, Fritz B, Wrobel G, Nessling M, Lampel S, Goettel D, Kuepper M, Joos S, Hopman T, Kokocinski F, Dohner H, Bentz M, Schwaenen C, Lichter P (2002) Automated screening for genomic imbalances using matrix-based comparative genomic hybridization. Lab Invest 82(1):47–60
6. Solinas-Toldo S, Lampel S, Stilgenbauer S, Nickolenko J, Benner A, Dohner H, Cremer T, Lichter P (1997) Matrix-based comparative genomic hybridization: biochips to screen for genomic imbalances. Genes Chromosomes Cancer 20(4):399–407
7. Niemeyer C, Stollmann-Gibbels B, Ebell W, Gaedicke G, Creutzig U (1992) Myelodysplastic diseases in childhood. Klin Padiatr 204(4):190–197
8. Clark DM, Lampert IA (1990) Apoptosis is a common histopathological finding in myelodysplasia: the correlate of ineffective haematopoiesis. Leuk Lymphoma 2(6):415–418
9. Kerbauy DB, Deeg HJ (2007) Apoptosis and antiapoptotic mechanisms in the progression of myelodysplastic syndrome. Exp Hematol 35(11):1739–1746
10. Gondek LP, Tiu R, O'Keefe CL, Sekeres MA, Theil KS, Maciejewski JP (2008) Chromosomal lesions and uniparental disomy detected by SNP arrays in MDS, MDS/MPD, and MDS-derived AML. Blood 111(3):1534–1542
11. Evers C, Beier M, Poelitz A, Hildebrandt B, Servan K, Drechsler M, Germing U, Royer HD, Royer-Pokora B (2007) Molecular definition of chromosome arm 5q deletion end points and detection of hidden aberrations in patients with myelodysplastic syndromes and isolated del(5q) using oligonucleotide array CGH. Genes Chromosomes Cancer 46(12):1119–1128
12. Praulich I, Tauscher M, Gohring G, Glaser S, Hofmann W, Feurstein S, Flotho C, Lichter P, Niemeyer CM, Schlegelberger B, Steinemann D (2010) Clonal heterogeneity in childhood myelodysplastic syndromes–challenge for the detection of chromosomal imbalances by array-CGH. Genes Chromosomes Cancer 49(10):885–900
13. Griesshammer M, Klippel S, Strunck E, Temerinac S, Mohr U, Heimpel H, Pahl HL (2004) PRV-1 mRNA expression discriminates two types of essential thrombocythemia. Ann Hematol 83(6):364–370
14. Porta G, Maserati E, Mattarucchi E, Minelli A, Pressato B, Valli R, Zecca M, Bernardo ME, Lo CF, Locatelli F, Danesino C, Pasquali F (2007) Monosomy 7 in myeloid malignancies: parental origin and monitoring by real-time quantitative PCR. Leukemia 21(8):1833–1835

Chapter 18

Nonfamilial Breast Cancer Subtypes

Markus Ringnér, Johan Staaf, and Göran Jönsson

Abstract

Over the last decade, our knowledge in somatic genetic events related to breast cancer has increased enormously. Through usage of various genome-wide molecular approaches, it has become increasingly clear that breast cancer is a vastly heterogeneous disease. Microarray-based gene expression profiling has divided breast cancer into five distinct intrinsic subtypes termed basal-like, HER2-enriched, normal-like, luminal A, and luminal B. Importantly, these subtypes are closely correlated to clinical variables as well as different outcomes, with luminal A tumors as the good prognostic group. Initial studies using genome-wide DNA copy number data broadly partitioned breast cancers into three types, complex, amplifier, and simple, and moreover associated distinct copy number changes with the intrinsic subtypes defined by gene expression profiles. More recently, this genomic classification was refined into six genomic subtypes demonstrating strong resemblance to the intrinsic gene expression classification. Additionally, inherited *BRCA1*- and *BRCA2*-mutated tumors were significantly correlated to specific subtypes. In this chapter, we will review the current status regarding genomic subtypes of nonfamilial breast cancer.

Key words: Breast cancer, aCGH, Subtypes, Prognosis, Gene expression, Amplification, Deletion

1. Introduction

Accumulation of genomic alterations and genomic instability is a hallmark of solid tumor development. Identification of target genes may potentially reveal underlying mechanisms crucial for tumor development. Breast cancer is characterized by a profound heterogeneous histopathology, biology, and clinical course. The etiology of breast cancer is complex and involves a multitude of risk factors such as race, diet, age, early menarche, late menopause, reproductive and environmental factors, as well as genetics. The clinical diversity can be partially unraveled by pathological markers such as estrogen receptor (ER) expression and amplification of the *HER2* oncogene. ER positivity defines a subset of patients eligible for

Diponkar Banerjee and Sohrab P. Shah (eds.), *Array Comparative Genomic Hybridization: Protocols and Applications*, Methods in Molecular Biology vol. 973, DOI 10.1007/978-1-62703-281-0_18, © Springer Science+Business Media, LLC 2013

hormone modulation therapy using, for example, tamoxifen and aromatase inhibitors. Likewise, patients with tumors harboring *HER2* amplification are eligible to HER2-targeted therapy including monoclonal antibodies or small tyrosine kinase inhibitors. Thus far, no specific treatment modalities are available for patients negative for both ER and HER2. Breast cancer has been extensively investigated using microarrays, revealing molecular subtypes and reinforcing the concept of heterogeneity of breast cancer. Heterogeneity unraveled by genome-wide DNA copy number data will be the focus of this chapter.

1.1. Gene Expression Profiling-Based Classification of Breast Cancer

Over the past decade, the number of studies utilizing gene expression profiling in breast cancer has increased dramatically. A pioneer study by Perou et al. utilized an intrinsic approach to define a set of genes that divided 65 tumors from 42 patients into four subtypes: (a) the basal-like, (b) HER2-enriched, (c) normal-like, and (d) the luminal epithelial/estrogen receptor (ER)-positive subtype (1). Importantly, these gene expression subtypes have been reproduced and further subdivided in subsequent studies, for example, dividing the luminal epithelial/ER-positive group into the luminal A and luminal B subtypes (2, 3). Luminal A and B subtypes both express markers of the luminal epithelial layer of normal breast ducts such as Keratin 8 and 18. Moreover, both subtypes generally express the ER and progesterone receptor (PR). The main distinction between luminal A and B tumors is based on proliferation-related genes since luminal B tumors display a higher expression of such genes, supported by higher Ki67 proliferation indices and S-phase fractions (4). The HER2-enriched group is characterized by overexpression of genes located within the HER2 amplicon on chromosome 17q12-q21, while the basal-like group expresses markers of the basal layer of the normal breast duct such as keratin 5/6 and is generally ER, PR, and HER2 negative. Interestingly, a strong association of specific subtypes with hereditary breast cancer has repeatedly been shown (2, 5). Patients carrying germ-line mutations in *BRCA1* most frequently develop basal-like tumors, while tumors arising in patients with germ-line *BRCA2* mutations are most often of the luminal B subtype (5). Moreover, the subtypes have also been found to have characteristic DNA methylation profiles (6). Importantly, subsequent studies demonstrated different prognosis of the subtypes and association with several established clinical variables (2, 3). The two predominantly ER-negative subtypes, basal-like and HER2-enriched, displayed the worst prognosis followed by luminal B, whereas luminal A was a good prognosis group. However, luminal B tumors often have the worst long-term prognosis, as late recurrences are more common in these tumors compared to basal-like and HER2-enriched tumors (7).

Besides the five intrinsic subtypes, additional subtypes have been reported. In a recent study of 537 breast tumors, six molecular subclasses were identified (8). In addition to the basal-like, normal-like, luminal A, and luminal B subtypes, two subtypes, the luminal C and the molecular apocrine type, were defined. Notably, in this study, no specific HER2-enriched subtype was identified. The luminal C type was highly proliferative and displayed borderline ER positivity and an increased frequency of *HER2*-amplified cases. The molecular apocrine type was similar to a previously described group comprising ER-negative and androgen receptor-positive breast cancers, frequently overexpressing *HER2* (9, 10). Moreover, by comparing expression profiles of mouse and human breast tumors, a claudin-low subtype has been identified (11). The claudin-low subtype is a rare subtype of human breast cancer characterized by low levels of tight junction genes and high levels of epithelial–mesenchymal transition genes (12). Claudin-low tumors have been shown to be enriched for gene expression signatures derived from tumor-initiating cells and human mammary stem cells and to be enriched in breast cancer cell populations surviving after conventional treatment (13).

The extensive confirmation of the intrinsic subtypes in breast cancer has lead to the development of a QPCR-based classifier consisting of 50 genes, which is also applicable to FFPE samples (7). IHC-based subtype classification using markers for ER, HER2, and proliferation has also shown considerable overlap with gene expression classification (14). Moreover, the first generation of gene signature-based diagnostic tests (Oncotype DX and MammaPrint) largely stratify patients on known clinical factors. Although, these first signatures show improvement in prognostication compared to clinical models based on ER, HER2, and histological grade, the improvements are relatively small (15). In addition, these gene signatures appear to have limited prognostic value in often more challenging breast cancer subgroups such as ER-negative, HER2-positive, or histological grade 3 tumors (16, 17). These clinical groups of tumors display heterogeneous patient outcomes, and gene expression signatures that stratify these tumor groups based on outcome are emerging (10, 18, 19). Such results suggest that the second generation of tests influenced from gene expression studies will provide a larger improvement on current clinical models.

1.2. Gene Amplifications in Breast Cancer

Chromosomal aberrations in breast cancer have been extensively analyzed using a wide range of techniques such as cytogenetics, conventional comparative genomic hybridization (CGH), and fluorescent in situ hybridization (FISH). Recurrent amplifications on chromosomes 8p12, 11q13, 17q12, and 20q13 including well-established oncogene drivers such as *CCND1* and *HER2* have been identified in several studies. With the development of array-based CGH (aCGH), the possibility to delineate DNA copy number changes on a genome-wide level has increased immensely.

Genome-wide aCGH was early applied to breast tumors (20, 21); however, despite the increased resolution from technology development, few recurrent novel oncogenic loci have been identified. Instead, the technique has allowed fine mapping and pinpointing of oncogenes, as well as providing a way of comparing global patterns of genomic alterations across large sets of tumors (5, 22–25).

The most frequently amplified region in breast cancer is located on chromosome 17q12-q21, includes the *HER2* oncogene, and occurs in 20–25% of cases (26, 27). Gene amplification of *HER2* coupled to protein overexpression defines a clinically important subgroup of breast cancer that is generally associated with high cell proliferation rate, DNA aneuploidy, high tumor grade, poor prognosis, and variable response to conventional systemic cytotoxic therapy (For a review see (28)). Strategies to therapeutically target the HER2 protein with monoclonal antibodies (e.g., trastuzumab) or tyrosine kinase inhibitors (e.g., lapatinib) have been successful in both the metastatic and adjuvant setting (29–31), representing one of the success stories of modern oncology. Using aCGH it has been shown that the 17q12 amplicon can vary considerably in size and include also other genes, such as *TOP2A* located ~700kbp telomeric of *HER2*, which may contribute to tumor progression and treatment effect (32, 33). In a comprehensive aCGH study of *HER2*-amplified breast cancer, the smallest region of amplification overlap for the 17q12 locus was narrowed to an 85.92-kbp region including the *TCAP*, *PNMT*, *PERLD1*, *HER2*, *C17orf37*, and *GRB7* genes. By integrating gene expression with aCGH data, the *HER2* gene was found to be most frequently upregulated followed by *GRB7* (33).

In contrast to amplification of *HER2*, which is more frequent in ER-negative tumors, amplification of other prevalent loci in breast cancer, such as 8p12, 11q13, and 20q13, are more frequent in ER-positive tumors (34). Chromosome 8p12 amplification occurs in approximately 10–15% of breast cancer cases, and several candidate genes, such as *FGFR1*, *BRF2*, and *RAB11FIP1*, have been suggested (22, 35–37). The role of *FGFR1* is elusive since some 8p12-amplified tumors do not overexpress *FGFR1*; however, it has been shown that *FGFR1* is frequently activated in lobular breast carcinomas (36). In an elegant study in which 1,172 breast tumors were analyzed by gene expression microarrays and genome-wide SNP microarrays, it was shown that the minimal 8p12 amplicon contained only the *ZNF703* gene (22). Specifically, amplification of *ZNF703* has been associated with increased gene expression, poor clinical outcome, and the luminal B subtype (22, 37). Amplification of 11q13 is observed in about 15% of breast cancers with *CCND1* as the most obvious candidate gene. However, several different amplified regions on 11q13 have been identified including other candidate genes such as *PAK1* and *EMSY* (38–40) albeit at a lower frequency as compared to *CCND1* (39). Moreover, co-amplification of 11q13 and 8p12 is relatively frequent, suggesting possible cooperation of genes in these regions during oncogenesis (5, 41, 42), although the exact mechanism remains to be determined. A third locus commonly amplified in ER-positive tumors maps to chromosome 20q13. Using aCGH, the amplicon was split in two peaks, each harboring candidate oncogenes, *ZNF217* and *CYP24D* (43). Overall, the development of aCGH has lead to better understanding and fine mapping of amplifications in breast cancer.

1.3. Characterization of Intrinsic Subtypes by aCGH

It is well established that breast cancer may be divided into subtypes based on distinct gene expression patterns. Since the discovery of these subtypes, they have been carefully characterized for a number of different traits including genome-wide copy number changes. Bergamaschi et al. used a cDNA microarray to characterize genome-wide copy number changes in 89 tumors that had been classified into the intrinsic subtypes (44). The authors found distinct copy number changes associated with the intrinsic subtypes and, additionally, found basal-like tumors to harbor a high frequency of low-level copy number changes and luminal B tumors to frequently harbor gene amplifications. Chin et al. also performed combined analyses of genome-wide copy number and gene expression in a cohort of 130 breast tumors (24). Statistical analyses demonstrated enrichment of low-level copy number changes on chromosomes 3q, 10p, 4, and 5q in basal-like tumors, which was found to be the subtype with the most unstable tumor genome. In contrast, luminal A tumors were characterized by stable tumor genomes and frequent copy number changes on chromosomes 1q and 16. These results were firmly corroborated in a large series of 359 breast tumors (Fig. 1) including also a notable number of *BRCA1*- and *BRCA2*-mutated tumors (5). Intriguingly, *BRCA1*-mutated tumors displayed almost indistinguishable genomic profiles from non-*BRCA1*-mutated basal-like tumors, suggesting that such tumors may be deficient for DNA repair as well.

2. Analysis Approaches for Genomic Copy Number Data

Studies aiming at genomic classification of breast cancer initially used noncommercial aCGH platforms consisting of spotted microarrays. These genomic microarrays were constructed from either cDNA libraries or, later on, predominantly from bacterial artificial chromosome (BAC) DNA libraries providing a genome-wide coverage. However, commercial aCGH and SNP platforms based on oligonucleotide sequences (20–60 bp) providing increased

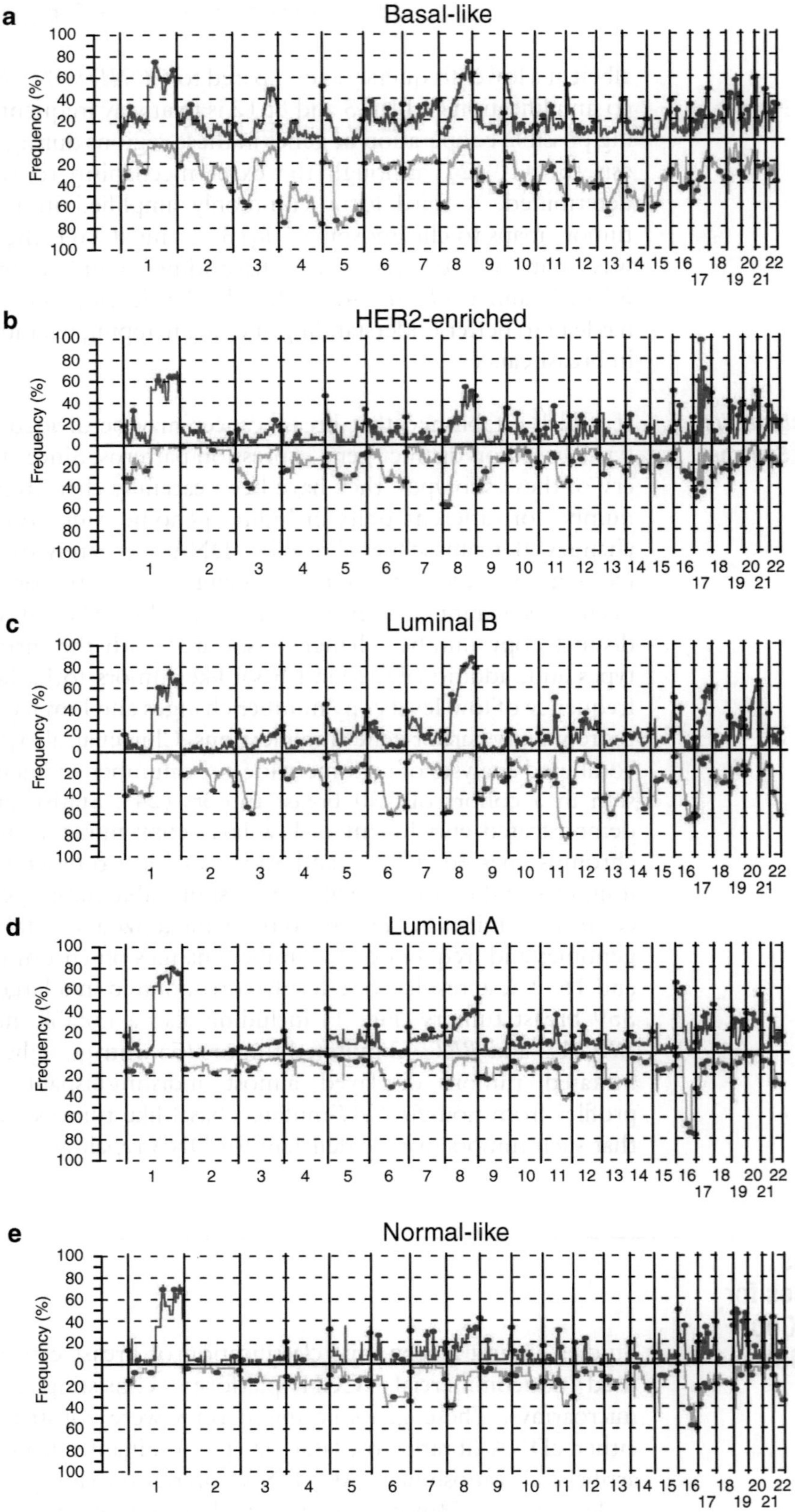

Fig. 1. Frequency plots of intrinsic subtypes in breast cancer from the study by Jönsson et al. (5). (**a**) Basal-like, (**b**) HER2-enriched, (**c**) luminal B, (**d**) luminal A, and (**e**) normal-like frequency plots indicate common gains and losses. Additionally, identified GISTIC peaks are marked by *dots*.

resolution have dominated in more recent studies. To determine copy number status, conventional aCGH platforms use test and reference DNA (labeled with different fluorescent dyes) in competitive hybridization to unique DNA fragments deposited on the microarray surface. The reference DNA is normally selected as DNA from normal cells, either from the test subject or a pool of unrelated individuals with an expected normal karyotype. Raw data from an aCGH experiment consist of intensity estimates for each fluorescent dye for each unique DNA fragment on the microarray. Combination of test and reference intensities into an intensity ratio creates a quantitative estimation of the DNA amount in the sample relative to the reference for each genomic position investigated by a DNA fragment on the microarray. By taking advantage of the unique genomic position of each DNA fragment on the microarray, an ordered genomic profile of copy number estimates can be generated from which genomic alterations may be identified. In addition to copy number status, single nucleotide polymorphism (SNP) platforms allow for simultaneous analysis of allelic imbalances, making SNP platforms an attractive alternative to aCGH. With SNP platforms only the sample DNA is analyzed, and copy number estimates for each SNP locus are formed by comparing test sample (allele-specific) signal intensities to those of an existing set of identically processed reference samples.

Raw intensity data obtained from the experimental analysis are commonly normalized and $\log_2$ transformed to calibrate and define copy number profiles. The purpose of normalization is to remove experimental variation and to shift $\log_2$ ratios toward a common baseline to facilitate, for example, comparisons between samples. Normalization of aCGH data can be performed in a variety of ways using different algorithms both specifically developed for aCGH (45) or adopted from analysis of gene expression data. In general, normalized copy number profiles from both aCGH and SNP microarrays are subsequently partitioned into genomic segments with the aim to identify genomic breakpoints and regions of common copy number. Numerous partitioning methods have been developed (46–48), all with the same underlying idea that the genome may be broken up into segments in which array probes between breakpoints share the same underlying copy number.

Since SNP arrays measure each allele separately, this allows for determination of both the total copy number at a given loci and its allelic composition. The latter may be viewed by, for example, B allele frequency plots (49) showing the proportions of each allele for SNPs ordered according to their genomic position in the genome. Moreover, integration of the total copy number with the allelic composition allows for estimation of both the fraction of non-aberrant cells and the ploidy of the sample (50–52). Thereby, SNP arrays can be used to generate allele-specific segmented copy number profiles adjusted for these cell characteristics (50–52).

Segmented copy number profiles reflect all detected copy number alterations that occurred during tumor development, as well as germ-line copy number variations if not corrected for. Solid tumors often harbor many altered regions, and each region may comprise thousands of genes. A focus of higher-level analysis methods for aCGH and SNP data is on identification of altered regions harboring genes with potential tumor-promoting capabilities. Such regions are commonly assumed to be affected in many tumors. Therefore, large tumor data sets are often investigated with the aim to identify regions that are frequently affected by copy number alterations. Moreover, large tumor data sets enable elimination of putative targets by defining boundaries based on the smallest region of alteration across the tumor set. A widely used method to identify frequently affected regions, their boundaries, and target genes in large tumor data sets is genomic identification of significant targets in cancer (GISTIC) (53, 54). Finally, copy number data is commonly integrated with gene expression data and other molecular assays to further pinpoint target genes.

To identify genomic subtypes of breast cancer, various unsupervised clustering algorithms have typically been used. Several studies have used hierarchical clustering, either using all regions (24, 25) or using only frequently altered regions identified by, for example, GISTIC (5). In addition, other clustering methods such as the nonnegative matrix factorization method have also been used (25). Breast tumors have also been grouped based on partly manual decisions, for example, by grouping tumors based on patterns of whole-arm or regional gains and losses (23), as well as overall appearance of genomic profiles (55).

3. Genomic Subtypes in Breast Cancer

3.1. Class Discovery in Breast Cancer Using aCGH

As with gene expression profiles genome-wide copy number changes have been used for class discovery purposes. Different analysis approaches exist that have resulted in five elegant studies summarized below. Chin et al. profiled 145 breast tumors for genome copy number changes using an aCGH platform consisting of 2,464 bacterial artificial chromosome (BAC) probes (24). Unsupervised hierarchical clustering using all BACs defined three major genomic subtypes referred to as the complex, amplifier, and 1q/16q subtypes. The complex type of tumors was characterized by a high frequency of copy number changes and frequent losses on chromosomes 3p, 4, and 5q. Moreover, the complex subtype was associated with aberrant expression of p53 and contained the majority of basal-like tumors. In contrast, the 1q/16q subtype displayed low frequency of copy number changes that were mainly found on chromosomes 1q and 16. Additionally, the majority of luminal A tumors were grouped in the 1q/16q group, and 1q/16q tumors tended to have Ki67 proliferation indices <10%. Amplifier tumors typically harbored gene amplifications targeting oncogenes such as *HER2* and *CCND1*. Importantly, these subtypes demonstrated different survival patterns with the 1q/16q subtype having a good outcome as compared to the complex and amplifier subtypes.

A slightly different approach was undertaken by Hicks et al., grouping tumors into three subtypes, the simplex, sawtooth, and firestorm, based on manual inspection of genome plots (55). Simplex tumors were predominantly diploid and harbored broad segments of duplication and deletion, while sawtooth tumors were characterized by a large number of narrow segments of duplication and deletion affecting all chromosomes. In sawtooth tumors, the majority of the genome was affected by copy number change; however, these tumors did not typically harbor gene amplifications. Firestorm tumors had at least one chromosomal region with multiple narrow amplification peaks (referred to as firestorms) but an otherwise less-rearranged genome compared to sawtooth tumors. Furthermore, Hicks and coworkers presented a mathematical measure, the firestorm index, to distinguish sawtooth and firestorm from simplex tumors, while simultaneously incorporating the total number of genomic events. The firestorm index was associated with amplification of *HER2*, *CCND1*, and *MYC* as well as age at diagnosis; nevertheless, the firestorm index demonstrated independent prognostic information when adjusting for these variables in a multivariate model for diploid tumors.

Andre et al. used a combined approach of genome copy number and expression data in 106 breast tumors (25). Based on non-negative matrix factorization, three classes were discovered. Class I was characterized by a basal-like intrinsic subtype and triple negativity (i.e., ER, PR, and HER2 negative). Class II tumors were mainly ER positive and classified as luminal B, whereas class III tumors were frequently classified as luminal A. Even though these classes were not investigated for correlation to clinical information, they will most likely relate to prognosis. Moreover, the authors proceeded by performing unsupervised hierarchical clustering. Again, three types were identified, of which two were overlapping with the previous approach. The third group identified consisted of tumors that almost exclusively lacked chromosomal alterations. This somewhat intriguing observation should be carefully investigated and could potentially also reflect the fraction of contaminating normal cells. Samples with extensive amount of contaminating non-tumor cells display very few if any chromosomal alterations.

More recently, Jönsson et al. approached the matter of genomic subtypes by using a combined analysis of genome copy number and expression data in 359 breast tumors (5). Here, the authors utilized GISTIC to define chromosomal regions of importance in breast cancer. GISTIC analysis identified 66 regions of gain and 67 regions of losses, which subsequently were used in hierarchical clustering algorithms. This approach defined six genomic subtypes, the 17q12, basal-complex, luminal-simple, luminal-complex, amplifier, and mixed. Four of the subtypes displayed striking similarity to the gene expression-derived intrinsic subtypes. The 17q12 genomic subtype included 88% of HER2-enriched classified cases suggesting an extensive overlap between classification approaches due to amplification of the 17q12 region. The basal-complex subtype included mainly basal-like cases and displayed higher frequency of copy number changes and higher S-phase fraction. In contrast, the luminal-simple type was characterized by frequent losses on 16q and gains on 1q and 16p. Furthermore, 67% of luminal A classified cases were grouped in luminal-simple and displayed lower frequency of copy number changes and lower S-phase fraction than other subtypes. The luminal-complex subtype included 69% of all luminal B classified cases but also 33% of luminal A classified cases. This subtype also displayed high frequency of copy number changes and was by far the largest genomic subtype in the cohort. Two additional genomic subtypes were defined, mixed and amplifier, both of which displayed mixed intrinsic subtype characteristics. As expected, there was a significant difference in survival with 17q12 subtype tumors having poor prognosis and luminal-simple tumors having good prognosis. An interesting finding of this study concerned the intrinsic luminal A subtype which was divided in luminal-simple and luminal-complex. Several differences were found, indicating that the intrinsic classification of luminal A tumors can be further refined. Moreover, this study confirmed that hereditary *BRCA1*- and *BRCA2*-mutated breast tumors are classified into distinct molecular classes.

The group of Borresen-Dale in Norway constructed platform-independent algorithms to determine whole-chromosome arm aberration index (WAAI) and complex arm aberration index (CAAI) (23). Four groups of tumors were initially defined: those with whole-arm gain of 1q and/or loss of 16q (group A), those with regional loss on 5q and/or gain on 10p (group B), those with both (group AB), and those with neither (group C). Based on the WAAI and CAAI indices, each group was subsequently split into two subgroups. Using a whole-genome sequencing approach, Russnes and coworkers found that these groups were associated with inter- and intrachromosomal rearrangements (23). Two simple groups (A1 and C1) with tumors that displayed low frequency of genomic changes, and also few chromosomal rearrangements demonstrated using whole-genome sequencing, were identified. The main difference between A1 and C1 tumors was alterations on chromosomes 1q and 16q found in A1 tumors and that A1 tumors showed strong correlation to the intrinsic luminal A subtype.

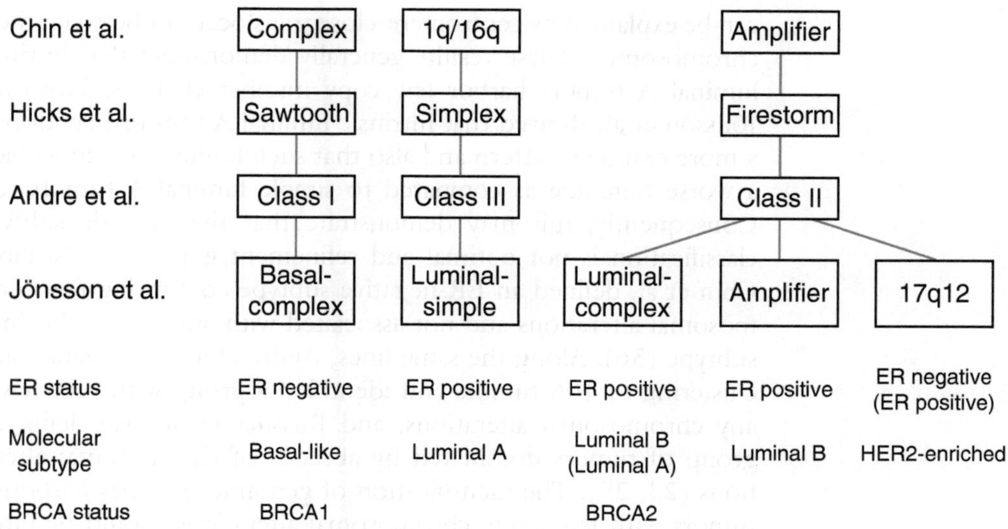

Fig. 2. An overview of aCGH classifications of breast cancer. Subtypes identified by Chin et al. (24), Hicks et al. (55), Andre et al. (25), and Jönsson et al. (5) are displayed together with molecular characteristics.

Six groups with a higher frequency of genomic imbalances were identified (A2, B1, B2, AB1, AB2, and C2). Differences were highlighted by whether altered regions were due to whole-arm aberrations or more complex rearrangements as quantified by WAAI and CAAI. To some extent, this classification also demonstrated overlap with the intrinsic subtype classification. Additionally, the data suggested that high-level CAAI is an independent prognostic marker in breast cancer.

Overall, these results clearly demonstrate that class discovery is possible using aCGH data and that it has correlation to clinical outcome and biological subtypes.

3.2. Overlap in aCGH Classification Systems

Results from studies using aCGH for class discovery purposes demonstrate extensive overlap (Fig. 2). From the work on gene expression and aCGH-based classification, it is obvious that the results of these two profiling techniques are closely connected and correlated. All studies included herein with the aim of defining a genomic classification of breast cancer identified a subtype harboring few chromosomal alterations. Chin et al. and Jönsson et al. found a type of breast tumors harboring frequent alterations on chromosomes 1q and 16 (5, 24). Such tumors were commonly ER positive and classified as luminal A by gene expression profiling. Russnes et al. verified tumors with a simple copy number profile (1q/16q) and subsequently performed a detailed map of chromosomal rearrangements using whole-genome DNA sequencing (23). Such analysis demonstrated low frequency of inter- and intrachromosomal changes. Presumably, tumors with 1q gain and 16q loss can be explained by centromere-close translocations between these chromosomes. These results generally demonstrate that intrinsic luminal A tumors harbor few copy number changes. However, Jönsson et al. showed that intrinsic luminal A tumors may display a more complex pattern and also that such luminal A tumors have a worse outcome as compared to simple luminal A tumors (5). Consequently, this may demonstrate that the intrinsic subtype classification is not optimal and refinement is possible. Notably, Chin et al. defined an ER-negative subtype comprising few chromosomal alterations and not associated with any distinct intrinsic subtype (56). Along the same lines, Andre et al. used hierarchical clustering of 106 tumors and identified a group with very few if any chromosomal alterations, and Russnes et al. also defined a group of tumors dominated by absence of chromosomal alterations (23, 25). The identification of genomic subtypes harboring tumors with few if any chromosomal alterations should be interpreted with caution since these could potentially include tumors with a high frequency of contaminating stromal cells and thus compress the ratios obtained in the analysis of aCGH data. Moreover, it is not clear that these subtypes identified in different studies share other characteristics.

Amplified DNA in cancer cells was originally observed by cytogenetic analysis and occurs either as homogeneously staining regions (HSR) or as double minutes. The origin of DNA amplifications has been thoroughly studied, and the most probable mechanism in breast tumors includes breakage–fusion–bridge cycles. The most common gene amplification in breast cancer includes the *HER2* oncogene, and by intrinsic subtyping, the majority of *HER2*-amplified cases are included in the HER2-enriched group. Likewise, the aCGH classification system proposed by Jönsson et al. grouped *HER2*-amplified tumors in a 17q12 subtype (5). The formation of the 17q12 group was very much the result of gene amplifications on chromosome 17q, predominantly found in *HER2*-amplified tumors compared to other subtypes (33). Other common gene amplifications in breast cancer include chromosomes 11q13, 8p12, 8q24, and 20q13, and tumors harboring such gene amplifications are associated with the intrinsic luminal B subtype (5, 24, 44). Despite different analyses approaches, most aCGH-based classification systems group all tumors harboring gene amplifications, including *HER2*-amplified tumors, into an amplifier/firestorm group (24, 25, 55).

The third major type of tumors includes those having a high frequency of low-level DNA copy number changes. The underlying mechanism for this extensive genomic distortion is not fully understood. A plausible hypothesis includes deficiencies in the DNA repair machinery including potential alterations in the

BRCA1 gene. Indeed, it has been shown that tumors from *BRCA1* mutation carriers have a high frequency of genomic changes (57, 58) and are genomically similar to basal-like classified tumors (5, 59), suggesting similar underlying mechanisms causing the genomic distortion found in such tumors. Furthermore, tumors from *BRCA1* mutation carriers are mainly classified as basal-like (2, 5) and sporadic basal-like, and tumors from *BRCA1* carriers exhibit similar histology with high Ki67 indices, lymphocytic infiltration, medullary features, and "pushing" margins (60–62). Further support for the notion that sporadic basal-like tumors may have defective BRCA1 function comes from BRCA1 promoter methylation studies showing that such tumors belong to the basal-like subtype (63, 64). Consequently, complex tumors that have a high frequency of low-level copy number changes display many similarities to tumors from BRCA1 mutation carriers. Another feature found in both sporadic basal-like and BRCA1 tumors is frequent loss of PTEN protein expression. Inactivation of PTEN in these tumors occurs through gross mutations such as inversions and deletions at the PTEN locus (65). These rearrangements are most likely a consequence of defective homologous recombination supporting the notion that sporadic basal-like tumors have a deficiency in the BRCA1 pathway. Taken together, this suggests that sporadic basal-like tumors or tumors demonstrating similar genome profile to BRCA1 tumors have a defective DNA repair machinery.

Overall, genomic classification demonstrates extensive overlap between studies suggesting that the dominant genomic subtypes of breast cancer are fairly robust since they have been identified by several different methods and parameter choices across multiple independent tumor sets (Fig. 2). In breast cancer, tumors with a simple, complex, or amplifier genome exist which can be further refined into additional subgroups. These subgroups display strong association to intrinsic subtyping with complex tumors classified mainly as basal-like, amplifier as luminal B, and simple as luminal A. However, much remains to be discovered, and the enormous development of next-generation sequencing tools will most likely contribute to this.

4. Clinical Implications: Prognosis and Prediction

Many studies have demonstrated the usage of genome DNA copy number profiling for prediction of prognosis and improving classification of breast tumors (5, 24, 25, 66, 67). Most directly, patients with amplifier tumors have a poor prognosis, and, furthermore, the number of amplifications is correlated to prognosis (24, 55). Corroborating these results are findings of specific amplifications associated with poor prognosis (24, 68). However, it appears that the underlying mechanism for poor prognosis in such tumors is reflected by a general amplification pattern and not by specific amplifications.

Genomic instability in general has been demonstrated to be a strong determinant of survival in cancer (69). The clinical significance of genomic instability in breast cancer was corroborated in a study where 313 lymph-node-negative breast cancers were analyzed (66). Here, a chromosomal instability score was significantly associated with prognosis in ER-positive, luminal B, and HER2-enriched subgroups but not in ER-negative patients. The clinical subgroup of patients with high genomic instability is obviously strongly correlated to the complex subtypes in different classification systems (basal-complex, complex, sawtooth, high CAAI). In contrast, it has been difficult to identify distinct genomic regions to be used as clinical markers. Encouragingly, Thompson et al. created a 19-region copy number classifier that identified a low-risk and a high-risk group by subjecting 971 breast cancers to genome profiling using molecular inversion probe-based (MIP) arrays (67). For the 19-region classifier, amplifications or gains in the regions, but not genomic instability, were associated with poor prognosis, suggesting a complex pattern of oncogene activation in aggressive tumors. In fact, the classifier could also divide ER-negative tumors according to outcome, which has not been observed in the different genomic subtype classifications.

Both gene amplification and genomic instability offer opportunities for treatment selection. A classical example of a gene amplification used in treatment prediction is the *HER2* oncogene for which targeted treatment now exists. An interesting feature of genomic instability includes therapeutically targeting chromosomal instability itself by taking advantage of "synthetic lethality." A distinct example is that tumors deficient for BRCA1 or BRCA2 respond markedly to PARP inhibitors (70). It is assumed that PARP inhibition leads to collapsed replication forks and double-strand break repair. This makes the treatment eligible only to patients having tumor deficiencies in the BRCA1 or BRCA2 pathway. Indeed, clinical trials support the efficacy of PARP inhibitors in patients with germ-line mutations in *BRCA1* or *BRCA2* (71). Others have showed that tumors with high genomic instability such as triple-negative tumors respond better to DNA damaging agents (e.g., cisplatinum) (72).

Future studies will most likely focus on sequence-based approaches, and it will be interesting to see how genomic subtypes in breast cancer will evolve.

References

1. Perou CM, Sorlie T, Eisen MB et al (2000) Molecular portraits of human breast tumours. Nature 406(6797):747–752
2. Sorlie T, Tibshirani R, Parker J et al (2003) Repeated observation of breast tumor subtypes in independent gene expression data sets. Proc Natl Acad Sci USA 100(14):8418–8423
3. Sorlie T, Perou CM, Tibshirani R et al (2001) Gene expression patterns of breast carcinomas distinguish tumor subclasses with clinical implications. Proc Natl Acad Sci USA 98(19):10869–10874
4. Cheang MC, Chia SK, Voduc D et al (2009) Ki67 index, her2 status, and prognosis of patients with luminal b breast cancer. J Natl Cancer Inst 101(10):736–750
5. Jonsson G, Staaf J, Vallon-Christersson J et al (2010) Genomic subtypes of breast cancer identified by array-comparative genomic hybridization display distinct molecular and clinical characteristics. Breast Cancer Res 12(3):R42
6. Holm K, Hegardt C, Staaf J et al (2010) Molecular subtypes of breast cancer are associated with characteristic DNA methylation patterns. Breast Cancer Res 12(3):R36
7. Parker JS, Mullins M, Cheang MC et al (2009) Supervised risk predictor of breast cancer based on intrinsic subtypes. J Clin Oncol 27(8):1160–1167
8. Guedj M, Marisa L, De Reynies A et al (2011) A refined molecular taxonomy of breast cancer. Oncogene 31(9):1196–1206
9. Farmer P, Bonnefoi H, Becette V et al (2005) Identification of molecular apocrine breast tumours by microarray analysis. Oncogene 24(29):4660–4671
10. Teschendorff AE, Miremadi A, Pinder SE et al (2007) An immune response gene expression module identifies a good prognosis subtype in estrogen receptor negative breast cancer. Genome Biol 8(8):R157
11. Herschkowitz JI, Simin K, Weigman VJ et al (2007) Identification of conserved gene expression features between murine mammary carcinoma models and human breast tumors. Genome Biol 8(5):R76
12. Prat A, Parker JS, Karginova O et al (2010) Phenotypic and molecular characterization of the claudin-low intrinsic subtype of breast cancer. Breast Cancer Res 12(5):R68
13. Creighton CJ, Li X, Landis M et al (2009) Residual breast cancers after conventional therapy display mesenchymal as well as tumor-initiating features. Proc Natl Acad Sci USA 106(33):13820–13825
14. Nielsen TO, Parker JS, Leung S et al (2010) A comparison of pam50 intrinsic subtyping with immunohistochemistry and clinical prognostic factors in tamoxifen-treated estrogen receptor-positive breast cancer. Clin Cancer Res 16(21):5222–5232
15. Sotiriou C, Pusztai L (2009) Gene-expression signatures in breast cancer. N Engl J Med 360(8):790–800
16. Desmedt C, Haibe-Kains B, Wirapati P et al (2008) Biological processes associated with breast cancer clinical outcome depend on the molecular subtypes. Clin Cancer Res 14(16):5158–5165
17. Reyal F, Van Vliet MH, Armstrong NJ et al (2008) A comprehensive analysis of prognostic signatures reveals the high predictive capacity of the proliferation, immune response and rna splicing modules in breast cancer. Breast Cancer Res 10(6):R93
18. Finak G, Bertos N, Pepin F et al (2008) Stromal gene expression predicts clinical outcome in breast cancer. Nat Med 14(5):518–527
19. Staaf J, Ringner M, Vallon-Christersson J et al (2010) Identification of subtypes in human epidermal growth factor receptor 2–positive breast cancer reveals a gene signature prognostic of outcome. J Clin Oncol 28(11):1813–1820
20. Pollack JR, Sorlie T, Perou CM et al (2002) Microarray analysis reveals a major direct role of DNA copy number alteration in the transcriptional program of human breast tumors. Proc Natl Acad Sci USA 99(20):12963–12968
21. Hyman E, Kauraniemi P, Hautaniemi S et al (2002) Impact of DNA amplification on gene expression patterns in breast cancer. Cancer Res 62(21):6240–6245
22. Holland DG, Burleigh A, Git A et al (2011) Znf703 is a common luminal b breast cancer oncogene that differentially regulates luminal and basal progenitors in human mammary epithelium. EMBO Mol Med 3(3):167–180
23. Russnes HG, Vollan HK, Lingjaerde OC et al (2010) Genomic architecture characterizes tumor progression paths and fate in breast cancer patients. Sci Transl Med 2(38):38ra47
24. Chin K, Devries S, Fridlyand J et al (2006) Genomic and transcriptional aberrations linked to breast cancer pathophysiologies. Cancer Cell 10(6):529–541
25. Andre F, Job B, Dessen P et al (2009) Molecular characterization of breast cancer with high-resolution oligonucleotide comparative genomic hybridization array. Clin Cancer Res 15(2):441–451

26. Owens MA, Horten BC, Da Silva MM (2004) Her2 amplification ratios by fluorescence in situ hybridization and correlation with immunohistochemistry in a cohort of 6556 breast cancer tissues. Clin Breast Cancer 5(1):63–69
27. Slamon DJ, Clark GM, Wong SG et al (1987) Human breast cancer: correlation of relapse and survival with amplification of the her-2/neu oncogene. Science 235(4785):177–182
28. Ross JS, Fletcher JA, Linette GP et al (2003) The her-2/neu gene and protein in breast cancer 2003: biomarker and target of therapy. Oncologist 8(4):307–325
29. Geyer CE, Forster J, Lindquist D et al (2006) Lapatinib plus capecitabine for her2-positive advanced breast cancer. N Engl J Med 355(26):2733–2743
30. Piccart-Gebhart MJ, Procter M, Leyland-Jones B et al (2005) Trastuzumab after adjuvant chemotherapy in her2-positive breast cancer. N Engl J Med 353(16):1659–1672
31. Slamon DJ, Leyland-Jones B, Shak S et al (2001) Use of chemotherapy plus a monoclonal antibody against her2 for metastatic breast cancer that overexpresses her2. N Engl J Med 344(11):783–792
32. Sircoulomb F, Bekhouche I, Finetti P et al (2010) Genome profiling of erbb2-amplified breast cancers. BMC Cancer 10:539
33. Staaf J, Jonsson G, Ringner M et al (2010) High-resolution genomic and expression analyses of copy number alterations in her2-amplified breast cancer. Breast Cancer Res 12(3):R25
34. Letessier A, Sircoulomb F, Ginestier C et al (2006) Frequency, prognostic impact, and subtype association of 8p12, 8q24, 11q13, 12p13, 17q12, and 20q13 amplifications in breast cancers. BMC Cancer 6:245
35. Garcia MJ, Pole JC, Chin SF et al (2005) A 1 mb minimal amplicon at 8p11-12 in breast cancer identifies new candidate oncogenes. Oncogene 24(33):5235–5245
36. Reis-Filho JS, Simpson PT, Turner NC et al (2006) Fgfr1 emerges as a potential therapeutic target for lobular breast carcinomas. Clin Cancer Res 12(22):6652–6662
37. Sircoulomb F, Nicolas N, Ferrari A et al (2011) Znf703 gene amplification at 8p12 specifies luminal b breast cancer. EMBO Mol Med 3(3):153–166
38. Bostner J, Ahnstrom Waltersson M, Fornander T et al (2007) Amplification of ccnd1 and pak1 as predictors of recurrence and tamoxifen resistance in postmenopausal breast cancer. Oncogene 26(49):6997–7005
39. Holm K, Staaf J, Jonsson G et al (2012) Characterisation of amplification patterns and target genes at chromosome 11q13 in ccnd1-amplified sporadic and familial breast tumours. Breast Cancer Res Treat 133(2):583–594
40. Hughes-Davies L, Huntsman D, Ruas M et al (2003) Emsy links the brca2 pathway to sporadic breast and ovarian cancer. Cell 115(5):523–535
41. Karlsson E, Waltersson MA, Bostner J et al (2011) High-resolution genomic analysis of the 11q13 amplicon in breast cancers identifies synergy with 8p12 amplification, involving the mtor targets s6k2 and 4ebp1. Genes Chromosomes Cancer 50(10):775–787
42. Kwek SS, Roy R, Zhou H et al (2009) Co-amplified genes at 8p12 and 11q13 in breast tumors cooperate with two major pathways in oncogenesis. Oncogene 28(17):1892–1903
43. Albertson DG, Ylstra B, Segraves R et al (2000) Quantitative mapping of amplicon structure by array cgh identifies cyp24 as a candidate oncogene. Nat Genet 25(2):144–146
44. Bergamaschi A, Kim YH, Wang P et al (2006) Distinct patterns of DNA copy number alteration are associated with different clinicopathological features and gene-expression subtypes of breast cancer. Genes Chromosomes Cancer 45(11):1033–1040
45. Staaf J, Jonsson G, Ringner M et al (2007) Normalization of array-cgh data: influence of copy number imbalances. BMC Genomics 8:382
46. Autio R, Hautaniemi S, Kauraniemi P et al (2003) Cgh-plotter: matlab toolbox for cgh-data analysis. Bioinformatics 19(13):1714–1715
47. Hupe P, Stransky N, Thiery JP et al (2004) Analysis of array cgh data: from signal ratio to gain and loss of DNA regions. Bioinformatics 20(18):3413–3422
48. Venkatraman ES, Olshen AB (2007) A faster circular binary segmentation algorithm for the analysis of array cgh data. Bioinformatics 23(6):657–663
49. Peiffer DA, Le JM, Steemers FJ et al (2006) High-resolution genomic profiling of chromosomal aberrations using infinium whole-genome genotyping. Genome Res 16(9):1136–1148
50. Assie G, Laframboise T, Platzer P et al (2008) Snp arrays in heterogeneous tissue: highly accurate collection of both germline and somatic genetic information from unpaired single tumor samples. Am J Hum Genet 82(4):903–915
51. Popova T, Manie E, Stoppa-Lyonnet D et al (2009) Genome alteration print (gap): a tool to visualize and mine complex cancer genomic profiles obtained by snp arrays. Genome Biol 10(11):R128

52. Van Loo P, Nordgard SH, Lingjaerde OC et al (2010) Allele-specific copy number analysis of tumors. Proc Natl Acad Sci USA 107(39): 16910–16915
53. Beroukhim R, Getz G, Nghiemphu L et al (2007) Assessing the significance of chromosomal aberrations in cancer: methodology and application to glioma. Proc Natl Acad Sci USA 104(50):20007–20012
54. Mermel CH, Schumacher SE, Hill B et al (2011) Gistic2.0 facilitates sensitive and confident localization of the targets of focal somatic copy-number alteration in human cancers. Genome Biol 12(4):R41
55. Hicks J, Krasnitz A, Lakshmi B et al (2006) Novel patterns of genome rearrangement and their association with survival in breast cancer. Genome Res 16(12):1465–1479
56. Chin SF, Teschendorff AE, Marioni JC et al (2007) High-resolution acgh and expression profiling identifies a novel genomic subtype of er negative breast cancer. Genome Biol 8(10):R215
57. Jonsson G, Naylor TL, Vallon-Christersson J et al (2005) Distinct genomic profiles in hereditary breast tumors identified by array-based comparative genomic hybridization. Cancer Res 65(17):7612–7621
58. Joosse SA, Van Beers EH, Tielen IH et al (2009) Prediction of brca1-association in hereditary non-brca1/2 breast carcinomas with array-cgh. Breast Cancer Res Treat 116(3):479–489
59. Holstege H, Horlings HM, Velds A et al (2010) Brca1-mutated and basal-like breast cancers have similar acgh profiles and a high incidence of protein truncating tp53 mutations. BMC Cancer 10:654
60. Cortesi L, Turchetti D, Bertoni C et al (2000) Comparison between genotype and phenotype identifies a high-risk population carrying brca1 mutations. Genes Chromosomes Cancer 27(2):130–135
61. Lakhani SR, Gusterson BA, Jacquemier J et al (2000) The pathology of familial breast cancer: histological features of cancers in families not attributable to mutations in brca1 or brca2. Clin Cancer Res 6(3):782–789
62. Lakhani SR, Jacquemier J, Sloane JP et al (1998) Multifactorial analysis of differences between sporadic breast cancers and cancers involving brca1 and brca2 mutations. J Natl Cancer Inst 90(15):1138–1145
63. Stefansson OA, Jonasson JG, Olafsdottir K et al (2011) Cpg island hypermethylation of brca1 and loss of prb as co-occurring events in basal/triple-negative breast cancer. Epigenetics 6(5):638–649
64. Joosse SA, Brandwijk KI, Mulder L et al (2011) Genomic signature of brca1 deficiency in sporadic basal-like breast tumors. Genes Chromosomes Cancer 50(2):71–81
65. Saal LH, Gruvberger-Saal SK, Persson C et al (2008) Recurrent gross mutations of the pten tumor suppressor gene in breast cancers with deficient dsb repair. Nat Genet 40(1):102–107
66. Smid M, Hoes M, Sieuwerts AM et al (2011) Patterns and incidence of chromosomal instability and their prognostic relevance in breast cancer subtypes. Breast Cancer Res Treat 128(1):23–30
67. Thompson PA, Brewster AM, Kim-Anh D et al (2011) Selective genomic copy number imbalances and probability of recurrence in early-stage breast cancer. PLoS One 6(8):e23543
68. Borg A, Tandon AK, Sigurdsson H et al (1990) Her-2/neu amplification predicts poor survival in node-positive breast cancer. Cancer Res 50(14):4332–4337
69. Carter SL, Eklund AC, Kohane IS et al (2006) A signature of chromosomal instability inferred from gene expression profiles predicts clinical outcome in multiple human cancers. Nat Genet 38(9):1043–1048
70. Farmer H, Mccabe N, Lord CJ et al (2005) Targeting the DNA repair defect in brca mutant cells as a therapeutic strategy. Nature 434(7035):917–921
71. Fong PC, Boss DS, Yap TA et al (2009) Inhibition of poly(adp-ribose) polymerase in tumors from brca mutation carriers. N Engl J Med 361(2):123–134
72. Silver DP, Richardson AL, Eklund AC et al (2010) Efficacy of neoadjuvant cisplatin in triple-negative breast cancer. J Clin Oncol 28(7):1145–1153

Chapter 19

Applications of Array-CGH for Lung Cancer

Kenneth J. Craddock, Wan L. Lam, and Ming-Sound Tsao

Abstract

This chapter summarizes the current knowledge on gene copy number changes found in lung tumors, and their application in the diagnosis, prognostication, and prediction of response to chemotherapy. Examples of the identification of specific "driver" oncogenes within amplified DNA segments are described. A model of how array-CGH could be integrated clinically into the routine workup of lung cancers in clinical laboratory is proposed.

Key words: Microarray, CGH, Lung cancer, Diagnosis, Prognosis, Predictive, Application

1. Introduction

Lung cancer is the leading cause of cancer-related death worldwide for men, and second leading cause for women (1). In 2008, lung cancer accounted for 13% (1.6 million) of all cancer deaths worldwide (1). Despite recent data showing decreasing incidence in developed countries and notable advancements in its treatment and early detection, lung cancer is projected to be the sixth leading cause of death by 2030, advancing from the ninth position in 2002 (2). In North America, the mortality rate of lung cancer is greater than the three next most common cancers combined, colorectal, breast, and prostate cancer (3). Smoking is well known as the most important etiology for lung cancer, but many other environmental carcinogens have also been implicated, including arsenic, nickel, and radon (4). It is estimated that the health benefits of smoking cessation would take approximately 20 years to realize, since as many as half of current new lung cancer patients are former smokers who have quit for more than 10 years (171) (5). Importantly,

Diponkar Banerjee and Sohrab P. Shah (eds.), *Array Comparative Genomic Hybridization: Protocols and Applications*, Methods in Molecular Biology vol. 973, DOI 10.1007/978-1-62703-281-0_19, © Springer Science+Business Media, LLC 2013

lung cancers of never-smokers demonstrate distinct clinical features compared to those of smokers, accounting for 25% of lung cancer cases worldwide (6). Thus, lung cancer will remain a major global health burden for decades to come.

Currently, the overall 5-year survival rate for all lung cancer patients is about 15% (3). This poor rate is largely due to diagnosis at a late stage of disease; about two-thirds of patients are initially diagnosed at a stage too advanced to be curable by surgery. The management of lung cancer patients is largely based on the histologic type and stage of the disease. Table 1 summarizes the major categories of malignant lung tumors in the WHO classification. Two major histological types with different biology and therapeutic responses are recognized: small cell lung cancer (SCLC) and non-small cell lung cancer (NSCLC) (4). SCLC is a highly aggressive cancer with almost all patients presenting with systemic metastases at diagnosis. Therefore, SCLC patients are treated primarily by chemotherapy, to which it is highly sensitive. Yet, despite high initial response rate, very few SCLC patients are cured. The survival of SCLC is correlated with the tumor stage at diagnosis, (7) and the incidence of SCLC has decreased significantly during the last 30 years (8). In contrast, approximately one third of NSCLC patients are diagnosed with the tumor being localized still to the lung, and amenable to surgical resection (9). Therefore, the greatest potential in reducing lung cancer mortality would come from an effective screening program, to discover the cancers in their earliest stage. A recent randomized lung cancer screening trial comparing chest X-ray and low dose spiral computed tomography (CT) has confirmed the effectiveness of the latter detection technique, which is capable of detecting sub-centimeter cancers (10).

Nevertheless, the risk of these "early stage" NSCLC patients to develop metastatic recurrences within 5 years after surgery is still 30–60% (9). While adjuvant chemotherapy can improve significantly the survival of these surgically treated patients, there remains a need for biomarkers to identify poorer prognosis patients, for whom adjuvant chemotherapy might improve survival (11).

For advanced or recurrent NSCLC patients, the standard first line treatment has been chemotherapy, with overall response rates of 30–40% (12, 13). More recently, however, it has been demonstrated that more effective therapies can be achieved by drugs that are designed to target specific molecule or pathway to which a tumor is "addicted." Two such targets have so far been identified, epidermal growth factor receptor (EGFR) (14–16) and anaplastic lymphoma kinase (ALK) (17, 18). The protein product of these genes become bona fide targets when they are activated constitutively, by activating mutations in the kinase domain, high level amplification of the gene copy number, or a translocation which may bring the gene under a new transcriptional control, or form a novel chimeric fusion gene. Clinical trials have demonstrated that

Table 1
Major WHO categories of malignant lung tumors

Squamous cell carcinoma (SCC)		
Small cell carcinoma (SCLC)		
Adenocarcinoma (ADC)	Mixed subtype	
	Acinar adenocarcinoma	
	Papillary adenocarcinoma	
	Bronchioloalveolar carcinoma	Nonmucinous
		Mucinous
		Mixed nonmucinous and mucinous or indeterminate
	Solid adenocarcinoma with mucin production	
Large cell carcinoma	Large cell neuroendocrine carcinoma (other undifferentiated large cell carcinomas)	
Carcinoid tumor		
Preinvasive lesions	Squamous carcinoma in situ	
	Atypical adenomatous hyperplasia	
	Diffuse idiopathic pulmonary neuroendocrine cell hyperplasia	

when patients with tumors expressing these aberrant genes are treated by drugs that are highly specific for blocking their kinase activities, response rates of >60% have been achieved. Targeted therapies can improve cancer treatment outcome only when they are used selectively on patients that harbor the targeted aberrant proteins. Thus, in order to tailor therapy, cancer diagnostics need to include profiling of these specific genetic abnormalities including gene copy number aberrations (CNAs) and mutations. This chapter reviews the findings to date on gene copy aberrations that have been identified in lung cancers, using the array comparative genomic hybridization (aCGH) technique.

2. Array CGH Platforms and Samples

The principle of array CGH and its application to investigate genetic alterations in human diseases has been well documented (19, 20). Typically, DNA from a tumor and a reference sample are differentially labelled and co-hybridized to a microarray containing

DNA probe elements representing multiple loci of the genome. Alternately, a single sample can be hybridized to individual arrays, and hybridization signals can be imaged and compared to other samples computationally. In the early aCGH studies of genetic alterations in lung cancer, array platforms focusing on specific regions of chromosomes were commonly used (21–23). As aCGH technology has advanced, arrays with genome wide coverage have become the platforms of choice, (24–27) and arrays with single nucleotide polymorphism (SNP) based oligonucleotide probe elements are now used to deduce allelic information in addition to DNA dosage. However, the resolution of array platforms is significantly influenced by the density and the distribution of probes throughout the genome (28). Most platforms have been applied to fresh or frozen tumor or cell line material, but occasional studies have successfully been carried out on formalin-fixed, paraffin-embedded (FFPE) tissue. Array CGH platforms that have been used for investigating lung cancer genomes are summarized in Table 2, including which platforms have successfully reported results from FFPE materials.

Comparison of tumor DNA against noncancerous tissue from the same patient yields somatic alterations in the tumor genome. However, matched normal DNA is not always available for comparison, especially for archival material. The genome profiles of the HapMap panel are often used as a common reference (29–31). Differences between a tumor and the reference profile may be attributed to copy number variation (CNV) between individuals rather than somatic alteration. Much effort has been applied to catalogue the remarkable variations in segmental DNA copy in the human population (32, 33).

3. Data Analysis and Interpretation: Tools, Software

A variety of bioinformatics resources for display and analysis of copy number data are available in the public domain (19, 32, 34–36). See GH was the first whole genome copy number profile display software developed, released in 2004 accompanying the advent of whole genome tiling path array technology (24, 37). Array CGH software development has continued to be an area of active research, especially in data processing, analysis and visualization (38–44). Fast and sensitive algorithms for segmentation and calling are needed as array platforms increase in resolution and CGH data becomes increasingly voluminous (34). Copy number variation data is rapidly accumulating, and human copy number variation databases are becoming more comprehensive. DECIPHER (Database of Chromosomal Imbalance and Phenotype in Humans Using Ensembl Resources) and DGV (Database of Genomic Variations) (35) are prominent examples. To date there are no public

Table 2
Platforms used in array-CGH publications of lung tumors

Platform	Type of array	No of loci	No of references	FFPE samples tested?
Stanford	cDNA	22,000–39,632	3	
Affymetrix CentXba and CentHind	SNP	115,000	1	
Affymetrix Mapping 250 K Sty2 SNP Array	SNP	238,000	5	
Affymetrix GeneChip 5.0	SNP	784,544	1	
Affymetrix GeneChip Human Mapping 50 K Array Hind 240	SNP	114,000	1	
Affymetrix Genome-Wide Human SNP 6.0 Array	SNP	1,800,000	2	
Affymetrix Mapping 250 K Nsp SNP Array	SNP	250,000	4	
Affymetrix_Mapping250K_Sty	SNP	238,000	1	
Agilent 11,367 cDNA clones	cDNA	11,367	1	
Agilent Human 1	cDNA	12,814	2	
Agilent Human 1 M	Oligonucleotide	1,000,000	1	
Agilent Human 244 k	Oligonucleotide	244,000	3	
Agilent Human 44 k	Oligonucleotide	44,000	1	
Agilent-014693 Human Genome CGH Microarray 244A	Oligonucleotide	244,000	3	
Agilent-014698 Human Genome CGH Microarray 105A	Oligonucleotide	99,000	1	Yes
Agilent-014850 Whole Human Genome Microarray 4×44K	Oligonucleotide	44,000	1	
Agilent-014950 Human Genome CGH Microarray 4×44K	Oligonucleotide	44,000	1	
BCCR, Vancouver, SMRT array	BAC	26,363	1	Yes
BCCR, Vancouver, SMRT array v.2	BAC	32,433	5	Yes
Chungnam University, Korea	BAC/P1	4,046	1	
CNIO H. sapiens 13.6 k Oncochip 1		13,600	1	
custom MCG Cancer Array-800, Tokyo	BAC/P1	800	1	Yes
GeneChip Human Mapping 10-K and 250-K		250,000	1	
Illumina BeadChip 600 K	SNP	620,901	1	

(continued)

Table 2 (continued)

Platform	Type of array	No of loci	No of references	FFPE samples tested?
Illumina HumanOmni1-Quad BeadChip	SNP	1,134,514	1	
Konyang-homo sapiens-Macarrray4.5 k-v1.0	BAC/P1	4,362		
MacArray Karyo1400, Macrogen, Inc., Seoul	BAC	1,440	2	
MacArray Karyo4000, Macrogen, Inc., Seoul	BAC	4,046	2	
MCG CancerArray-800 ver.2	BAC	800	1	
MDACC, Houston, Research Genetics, Invitrogen, Huntsville	cDNA	8,000	1	
NimbleGen Human CGH 385 K Whole-Genome Tiling Array v1.0		385,000	1	
Peter Mac	BAC	2,400	1	
UCSF, San Francisco	BAC/P1	800–2,464	2	Yes
UMCG_6.5KBAC_ARRAY	BAC	6,500	2	
VUMC MACF human 30 K oligo	Oligonucleotide	25,000	2	

copy number databases developed specifically catering to the analysis of cancer, although such a resource would be invaluable for interpreting copy number data in the clinical setting.

4. Gene Copy Number Aberrations Detected by Array-CGH in Lung Cancer

Similar to other epithelial malignancies, the karyotypes of NSCLC invariably show polyploidy (in the range of 58–102 chromosomes per cell) and multiple complex chromosomal aberrations, resulting in net gain or loss of genetic material, indicative of genomic instability (45). Lung cancer is a very heterogeneous disease, yet the CNA profiles of the major histologic subtypes of NSCLC are remarkably similar, with frequent gains involving 5p, 8q, 3q, and 1q, frequent losses at 3p, 8p, 9p, 13q, and 17p (46). Amplifications are commonly observed in the form of double minutes. Knowledge of the order or progression of these aberrations remains very preliminary; some investigators have speculated that early events include trisomy 7, loss at 3p, and trisomy 12.

Array-CGH has allowed researchers to identify and refine CNAs in lung cancer in great detail. The high resolution of this technique is clarifying genomic amplification and deletion to regions often containing only a few genes, as well as identifying small, previously undetected aberrations. As a result, the list of genes implicated in NSCLC pathobiology is growing rapidly. Tonon et al. (47) identified 93 minimal common regions (MCRs) of aberration in 44 NSCLC tumors and cell lines, 21 of which spanned less than 0.5 Mb with a median of 5 genes in each, with virtually all genes previously implicated in NSCLC pathogenesis present within these regions, as well as many novel candidate genes. In a large study of 371 adenocarcinomas using SNP array-CGH, Weir et al. (26) identified 26 recurrent large-scale events involving gain or loss of at least half of a chromosome, together comprising more than half of the human genome. In addition, 31 focal amplifications and homozygous deletions were identified, including multiple novel candidate genes. The most commonly amplified regions in lung cancer include *MYC*, *TERT*, *CCND1*, and *EGFR*, but many other targets of amplification have been identified in smaller percentages of cases. Table 3 lists the major amplifications and deletions found in lung cancer, along with any associations with histology, and their associated oncogenes and tumor suppressor genes (26, 47–49). Figure 1 shows the frequency of gains and losses across the genome of NSCLC.

5. Squamous Cell Carcinoma (SCC) and Adenocarcinoma (ADC)

Many aCGH studies have focused their analysis on detecting differences in genomic copy number profile between different histologic types of lung cancer. A majority have examined specifically SCC and ADC, the two most common histologic types of NSCLC. Overall the patterns of aberrations were similar between adenocarcinoma (ADC) and squamous cell carcinoma (SCC); supervised or unsupervised clustering methods have for the most part been unable to differentiate the two. However, a number of distinct associations have been identified. Figure 2 shows the copy number changes across the genome, with comparison of SCC to ADC.

By far the most striking difference has been the gains of chromosome 3 long arm material spanning a large 30 Mb region including bands q25 to q29 (47, 50, 51). As well as being quite specific for SCC, gains at 3q occur very likely in early stages of the cancer (51, 52).

The association of 3q gain with SCC was previously recognized by classic chromosomal CGH and other molecular cytogenetic experiments. Microarray studies have confirmed these findings, and furthermore, they have allowed researchers to define

Table 3
Major amplifications and deletions in lung cancer

Amplification (A) or deletion (D)	Cytoband	Histology	Candidate genes
A	1p34.3-p34.2	NSCLC	BMP8B, COL9A2, MYCL1, PPT1, RLF, ZMPSTE24, PPIE, CAP1, HEYL, HPCAL4, TRIT1, OXCT2, NT5C1A, MFSD2, TMCO2, FLJ31434, YRDC, FLJ45459
A	1q32.2-q21.2	All	ARNT, LAMB3
A	2q14.2	SCC	GLI2, TFCP2L1
A	3q11.2-q29	SCC	PIK3CA, TP63, SOX2, DCUN1D1, EVI1, MDS1, TP73L, FGF12, SST, TGFA, and GLUT2
A	5p15.33-p14.3	All	BASP1, DAP, CTNND2, TERT, BRD9, IRX2, IRX1, ADAMTS16, MED10, SRD5A1, POLS, FASTKD3
A	6p21.1-p12.1	NSCLC	TINAG, C6orf142
A	7p11.2	NSCLC	**EGFR**, SEC61G, VSTM2A
A	7q11.21-q11.22	SCLC	PMS2L3, HIP1, ELN, FZD9, LIMK1, BAZ1B, GTF2IRD1, GTF2I, GTF2IRD2, GTF2IRD2B
A	8p12	SCC	FGFR1, FLJ43582, WHSC1L1, LETM2, BRF2
A	8q24.21	NSCLC	**MYC**
A	11q13.2-q13.3	NSCLC	CCND1
A	12p12.1	NSCLC	**KRAS**, BCAT1, LRMP, SOX5, FLJ32894, LOC196415
A	12q13.2-q24.23	All	ERBB3, CDK4, MDM2, DYRK2
A	14q13.3	ADC	**NKX2-1**, MBIP
A	14q32.13–14q32.31	All	KIAA1622, RTL1
A	17q24.3-q25.3	SCLC	SOX9, CBX2
A	18q11.2-q12.3	NSCLC	hsa-mir-1-2, SNRPD1, MIB1, ESCO1, ABHD3, DSC1, DSC2, DSG2, PIK3C3, RIT2, SYT4
A	19q12-q13.33	NSCLC	CCNE1, C19orf12, LTBP4, NUMBL, SPTBN4, ADCK4, ITPKC, SHKBP1
A	20q11.21-q13.32	NSCLC	BCL2L1, TPX2
A	22q11.21-q12.2	NSCLC	MTMR3, ASCC2, HORMAD2
A	3p14.1-p12.3	SCC	GBE1, ROBO1, ROBO2, MAGI1
D	4q21.3-q32.2	SCC	MAPK10, GPR103, PCDH18, RAPGEF2, FSTL5
D	7q11.22-q34	NSCLC	TAC1, LOC154761

(continued)

Table 3 (continued)

Amplification (A) or deletion (D)	Cytoband	Histology	Candidate genes
D	9p21.3	NSCLC	**CDKN2A**, **CDKN2B**, DMRTA1
D	10q23.2-q23.31	NSCLC	**PTEN**, ATAD1
D	11q11-11q21	All	OR4C11, OR4C6, OR4V1P†, MAML2
D	13q12.11-13q32.3	All	**RB1**, HSMPP8, PSPC1, FOXO1, ELF1, ITM2B, FARP1

Adapted from (26, 47–49)

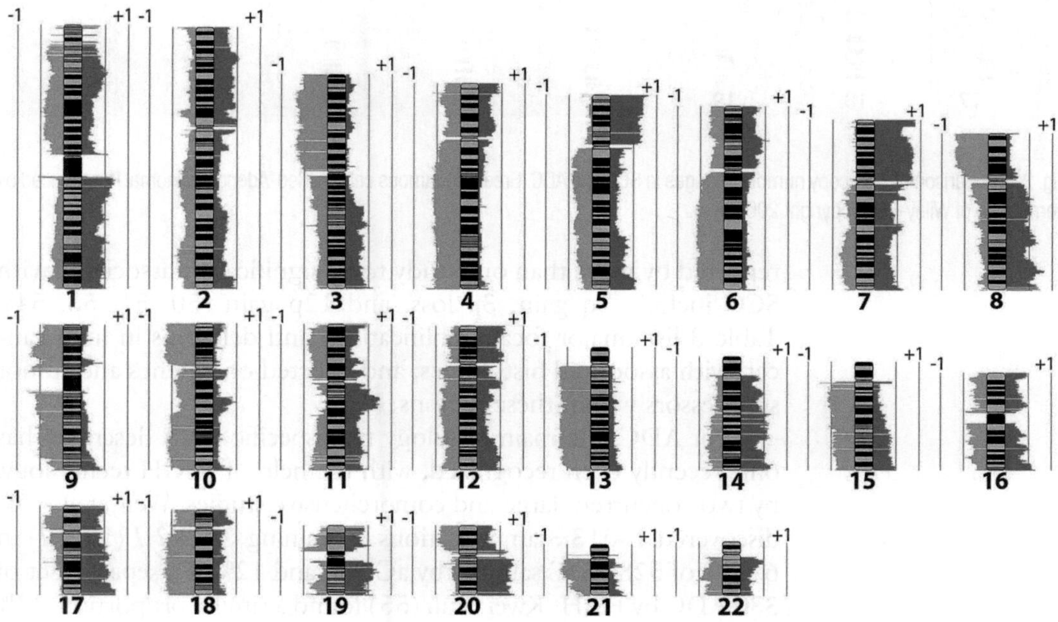

Fig. 1. Common DNA copy number changes in NSCLC. Frequency of gains and losses are plotted to the right and left of the chromosome location, respectively. Reproduced by permission of Wiley-Liss copyright 2005 (54).

the involved segments at a sufficiently high resolution to begin identifying with confidence, individual oncogenes that are the likely target(s) of amplification. Candidate oncogenes at this region have included *PIK3CA, TP63, SOX2, DCUN1D1, EVI1, MDS1, TP73L, FGF12, SST, TGFA,* and *GLUT2*. However, for most of these, the oncogenic role remains to be determined.

A region of 8p12 has been reported to be amplified in 10–15% of SCC, while ADC samples in contrast often show losses involving 8p (52). The identification of potential oncogenes at this location, including FGFR1, BRF2, and WHSC1L1, is discussed in later sections of this chapter. Other copy number alterations that have been

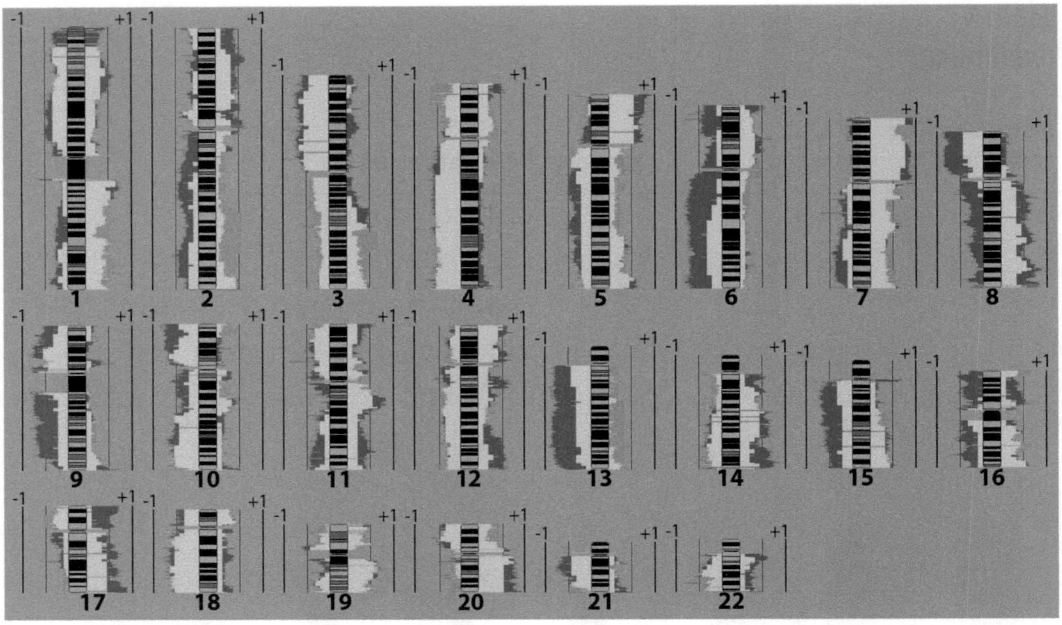

Fig. 2. Comparison of DNA copy number changes in SCC and ADC. Green-squamous cell CA, Red-Adenocarcinoma. Reproduced by permission of Wiley-Liss copyright 2005 (54).

reported by more than one study to be significantly associated with SCC include 2q gain, 3p loss, and 12p gain (50, 51, 53, 54). Table 3 lists major focal amplifications and deletions in lung cancer, with associated histologies, and selected oncogenes and tumor suppressors within these regions.

For ADC, the main histology type-specific CNA described has only recently been recognized, with the help of aCGH technology, by two concurrent large and comprehensive studies. Weir et al. (26) discovered 14q13.3 amplifications containing *NKX2-1* (*TITF1*) in 6–12% of 528 ADC samples by aCGH, and 12% of a separate set of 330 ADC by FISH. Kwei et al. (55) found a similar proportion, 11% of their 36 ADC, while only 1 of 40 SCC showed this amplification by aCGH. Notably, this amplification was not identified in a total of 385 cancers of the breast, prostate, colon, and pancreas, providing further evidence for its specificity for lung ADC.

Kang et al. (51) applied a targeted BAC array to 36 primary NSCLC and found alterations at the following loci to be significantly associated with ADC histology: gains at 1q, 6p, and 7q; and losses at 9q, 10p, and 15q. Garnis et al. (54) used the SMRT tiling-path 32 k bacterial artificial chromosome (BAC) arrays produced by the British Columbia Cancer Research Centre (BCCRC) to study 28 NSCLC cell lines. Alterations that were identified specific to ADC-derived lines included gains at 3q22 (proximal to the SCC-associated region at 3q26-29), 12, and 14; and losses at 2q and 13q. Newnham et al. (50) studied 69 frozen NSCLC samples from stage I–IIIA lesions excised with curative intent, using the BAC array generated at the Peter McCallum Cancer Centre. The copy number alterations found to be significantly associated with ADC in their dataset were gains of 6p21.31-25.1, and losses of 6q13-27. In addition, they found a significant tendency to greater magnitude of change in several regions of shared alterations with SCC: gains at 7p, 7q, and 8q; and losses at 9p, 13q, 17p, 18q, 19p, 19q, and 22q.

Advanced statistical methods have also been employed in the analysis of copy number changes between histologic subtypes. Massion et al. (53) applied agglomerative hierarchical clustering using the 100 most informative array elements to accurately discriminate between ADC and SCC. Shibata et al. (56) applied unsupervised hierarchical clustering using the Impressionist and GeneMaths software packages to aCGH data from a series of 55 microdissected, methanol-fixed, paraffin-embedded ADC samples of various stages (I–III), and found three distinct subcategories of ADC. One of these exhibited significantly fewer chromosomal number aberrations (CNAs) overall, but more frequent gains at 19q13.1 and losses on 22q12.2. The other two clusters showed many overlapping CNAs, but one showed more frequent gains at 1p25-32, 4p16.3, 11p15, 12q13-14, 16p11.2-13.3, 17q11.1-25, 19q13.2, 20p11, 20q11.2,and 22q12.2, and losses on 1p22, 6q26, 10q24-26, 13q22.1-34, 15q21-25, and 18p11.2; and the other had more frequent gains of 5p12-14.3, 7p12.3-21.1, 7q22, 7q31, 8q12-21, and 14q11-24, and losses at 1q23.3-41, 10q22.1, and Xq. These clusters were significantly associated with smoking habits, but not with any other clinicopathologic features, including histology, stage, or survival. Validation of these clusters on a separate dataset has not been performed in either of the above-mentioned studies.

6. Small Cell Lung Carcinoma

A major obstacle in studying the genetic lesions in SCLC has been lack of adequate tissue for study, since these tumors are rarely surgically resected. As a result, aCGH studies conducted using primary SCLC samples are relatively scarce, and instead many researchers have focused on examining cell lines. However, there are a few aCGH studies that have tested a number of primary tumor samples. Refer to Table 3. for common amplifications and deletions that have been associated with SCLC.

Voortman et al. (49) used the Agilent-014698 Human Genome CGH Microarray 105A (G4412A) and Agilent-022060 SurePrint G3 Human CGH Microarray 4 × 180 K platforms to examine the CNAs of 33 primary SCLC tumors and 13 SCLC cell lines, and found extensive sharing of genomic alterations, except for greater

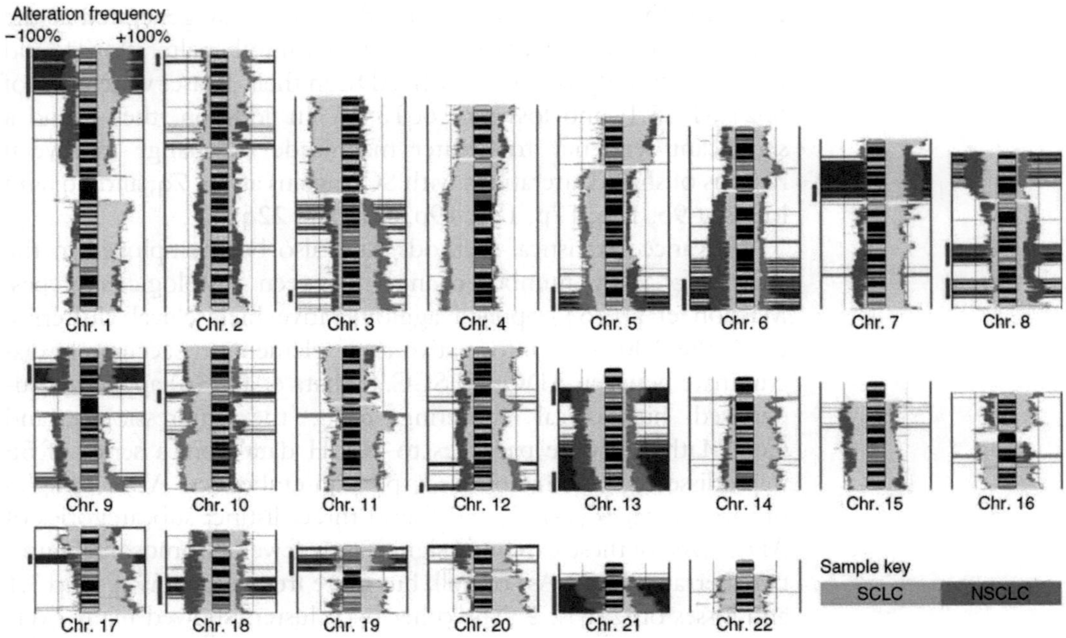

Fig. 3. Comparison of copy number changes in SCLC and NSCLC. Reproduced with permission of Nature Publishing Group (57).

copy number (CN) losses on chromosome 2 in cell lines. Recurrent CN gains were observed on chromosomes 1, 3q, 5p, 6p, 12, 14, 17a, 18, 19, and 20, while recurrent CN losses were noted on chromosomes 3p, 4, 5q, 10, 13, 16q, and 17p. Among ~20,000 genes analyzed, only ~4% differed in terms of presence of recurrent CNAs between primary and cell lines. Importantly among the tumor samples, primary ($n = 19$) and metastatic ($n = 14$) tumors demonstrated similar CNAs.

There have been very few studies that compared the CNAs of SCLC and NSCLC by aCGH. Coe et al. (57) examined a large group of 41 cell lines of SCLC and NSCLC using the BCCRC SMRT tiling 32 k BAC array. They found some striking similarities including gains at 5p, 7p, and 11q and losses at 3p, 4q, but also differences at multiple loci on 1p, 2p, 3q, 5q, 6q, 7p, 8q, 9p, 10q, 12q, 13q, 17q, 18p, 18q, 19p, and 21q (Fig. 3). The most striking among these were very frequent gains at 21q11.2-22.3 in SCLC, while the same region was frequently deleted in NSCLC lines. By expression microarray profiling, the genes TIAM1 and DSCAM, involved in MAPK signaling pathways, were found to be differentially upregulated in SCLC as compared to NSCLC in this region. The lack of 8q21.2-22.3 gains in SCLC lines also stood out compared to the very frequent gains in NSCLC at this region. They concluded that cell cycle upregulation in SCLC and NSCLC occurs through drastically different mechanisms, suggesting a need for differential therapeutic target selection in these two malignancies.

Peng et al. (58) examined 10 laser microdissected primary formalin-fixed and paraffin embedded (FFPE) SLCL samples using the custom targeted MCG Cancer Array-800, and found that gains at 18q11-22 and losses at 1p and 16q were infrequently reported in NSCLC, suggesting that these regions might play a pathogenetic role specific to high-grade neuroendocrine tumors. These findings are similar to those reported with conventional CGH analyses (45, 59).

7. Large Cell Neuroendocrine Carcinoma (LCNEC)

Peng et al. (58) also examined 31 primary LCNEC samples in their study and observed similar frequencies of most frequent chromosomal alterations in both SCLC and LCNEC, consistent with results of previous allelotyping, cCGH, and karyotyping analyses. However, losses at 3p and 4q were significantly associated with SCLC, while 2q gains and 6p losses were associated with LCNEC. These CNAs were present even in early stage I tumors, suggesting that they may be early events in tumor development and implicating them in the histologic difference between these two subtypes.

8. Bronchial Carcinoid

Aside from SCLC, Voortman et al. (49) also studied 19 carcinoid tumors of bronchial and 9 of gastrointestinal origin. When comparing to SCLC, the carcinoids revealed significantly less complex alterations, which also tended to be localized to narrower gene regions. CNAs that were very infrequently observed in carcinoids included 17p loss involving *TP53*, and amplifications involving *MYC* gene family members. High copy number gains that were found mainly in four distinct cytobands in SCLC were not observed in carcinoids. Deletion of 11q13 involving the *MEN1* gene, on the other hand, was relatively specific for bronchial carcinoids, and much less frequent in carcinoids of the GI tract. A number of other alterations, involving 203 genes and 59 microRNAs, were common to both tumor types, including most notably loss of *RB1* and gain of *DLK1-DIO3* domain. The latter has been postulated to be specific for neuroendocrine tumors.

9. Pleural Mesothelioma

A few studies have applied aCGH technology to pleural mesothelioma, (60–63) which have identified multiple recurrent gains and losses, including but not limited to losses at 1p36, qp21.3, 3p21.3, 4q22, 6q25, 9p21.3, 13q, and 22q, and gains at 1q and 8q. The most frequent amongst these are 9p losses involving CDKN2A and CDKN2B, which was reported in 40–70% of primary mesotheliomas by aCGH, (60, 61) and 70–90% by FISH (64). These changes have been noted to be marginally more frequent in sarcomatoid histology than in epithelioid histology, (64) but none of these have been shown to be specific for mesothelioma, as compared to NSCLC.

Some studies have examined for the presence of these abnormalities, and in particular 9p deletions, between reactive or benign mesothelial proliferations and mesothelioma; these studies have found that that these abnormalities are not found in reactive or benign lesions including mesothelial hyperplasia, pleuritis, adenomatoid tumor, and benign mesothelial multicystic tumor (64, 65).

10. Neoplastic and Malignant Progression in Lung Cancer Development

SCC is thought to arise by multistep carcinogenesis process. Evaluations of heavy smokers by serial bronchoscopies and biopsies have established the histologically defined epithelial lesions that epitomize neoplastic progression leading to SCC. These lesions include squamous metaplasia (SM), increasing grades of dysplasia and carcinoma-in-situ (21, 22, 52, 66, 67) van Boerdonk et al. (68) have determined that copy number alterations at 3p, 3q, 9p, and 17p in SM were predictive of subsequent development of invasive SCC. Along the same vein, Schneider et al. (69) have proposed *SOX2* amplification on 3q as a sensitive marker for screening or early detection of SCC in bronchial washings or possibly sputum samples.

Atypical adenomatous hyperplasia (AAH) and bronchioloalveolar carcinoma are regarded as putative precursor lesions of peripheral invasive lung adenocarcinoma (70, 71) bronchioloalveolar carcinoma is defined as ADC with tumor cells growing in a lepidic pattern along preexisting lung alveolar septae, without histological evidence of invasion (4). In a more recent proposal for a new classification of lung ADC, bronchioloalveolar carcinoma has been replaced by the term lung adenocarcinoma in situ (AIS), recognizing the pre-invasive nature of these tumors and the 100% survival of patients following complete surgical resection (72).

To identify CNAs that may be marker of invasion, Aviel-Ronen et al. (25) used the BCCRC tiling path BAC arrays to compare the CNAs of bronchioloalveolar carcinoma to ADC with bronchioloalveolar carcinoma-like features (AWBF). They were able to demonstrate a progression with respect to numbers of CNAs between bronchioloalveolar carcinoma and AWBF, but not between pure bronchiolocarcinoma and bronchioloalveolar carcinoma with focal areas of invasion. They generated a list of 113 differentially gained genes in AWBF compared to bronchioloalveolar carcinoma, and furthermore found that a fair proportion of these genes were prognostic markers for early-stage ADC using databases of publically available expression microarray data.

11. Tumor Lineage and Origin

Array-CGH is potentially a powerful tool for establishing the presence or absence of a clonal relationship between two physically separate tumors or neoplastic populations. Cytogenetic abnormalities by karyotype or FISH are used routinely in leukemia diagnosis and follow-up to determine whether a recurrence is related to the patient's previously diagnosed leukemia, or whether it is a malignancy secondary to previous therapy.

Array-CGH is capable of defining breakpoints related to copy number changes at a high resolution, and aside from some fragile sites to keep in mind, these breakpoints can serve as a "fingerprint" for a given neoplasm. Even for deletions or amplifications that are recurrently seen in a given type of cancer, these abnormalities can occur via a variety of methods and locations, and therefore, the associated breakpoints of these recurrent deletions and amplifications are quite variable from one tumor to another.

As a proof of principle, Buys et al. (73) used the BCCR SMRT tiling BAC-array to demonstrate that a rare 3-component tumor (SCLC, LCNEC, and ADC) was most likely to be a "collision" of 3 separately occurring malignancies, arising in the lungs of a 30 pack-year smoker. By contrast, when they tested two different areas of the SCLC component, distinguishable by divergent neuroendocrine marker expression (CD56 and chromogranin-A), they were able to demonstrate a clear clonal relationship according to multiple shared breakpoints.

Girard et al. (74) tested 24 pairs of NSCLCs, of separate synchronous and metachronous tumors, from 20 patients, using Agilent Human 244 K oligo-arrays. They found that a diagnosis of metastases or multiple primaries could be confidently made in 18 of the 22 paired comparisons. The remaining four cases were considered equivocal as they shared a lower proportion of breakpoints with their partner; however, when these cases were examined further by mutational profiling, the combined information was able to confidently classify these as metastases as well. Importantly, the use of this technique identified four cases (18%) that had been misclassified by the widely used Martini–Melamed guidelines, which are based on tumor histology and location.

12. Gene Copy Number Variation as Prognostic Markers

Relatively little conclusive information is available regarding the application of CNAs detected by aCGH for prognosis in lung cancer. Many studies do not include patient follow-up in their analysis, and among those that do, the patient numbers are often too small to make strong conclusions. Of the associations that have been made, most have not been validated on separate patient cohorts and thus could represent chance associations within their particular dataset.

A few survival associations have been made independently by more than one study. Gains of 8q and loss of 13q material have both been associated with poorer disease-free survival (DFS) in ADC, and with distant metastases in SCC (56, 75). In addition, 8q gains have been associated with prognosis in LCNEC patients (58). Another potential prognostic marker is a deletion at 4q which were found to be associated with lymph node metastases in SCC in one study, as well as presence of disseminated tumor cells in the bone marrow and metastasis to the brain in another study (75, 76).

Shibata et al. (56) studied 55 laser microdissected, methanol-fixed paraffin-embedded ADC samples from stage I to III patients using a targeted BAC array, and found two chromosomal alterations associated with disease-free survival: 8q24.2 gain ($P = 0.02$, hazard ratio, 2.92; 95% confidence interval, 1.16–7.37), and 13q14.1 loss. Suggested candidate genes were *GLI4* and *RECQL4* on 8q and *LCP1* at 13q.

Boelens et al. (75) interrogated laser-microdissected tumor samples from 34 patients who presented with centrally located SCC, with follow-up of at least 5 years, including 15 patients without any metastases, 8 with metastases in regional lymph nodes only, and 11 with metastases exclusively in distant organs within 2 years after surgery. Using a UMCG 6.5 K BAC array, they found significant associations with distant metastases for 8q gains, 8p losses, and 13q losses. Gains at various loci in the 8q22-q24 region ranged between 50 and 60% in patients with distant metastases, and only 7–20% of patients without metastases. Candidate genes at 8q commented by the authors, based on previously reported over-expressed genes included *CDH17, SPAG1, RAD21, MTBP, HAS2*, and *ANGPT1*, while *MYC* was mentioned to be located just outside of the significant region. The presence of lymph node metastases was associated with gains at 7q, 8p, 10q, and 12p, and losses at 4p and 4q. Tumors without metastases at 5 years showed association with gains at 2p, and losses at 11p.

Providing further support for the prognostic value of 8q gains, Iwakawa et al. (77) recently demonstrated an association of an 8q amplicon including *MYC* and *PVT1* with poor prognosis in 105 patients with small-sized ADCs of stages I to III. They went on to validate this finding on another large set of 162 stage I lung ADC, finding an association between 8q gain and poor recurrence-free survival.

Peng et al. (58) used a custom MCG targeted BAC array with 800 clones to study 41 stage I to III high-grade neuroendocrine carcinomas (10 SCLC and 31 LCNEC), and found 10 loci whose aberrations showed significant association with patient outcome, including gains at 8q. The other alterations with significant associations were located at 10q, 18q, 20p, and 20q. A multivariate analysis including sex, age, tumor size, lymph node metastasis, and stage found that loss of 10q was associated with poor prognosis in these tumors.

It is noteable that three of the four above studies showing association of 8q gains with poor patient outcome were in the Japanese population. Given the known geographic and ethnic differences in the occurrence of other genetic aberrations in lung cancer, most notably EGFR mutation, it is possible that these prognostic associations may be ethnicity or region specific.

In agreement with the finding of an association between 4q losses and lymph node metastasis in SCC, Wrage et al. (76) reported supporting evidence for the prognostic utility of losses at 4q by studying primary NSCLC tumors from 30 patients using VUMC MACF human 30 K oligo arrays. They found association between losses at 4q21, 10p, 10q and gains at 17q and 20q with the presence of disseminated tumor cells in the bone marrow, and validated the association of 4q losses by FISH on 29 additional NSCLC tumors, and 36 brain metastases. Loss of 4q was found by FISH in 50% of bone-marrow positive patients compared to 6% of bone marrow-negative patients. Losses at 4q were associated with SCC histology and tumor stage, and in multivariate analysis only bone marrow status and tumor stage remained as independent factors. The study did not identify a specific tumor suppressor gene at this location, but noted that differentially expressed genes clustered within this region into five separate hot spots in the region. Candidate genes in this region noted by Weiss et al. (48) included MAPK10, GPR103, PCDH18, RAPGEF2, and FSTL5.

Some other prognostic associations of aCGH findings in lung cancer have been made by individual studies that require further validation. In their study comparing CNAs of BAC and AWBF, Aviel-Ronen et al. (25) found 24% of 113 genes within the differentially aberrant regions being associated with poor prognosis in early stage ADC patients. In particular, PDCD6 at 5p was found to be an independent prognostic marker for early-stage ADC.

Kang et al. (78) used a MacArray™ Karyo4000K BAC chip to profile DNA extracted from touch preparations of 22 SCC of stages I to III, and found loss of 9p, found in ~35% of tumors, to be strongly associated with lymph node metastases, poorly differentiated histology, and high stage. The array platform allowed them to define two distinct recurrent regions of homozygous deletion on 9p: one spanning a 128 kbp region on 9p21.1, containing putative tumor suppressor genes *CDKN2A/p16*, *MTAP*, and *TEK*, and the other spanning a 200 kbp region on 9p24.3, containing 18 possible targets including putative tumor suppressor genes *DOCK8*, *DMRT1*, and *DMRT3*.

13. Gene Copy Number Variation as Predictive Markers

Applications of aCGH as predictive markers for lung cancer are currently very limited; however, this will likely change in the near future, as more and more molecularly targeted therapeutics are developed for their high efficacy against specific genetic lesions. The latter include activating mutations and/or amplification involving oncogenes, which render tumor cell dependence on these aberrations. The following examples illustrate how focal amplifications can expose specific oncogenes that are driving the growth of a given lung cancer, and how removing this driver using an appropriately directed therapy can result in dramatic treatment response.

EGFR mutations and amplifications in lung ADC have been established as predictive of higher response rate to treatment by EGFR tyrosine kinase inhibitors (TKIs) (14–16, 79, 80). Whether mutation or amplification is a better predictor of response and/or survival outcome has remained controversial, but it is clear that there is a strong correlation between the two aberrations (81, 82). Both are associated with a specific demographic: East Asian, female, never smokers, with adenocarcinomas often showing a bronchioloalveolar histologic pattern (80, 83). Most copy number studies have employed FISH to detect the presence of amplification. In practice most centers have chosen to test for common mutations in order to select patients for EGFR TKI therapy.

Li et al. (82) tested the ability of aCGH to detect *EGFR* amplifications in 77 consecutive patients with ADC, using Affymetrix Genome-Wide Human SNP array 6.0 microarrays. They reported that both amplification by FISH and by aCGH were strongly correlated with presence of *EGFR* mutation in the tyrosine kinase domain (exons 18–21) by direct sequencing ($p=0.001$ and $p<0.001$, respectively), demonstrating that aCGH is a valid method of detecting *EGFR* amplifications. Copy number gain by aCGH was detected in 42/50 (84%) of *EGFR*-mutated cases, and 7/24 (29%) of wild-type cases (42 cases were EGFR-gained and EGFR-mutated; 17 cases had normal copy number and EGFR-wildtype; 7 cases were EGFR-gained and EGFR-wildtype, and 8 cases were normal copy number and EGFR-mutant), suggesting that the detection of common *EGFR* mutations alone may miss a subset of patients who may respond to TKI therapy.

More recently, Weiss et al. (48) have demonstrated a common recurrent amplification in SCC containing *FGFR1* and *FLJ43582*, in 10% of 155 samples studied by the Affymetrix Genome-Wide Human SNP 6.0 Array. *FGFR1* amplifications were significantly associated with current smoking status, and second in frequency only to 3q amplifications involving *SOX2*, which were found to be almost completely mutually exclusive of each other. This lesion was deemed to be specific for SCC, as no significant (4 copies) amplifications were detected by FISH on a separate cohort of 77 ADC. The authors proceeded to build a compelling case for a role of an FGFR inhibitor, PD173074, in tumors with amplifications of this gene. In a series of experiments, they found that 3 of 4 lung cancer cell lines that were sensitive to PD173074 bore amplifications of *FGFR1*, and found an enrichment of *FGFR1* amplification in cell lines that were induced to apoptosis by PD173074. They further demonstrated decreased colony formation in soft agar, specific to amplified cell lines, in response to PD173074, and tumor shrinkage in mice engrafted with *FGFR1*-amplified cell lines.

Olejniczak et al. (84) also provided evidence that SCLC carrying *BCL2* amplifications at 18q are sensitive to Bcl-2 antagonist ABT-737. The group tested this compound against 23 SCLC cell lines, and found that the sensitivity of the cell lines to the drug was significantly associated with BCL2 copy number as assayed by Affymetrix GeneChip Human Mapping 50 K Array Hind 240 platform. The most sensitive cell lines showed the highest BCL2 copy number (4–5 copies). FISH on a tissue microarray of 62 SCLC tumors demonstrated low-level BCL2 gains in 40%, and high level gains in 8%, suggesting that a significant subset of tumors have the potential to respond to this new agent.

These experiments are likely just the tip of the iceberg. For example, based on the results of their aCGH study, Voortman et al. (49) suggested frequent high level amplifications of many genes for which putative inhibitors are currently available and thus could be tested for their activity in SCLC, including *FGFR1*, *JAK2*, and *PIK3CA*.

14. Identification of Oncogenes by Array-CGH

One of the unique strengths or benefits of aCGH studies has been the identification of many important oncogenes and tumor suppressors. Particularly with the most recent set of platforms, focal amplifications and deletions that are recurrent across multiple tumors has often allowed researchers to prioritize and fast track the list of potential gene candidates for time-consuming and costly functional studies required to demonstrate their roles in lung cancer carcinogenesis or biology. In combination with other genome-level studies (expression, methylation, and more recently next-generation sequencing), and with the use of bioinformatics tools and databases, researchers have exploited aCGH to assist in identifying the most important pathways that drive the malignancy of lung cancer.

14.1. 3q Amplicon in Squamous Cell Carcinoma

The task of identifying the oncogene(s) that reside on 3q gains in SCC has proven to be challenging but it has been facilitated by aCGH analyses. Kang et al. (51) were able to narrow the candidate genes down to a shortlist of 13 including *SST* and *FGF12*. Massion et al. (53) focused their attention to *PIK3CA*, demonstrating that PKB activity was increased in SCC as compared to ADC. They went on to suggest that *PIK3CA* gains had an oncogenic effect by increasing PI3K signaling, and postulated that this could make SCC susceptible to PI3K inhibitors. Another study showed that siRNA-mediated TP73L inhibition resulted in reduced cell survival but with maintenance of squamous characteristics (85).

More recently, two concurrent studies identified *SOX2* as an important oncogene amplified within the 3q region in SCC. Bass et al. (27) applied a high-resolution SNP array to 47 SCC of the lung and identified only four genes at the peak of the most significant region on 3q as defined by GISTIC analysis: *SOX2*, *ATP11B*, *DCUN1D1*, and *MCCC1*. Testing of multiple ADCs failed to identify significant *SOX2* amplification. The group was able to demonstrate increased SOX2 mRNA expression in tumors with 3q gains, and proceeded to perform functional studies that supported the important role of *SOX2* in SCC carcinogenesis. Suppression of *SOX2* using shRNA constructs in SCC cell lines showed greater anti-proliferative effect in 3q26.33 amplified cell lines as compared to a number of neighboring gene products that were also tested. Furthermore, *SOX2* suppression decreased colony formation, and induced expression of wild-type *SOX2* restored anchorage-independent growth. In contrast, non-amplified cell lines showed no effect of *SOX2* suppression.

Independently, Hussenet et al. (86) also identified *SOX2* as the most consistently overexpressed gene in the peak of the recurrent 3q amplicon in SCC using a dedicated chromosome 3 tiling BAC/

PAC array. They then demonstrated the ability of induced *SOX2* expression to convert non-tumorigenic human lung bronchial epithelial cells to a tumorigenic phenotype that give rise to poorly differentiated squamous tumors in immunocompromised mice. Furthermore, they showed that silencing of *SOX2* using shRNA induced massive tumor cell death within 10–20 days.

SOX2 has critical roles in foregut development and is hypothesized to be a lineage-survival oncogene for squamous cell differentiation and proliferation. SOX2 is considered one of the "magical four" crucial transcription factors capable of cooperating to reprogram differentiated cells into an induced pluriopotent stem cell-like phenotype.

Evidence is building that multiple genes are likely contributing to oncogenesis at the 3q26-29 region in SCC, which most likely has contributed to the difficulty in fine mapping of 3q alterations in this disease (53). TP63 is another notable gene in this amplified region, because immunohistochemistry for p63 protein is used commonly in clinical laboratories as a sensitive and specific marker for SCC at multiple body sites (87). Massion et al. (88) applied FISH and immunohistochemistry to detect TP63 gene amplification and p63 protein expression in tissue microarrays containing 217 NSCLC samples. They found TP63 copy number ≥3 in 88% of squamous carcinomas, 42% of large cell carcinomas, and 11% of adenocarcinomas of the lung. In their cohort, they found that increased copy number and increased immunostaining intensity were both significantly associated with a better survival in NSCLC, and furthermore they found a significant increase in CN in preinvasive lesions graded severe dysplasia or higher.

14.2. 8p Amplicon in Squamous Cell Carcinoma

A region at cytoband 8p12, including potential targeted oncogenes FGFR1, BRF2, and WHSC1L1, has been demonstrated to be amplified in 10–15% of SCC. A description of the functional experiments by Weiss et al. (48) demonstrating FGFR1 as an important gene in this amplicon which is potentially responsive to targeted therapy can be found earlier in this chapter.

Lockwood et al. (52) also examined this region closely using the BCCR SMRT array v.2, with 181 NSCLC samples and cell lines. High-throughput RNA expression microarrays were used to closely examine expression levels of genes in this region, compare between cancerous and noncancerous tissues, as well as correlate expression findings with the presence of concordant amplification. This analysis favored BRF2 over FGFR1 as the target of amplification. Further experiments demonstrated that knockdown of BRF2 expression in amplified cell lines significantly reduced anchorage-independent growth and cell proliferation; in contrast, siRNA knockdown of three cell lines without BRF2 amplification did not decrease proliferation.

These experiments demonstrate that possibly more than one gene within an amplicon, in this example FGFR1 and BRF2, could potentially be playing oncogenic roles simultaneously, or at least in separate subgroups of tumors.

14.3. Lineage-Specific Oncogene in Adenocarcinoma

Weir et al. (26) used an Affymetrix 250 K *Sty* SNP array to profile 528 frozen-banked primary ADC representing all stages of disease. Surprisingly, they found that the most common focal amplification, found at 14q13, contained no known proto-oncogene. The high resolution and large sample size in their study allowed them to narrow down the minimal common region of amplification to include just two known genes, *MBIP* and *NKX2*-1 (*TITF1,* encoding for TTF-1 protein). They confirmed the finding by FISH on an independent set of 330 ADC from tissue microarrays (TMAs), with high-level amplifications detected in 12% of these tumors. Whereas RNAi-based *MBIP* knockdown had no effect on colony formation of amplified cell lines, knockdown of *NKX2-1* by the same method resulted in substantial reduction of anchorage independent growth, cell viability, and colony formation of *NKX2-1* amplified and/or overexpressing cell lines. Kwei et al. (55) also investigated 76 NSCLC tumors and 52 NSCLC cell lines using cDNA microarrays from the Stanford Functional Genomics Facility covering ~22,000 genes. Among the eight genes within their minimally amplified region, *NKX2-1* showed promise as an oncogene due to increased expression at both the RNA and protein level in NSCLC cell lines with amplification. The common amplification of *NKX2-1* in lung cancer was recently confirmed by Tang et al. (89) who employed FISH to show high level copy gains (>10 gene copies) in 8.3% of ADC and 3.0% of SCC. *NKX2-1* gene copy gains (>4 copies/nuclei) by FISH was correlated with poorer survival in the entire cohort of 269 NSCLC patients. Several investigators have also confirmed that reducing *NKX2-1* expression using RNA interference resulted in decreased cell proliferation and increased apoptosis, whereas a non-amplified cell line showed no reduction in proliferation (26, 55, 90).

NKX2-1 has been determined to be a tissue-specific transcription factor, required during normal lung development and differentiation of pulmonary epithelial cells. It has been found to be expressed in approximately ~80% of ADC, but rarely in SCC (91). Its lineage-specific expression is routinely employed by pathologists to support the pulmonary origin of an adenocarcinoma, despite its role as a lineage marker of thyroid follicular cells and carcinoma.

15. aCGH as Molecular Diagnostic Assay in Clinical Cytogenetics Laboratory

Given the emerging diagnostic, prognostic, and predictive utility of genomic copy number determinations we discussed, a number of practical uses of aCGH can be envisaged for future cancer genetics laboratories. The decreasing cost of arrays combined with the increasing ability to batch multiple samples on the same slide has brought the cost of running an array to a similar level as performing a FISH panel assay.

Genomic techniques including aCGH will likely be instrumental in determining aberrant tyrosine kinases and pathways that are driving a given cancer, in order to customize molecularly directed therapy for the patient. EGFR is already established in this regard, and other players such as FGFR1 and BCL2 are already emerging. A combination of copy number, mutation status, and epigenetic (e.g., DNA methylation) genomic assays will need to be employed and integrated to provide the most accurate and complete information to guide personalized therapy. As costs decrease and analysis becomes more user-friendly, genomic analysis in at least a selected group of cancers (those that will not be cured by surgery alone) could be employed routinely in the diagnostic lab.

For early stage lung cancers, an initial targeted analysis of the copy number findings could be performed that would identify poor prognostic aberrations, for example presence of 8q gain or 13q loss, or relatively high number of abnormalities in a lepidic (bronchioloalveolar carcinoma) -predominant histology, or detection of a 4q loss (associated with metastasis), could provide the oncologist with additional information to decide whether or not to administer adjuvant chemotherapy or refer for radiation treatment after surgical resection. The presence of one of these poor prognostic markers could for example trigger an in-depth analysis for disrupted "druggable" targets by multiplex mutational analysis and targeted copy number analysis. It remains to be determined whether gene copy number and mutation analyses could be performed using the same multiplex platforms.

Common diagnostic scenarios in which aCGH would be informative can be easily imagined. Often it is difficult to determine whether a NSCLC is showing glandular or squamous differentiation; emerging histology-specific abnormalities such as *SOX2* and *NKX2-1* could be helpful in placing the cancer into one of these categories in order to help direct further testing (ex. ADC mutation panel or SCC mutation panel) and personalize therapy. Similarly, with small biopsies it can be difficult to distinguish SCLC and NSCLC; as discussed earlier, aCGH has the potential to address this question as well.

The question of second primary versus metastasis is a frequently debated question at lung tumor board meetings that can be unequivocally addressed by the demonstration of multiple common breakpoints or lack thereof in two separate lung tumors of similar histology. The correct answer to such a question has a significant impact on the staging and therefore proper treatment of the patient.

The differential diagnosis between a reactive pleural fibrosis and mesothelioma is a not infrequent diagnostic dilemma that confronts a pulmonary pathologist; the detection of any abnormality by aCGH, particularly the commonly recurrent 9p deletion, would provide strong evidence for a malignant diagnosis. Other benign versus malignant scenarios could include screening cytology samples of sputum or bronchial brushes for the most common gains or deletions associated with squamous cell carcinoma in smokers.

16. Conclusion

Array-CGH has been a very useful tool to understanding the genetic imbalances and resulting pathway disruptions in lung cancer. A vast amount of information, addressing questions from tumor classification to prognosis and predicting response to therapy, can be gained from this technique. Copy number information will be an important component in the comprehensive genetic workup that will be required to design the most potent therapy in the era of personalized cancer treatment.

Acknowledgments

This work is supported in parts by the Canadian Cancer Society Research Institute grant #020527 and the Ontario Ministry of Health and Long Term Care. Dr. Tsao is the M. Qasim Choksi Chair in Lung Cancer Translational Research.

References

1. Jemal A, Bray F, Center MM et al (2011) Global cancer statistics. CA Cancer J Clin 61(2):69–90
2. Mathers CD, Loncar D (2006) Projections of global mortality and burden of disease from 2002 to 2030. PLoS Med 3(11):e442–e442
3. Jemal A, Siegel R, Xu J et al (2010) Cancer statistics, 2010. CA Cancer J Clin 60(5):277–300
4. Ag TI (2004) Pathology and genetics of tumours of the lung, pleura, thymus and heart. WHO Publications, France
5. Peto R, Darby S, Deo H et al (2000) Smoking, smoking cessation, and lung cancer in the uk since 1950: combination of national statistics with two case–control studies. BMJ (Clin Res Ed) 321(7257):323–329

6. Sun S, Schiller JH, Gazdar AF (2007) Lung cancer in never smokers–a different disease. Nat Rev Cancer 7(10):778–790
7. Valliã Res E, Shepherd FA, Crowley J et al (2009) The iaslc lung cancer staging project: proposals regarding the relevance of tnm in the pathologic staging of small cell lung cancer in the forthcoming (seventh) edition of the tnm classification for lung cancer. J Thorac Oncol 4(9):1049–1059
8. Govindan R, Page N, Morgensztern D et al (2006) Changing epidemiology of small-cell lung cancer in the united states over the last 30 years: analysis of the surveillance, epidemiologic, and end results database. J Clin Oncol 24(28):4539–4544
9. Goldstraw P, Crowley J, Chansky K et al (2007) The iaslc lung cancer staging project: proposals for the revision of the tnm stage groupings in the forthcoming (seventh) edition of the tnm classification of malignant tumours. J Thorac Oncol 2(8):706–714
10. Reduced lung-cancer mortality with low-dose computed tomographic screening. N Engl J Med 365 (5):395–409
11. Goldstraw P, Ball D, Jett JR et al (2011) Non-small-cell lung cancer. Lancet 378(9804):1727–1740
12. Chemotherapy in addition to supportive care improves survival in advanced non-small-cell lung cancer: a systematic review and meta-analysis of individual patient data from 16 randomized controlled trials (2008). J Clin Oncol 26 (28):4617–4625
13. Visbal AL, Leighl NB, Feld R et al (2005) Adjuvant chemotherapy for early-stage non-small cell lung cancer. Chest 128(4):2933–2943. doi:128/4/2933 (pii) 10.1378/chest.128.4.2933
14. Cappuzzo F, Hirsch FR, Rossi E et al (2005) Epidermal growth factor receptor gene and protein and gefitinib sensitivity in non-small-cell lung cancer. J Natl Cancer Inst 97(9):643–655. doi:97/9/643 (pii) 10.1093/jnci/dji112
15. Takano T, Ohe Y, Sakamoto H et al (2005) Epidermal growth factor receptor gene mutations and increased copy numbers predict gefitinib sensitivity in patients with recurrent non-small-cell lung cancer. J Clin Oncol 23(28):6829–6837. doi:JCO.2005.01.0793 (pii) 10.1200/JCO.2005.01.0793
16. Hirsch FR, Varella-Garcia M, Cappuzzo F et al (2007) Combination of egfr gene copy number and protein expression predicts outcome for advanced non-small-cell lung cancer patients treated with gefitinib. Ann Oncol 18(4):752–760. doi:mdm003 (pii) 10.1093/annonc/mdm003
17. Koivunen JP, Mermel C, Zejnullahu K et al (2008) Eml4-alk fusion gene and efficacy of an alk kinase inhibitor in lung cancer. Clin Cancer Res 14(13):4275–4283
18. Camidge DR, Kono SA, Flacco A et al (2010) Optimizing the detection of lung cancer patients harboring anaplastic lymphoma kinase (alk) gene rearrangements potentially suitable for alk inhibitor treatment. Clin Cancer Res 16(22):5581–5590
19. Lockwood WW, Chari R, Chi B et al (2006) Recent advances in array comparative genomic hybridization technologies and their applications in human genetics. Euro J Hum Genet: EJHG 14(2):139–148
20. Costa JL, Meijer G, Ylstra B et al (2008) Array comparative genomic hybridization copy number profiling: a new tool for translational research in solid malignancies. Semin Radiat Oncol 18(2):98–104
21. Garnis C, Campbell J, Davies JJ et al (2005) Involvement of multiple developmental genes on chromosome 1p in lung tumorigenesis. Hum Mol Genet 14(4):475–482
22. Garnis C, Davies JJ, Buys TPH et al (2005) Chromosome 5p aberrations are early events in lung cancer: implication of glial cell line-derived neurotrophic factor in disease progression. Oncogene 24(30):4806–4812
23. Coe BP, Henderson LJ, Garnis C et al (2005) High-resolution chromosome arm 5p array cgh analysis of small cell lung carcinoma cell lines. Genes Chromosomes Cancer 42(3):308–313. doi:10.1002/gcc.20137
24. Ishkanian AS, Malloff CA, Watson SK et al (2004) A tiling resolution DNA microarray with complete coverage of the human genome. Nat Genet 36(3):299–303
25. Aviel-Ronen S, Coe BP, Lau SK et al (2008) Genomic markers for malignant progression in pulmonary adenocarcinoma with bronchioloalveolar features. Proc Natl Acad Sci USA 105(29):10155–10160. doi:0709618105 (pii) 10.1073/pnas.0709618105
26. Weir BA, Woo MS, Getz G et al (2007) Characterizing the cancer genome in lung adenocarcinoma. Nature 450(7171):893–898. doi:nature06358 (pii) 10.1038/nature06358
27. Bass AJ, Watanabe H, Mermel CH et al (2009) Sox2 is an amplified lineage-survival oncogene in lung and esophageal squamous cell carcinomas. Nat Genet 41(11):1238–1242. doi:ng.465 (pii) 10.1038/ng.465

28. Coe BP, Ylstra B, Carvalho B et al (2007) Resolving the resolution of array cgh. Genomics 89(5):647–653
29. Taylor BS, Barretina J, Socci ND et al (2008) Functional copy-number alterations in cancer. PLoS One 3(9):e3179–e3179
30. Yim S-H, Kim T-M, Hu H-J et al (2010) Copy number variations in east-asian population and their evolutionary and functional implications. Hum Mol Genet 19(6):1001–1008
31. Gunnarsson R, Staaf J, Jansson M et al (2008) Screening for copy-number alterations and loss of heterozygosity in chronic lymphocytic leukemia–a comparative study of four differently designed, high resolution microarray platforms. Genes Chromosomes Cancer 47(8):697–711
32. Vucic EA, Thu KL, Williams AC et al (2010) Copy number variations in the human genome and strategies for analysis. Methods Mol Biol 628:103–117
33. Alkan C, Coe BP, Eichler EE (2011) Genome structural variation discovery and genotyping. Nat Rev Genet 12(5):363–376
34. Coe BP, Chari R, Macaulay C et al (2010) Facade: a fast and sensitive algorithm for the segmentation and calling of high resolution array cgh data. Nucleic Acids Res 38(15):e157–e157
35. Zhang J, Feuk L, Duggan GE et al (2006) Development of bioinformatics resources for display and analysis of copy number and other structural variants in the human genome. Cytogenet Genome Res 115(3–4):205–214
36. Chari R, Lockwood WW, Lam WL (2006) Computational methods for the analysis of array comparative genomic hybridization. Cancer Informatics 2:48–58
37. Chi B, Deleeuw RJ, Coe BP et al (2004) Seegh–a software tool for visualization of whole genome array comparative genomic hybridization data. BMC Bioinformatics 5:13–13
38. Hsu F-H, Chen H-IH, Tsai M-H et al (2011) A model-based circular binary segmentation algorithm for the analysis of array cgh data. BMC Res Notes 4(1):394–394
39. Carro A, Rico D, Rueda OM et al (2010) Wavicgh: a web application for the analysis and visualization of genomic copy number alterations. Nucleic Acids Res 38:W182–W187
40. Van Houte BPP, Binsl TW, Hettling H et al (2010) Cghnormaliter: a bioconductor package for normalization of array cgh data with many cnas. Bioinformatics 26(10):1366–1367
41. Van De Wiel MA, Brosens R, Eilers PHC et al (2009) Smoothing waves in array cgh tumor profiles. Bioinformatics 25(9):1099–1104
42. Chari R, Coe BP, Wedseltoft C et al (2008) Sigma2: a system for the integrative genomic multi-dimensional analysis of cancer genomes, epigenomes, and transcriptomes. BMC Bioinformatics 9:422–422
43. Shah SP, Lam WL, Ng RT et al (2007) Modeling recurrent DNA copy number alterations in array cgh data. Bioinformatics 23(13):i450–i458
44. Chi B, Deleeuw RJ, Coe BP et al (2008) Md-seegh: a platform for integrative analysis of multi-dimensional genomic data. BMC Bioinformatics 9:243–243
45. Balsara BR, Testa JR (2002) Chromosomal imbalances in human lung cancer. Oncogene 21(45):6877–6883
46. Hoglund M, Gisselsson D, Hansen GB et al (2004) Statistical dissection of cytogenetic patterns in lung cancer reveals multiple modes of karyotypic evolution independent of histological classification. Cancer Genet Cytogenet 154(2):99–109. doi:S0165-4608(04)00067-6 (pii) 10.1016/j.cancergencyto.2004.01.030
47. Tonon G, Wong KK, Maulik G et al (2005) High-resolution genomic profiles of human lung cancer. Proc Natl Acad Sci USA 102(27):9625–9630. doi:0504126102 (pii) 10.1073/pnas.0504126102
48. Weiss J, Sos ML, Seidel D et al (2010) Frequent and focal FGFR1 amplification associates with therapeutically tractable FGFR1 dependency in squamous cell lung cancer. Sci Transl Med 2(62):62ra93. doi:2/62/62ra93 (pii) 10.1126/scitranslmed.3001451
49. Voortman J, Lee JH, Killian JK et al (2010) Array comparative genomic hybridization-based characterization of genetic alterations in pulmonary neuroendocrine tumors. Proc Natl Acad Sci USA 107(29):13040–13045. doi:1008132107 (pii) 10.1073/pnas.1008132107
50. Newnham GM, Conron M, Mclachlan S et al (2011) Integrated mutation, copy number and expression profiling in resectable non-small cell lung cancer. BMC Cancer 11:93–93
51. Kang JU, Koo SH, Kwon KC et al (2009) Identification of novel candidate target genes, including ephb3, masp1 and sst at 3q26.2-q29 in squamous cell carcinoma of the lung. BMC Cancer 9:237. doi:1471-2407-9-237 (pii) 10.1186/1471-2407-9-237
52. Lockwood WW, Chari R, Coe BP et al (2010) Integrative genomic analyses identify brf2 as a novel lineage-specific oncogene in lung squamous cell carcinoma. PLoS Med 7(7):e1000315–e1000315
53. Massion PP, Kuo WL, Stokoe D et al (2002) Genomic copy number analysis of non-small cell lung cancer using array comparative genomic hybridization: implications of the

phosphatidylinositol 3-kinase pathway. Cancer Res 62(13):3636–3640
54. Garnis C, Lockwood WW, Vucic E et al (2006) High resolution analysis of non-small cell lung cancer cell lines by whole genome tiling path array cgh. Int J Cancer 118(6):1556–1564. doi:10.1002/ijc.21491
55. Kwei KA, Kim YH, Girard L et al (2008) Genomic profiling identifies titf1 as a lineage-specific oncogene amplified in lung cancer. Oncogene 27(25):3635–3640. doi:1211012 (pii) 10.1038/sj.onc.1211012
56. Shibata T, Uryu S, Kokubu A et al (2005) Genetic classification of lung adenocarcinoma based on array-based comparative genomic hybridization analysis: its association with clinicopathologic features. Clin Cancer Res 11(17):6177–6185. doi:11/17/6177 (pii) 10.1158/1078-0432.CCR-05-0293
57. Coe BP, Lockwood WW, Girard L et al (2006) Differential disruption of cell cycle pathways in small cell and non-small cell lung cancer. Br J Cancer 94(12):1927–1935
58. Peng W-X, Shibata T, Katoh H et al (2005) Array-based comparative genomic hybridization analysis of high-grade neuroendocrine tumors of the lung. Cancer Sci 96(10):661–667
59. Ullmann R, Schwendel A, Klemen H et al (1998) Unbalanced chromosomal aberrations in neuroendocrine lung tumors as detected by comparative genomic hybridization. Hum Pathol 29(10):1145–1149. doi:S0046-8177(98)90428-2 (pii)
60. Taniguchi T, Karnan S, Fukui T et al (2007) Genomic profiling of malignant pleural mesothelioma with array-based comparative genomic hybridization shows frequent non-random chromosomal alteration regions including jun amplification on 1p32. Cancer Sci 98(3):438–446
61. Christensen BC, Houseman EA, Poage GM et al (2010) Integrated profiling reveals a global correlation between epigenetic and genetic alterations in mesothelioma. Cancer Res 70(14):5686–5694. doi:0008-5472.CAN-10-0190 (pii) 10.1158/0008-5472.CAN-10-0190
62. Bott M, Brevet M, Taylor BS et al (2011) The nuclear deubiquitinase bap1 is commonly inactivated by somatic mutations and 3p21.1 losses in malignant pleural mesothelioma. Nat Genet 43(7):668–672. doi:ng.855 (pii) 10.1038/ng.855
63. Murakami H, Mizuno T, Taniguchi T et al (2011) Lats2 is a tumor suppressor gene of malignant mesothelioma. Cancer Res 71(3):873–883. doi:0008-5472.CAN-10-2164 (pii) 10.1158/0008-5472.CAN-10-2164
64. Takeda M, Kasai T, Enomoto Y et al (2012) Genomic gains and losses in malignant mesothelioma demonstrated by fish analysis of paraffin-embedded tissues. J Clin Pathol 65(1):77–82
65. Chiosea S, Krasinskas A, Cagle PT et al (2008) Diagnostic importance of 9p21 homozygous deletion in malignant mesotheliomas. Modern Pathol 21(6):742–747
66. Lonergan KM, Chari R, Coe BP et al (2010) Transcriptome profiles of carcinoma-in-situ and invasive non-small cell lung cancer as revealed by sage. PLoS One 5(2):e9162–e9162
67. Garnis C, Macaulay C, Lam S et al (2004) Genetic alteration on 8q distinct from myc in bronchial carcinoma in situ lesions. Lung Cancer 44(3):403–404
68. Van Boerdonk RA, Sutedja TG, Snijders PJ et al (2011) DNA copy number alterations in endobronchial squamous metaplastic lesions predict lung cancer. Am J Respir Crit Care Med 184(8):948–956. doi:201102-0218OC (pii) 10.1164/rccm.201102-0218OC
69. Schneider KU, Dietrich D, Fleischhacker M et al (2011) Correlation of shox2 gene amplification and DNA methylation in lung cancer tumors. BMC Cancer 11:102–102
70. Kerr KM (2001) Pulmonary preinvasive neoplasia. J Clin Pathol 54(4):257–271
71. Lantuéjoul S, Salameire D, Salon C et al (2009) Pulmonary preneoplasia–sequential molecular carcinogenetic events. Histopathology 54(1):43–54
72. Travis WD, Brambilla E, Noguchi M et al (2011) International association for the study of lung cancer/american thoracic society/european respiratory society international multidisciplinary classification of lung adenocarcinoma. J Thorac Oncol 6(2):244–285
73. Buys TPH, Aviel-Ronen S, Waddell TK et al (2009) Defining genomic alteration boundaries for a combined small cell and non-small cell lung carcinoma. J Thorac Oncol 4(2):227–239
74. Girard N, Ostrovnaya I, Lau C et al (2009) Genomic and mutational profiling to assess clonal relationships between multiple non-small cell lung cancers. Clin Cancer Res 15(16):5184–5190
75. Boelens MC, Kok K, Van Der Vlies P et al (2009) Genomic aberrations in squamous cell lung carcinoma related to lymph node or distant metastasis. Lung Cancer 66(3):372–378. doi:S0169-5002(09)00115-9 (pii) 10.1016/j.lungcan.2009.02.017
76. Wrage M, Ruosaari S, Eijk PP et al (2009) Genomic profiles associated with early micrometastasis in lung cancer: relevance of 4q deletion. Clin Cancer Res 15(5):1566–1574.

doi:1078-0432.CCR-08-2188 (pii) 10.1158/1078-0432.CCR-08-2188

77. Iwakawa et al (2011) MYC amplification as a prognostic markers of early-stage lung adenocarcinoma identified by whole genome copy numbers analysis. Clin Cancer Res 17: 1481–1489.
78. Kang JU, Koo SH, Kwon KC et al (2010) Frequent silence of chromosome 9p, homozygous dock8, dmrt1 and dmrt3 deletion at 9p24.3 in squamous cell carcinoma of the lung. Int J Oncol 37(2):327–335
79. Lynch TJ, Bell DW, Sordella R et al (2004) Activating mutations in the epidermal growth factor receptor underlying responsiveness of non-small-cell lung cancer to gefitinib. N Engl J Med 350(21):2129–2139. doi: 10.1056/NEJMoa040938 NEJMoa040938 (pii)
80. Sequist LV, Bell DW, Lynch TJ et al (2007) Molecular predictors of response to epidermal growth factor receptor antagonists in non-small-cell lung cancer. J Clin Oncol 25(5): 587–595. doi:25/5/587 (pii) 10.1200/JCO.2006.07.3585
81. Thomas RK, Weir B, Meyerson M (2006) Genomic approaches to lung cancer. Clin Cancer Res12(14Pt2):4384s–4391s.doi:12/14/4384s (pii) 10.1158/1078-0432.CCR-06-0098
82. Azzoli CG, Park BJ, Pao W et al (2008) Molecularly tailored adjuvant chemotherapy for resected non-small cell lung cancer: a time for excitement and equipoise. J Thorac Oncol 3(1):84–93. doi:10.1097/JTO.0b013e31815efe2401243894-200801000-00015 (pii)
83. Li C, Sun Y, Fang Z et al (2011) Comprehensive analysis of epidermal growth factor receptor gene status in lung adenocarcinoma. J Thorac Oncol 6(6):1016–1021
84. Olejniczak ET, Van Sant C, Anderson MG et al (2007) Integrative genomic analysis of small-cell lung carcinoma reveals correlates of sensitivity to bcl-2 antagonists and uncovers novel chromosomal gains. Mole Cancer Res: MCR 5(4):331–339
85. Au NHC, Gown AM, Cheang M et al (2004) P63 expression in lung carcinoma: a tissue microarray study of 408 cases. Appl Immunohistochem Mol Morphol 12(3):240–247
86. Hussenet T, Dali S, Exinger J et al (2010) Sox2 is an oncogene activated by recurrent 3q26.3 amplifications in human lung squamous cell carcinomas. PLoS One 5(1):e8960. doi:10.1371/journal.pone.0008960
87. Rekhtman N, Ang DC, Sima CS et al (2011) Immunohistochemical algorithm for differentiation of lung adenocarcinoma and squamous cell carcinoma based on large series of whole-tissue sections with validation in small specimens. Mod Pathol 24(10):1348–1359
88. Massion PP, Taflan PM, Jamshedur Rahman SM et al (2003) Significance of p63 amplification and overexpression in lung cancer development and prognosis. Cancer Res 63(21):7113–7121
89. Tang X, Kadara H, Behrens C et al (2011) Abnormalities of the titf-1 lineage-specific oncogene in nsclc: implications in lung cancer pathogenesis and prognosis. Clin Cancer Res 17(8):2434–2443
90. Tanaka H, Yanagisawa K, Shinjo K et al (2007) Lineage-specific dependency of lung adenocarcinomas on the lung development regulator ttf-1. Cancer Res 67(13):6007–6011
91. Loo PS, Thomas SC, Nicolson MC et al (2010) Subtyping of undifferentiated non-small cell carcinomas in bronchial biopsy specimens. J Thorac Oncol 5(4):442–447

Chapter 20

Array CGH in Brain Tumors

Gayatry Mohapatra, Julia Sharma, and Stephen Yip

Abstract

Alterations in the copy number of the cancer genome are frequently observed in brain tumors especially gliomas. Some pertinent examples include amplification of the *EGFR* locus in chromosome 7p and loss of the *PTEN* locus in 10q in glioblastoma. Meningiomas are often associated with loss of the *NF2* locus in 22q. Array CGH or aCGH probes provide a reliable, consistent, and economical method of profiling genome-wide copy number alterations (CNAs) of cancer specimens at fairly robust resolution. This has allowed for the systematic assessment of brain tumors for recurrent genomic CNAs. In addition, recent technical advancements have increased the robustness of this technique to accommodate DNA derived from formalin-fixed paraffin-embedded (FFPE) tissue. Lastly, novel technologies such as next-generation sequencing and multiplex digital gene counting technology such as NanoString will expand the repertoire of techniques for assessing CNAs in brain tumors.

Key words: Brain tumors, Glioma, Meningioma, Array CGH, Copy number alterations, Oligonucleotide microarray

1. Introduction

The introduction, technical evolution, and widespread adoption of array-based comparative genomic hybridization (aCGH) technology have significantly increased our appreciation of the underlying copy number variability of the human genome, even in otherwise healthy individuals (1, 2). In addition, these copy number variations (CNVs) account for many phenotypic differences among individuals and are a significant but poorly studied cause of complex human diseases such as autism and schizophrenia (3–5). In the field of oncology, copy number alterations (CNAs) or changes of somatic chromosomal copy numbers often lead to growth advantage of the malignant clone and define the type and subtype of cancer especially brain tumors (6). Constantly evolving technology in the past

Diponkar Banerjee and Sohrab P. Shah (eds.), *Array Comparative Genomic Hybridization: Protocols and Applications*, Methods in Molecular Biology vol. 973, DOI 10.1007/978-1-62703-281-0_20, © Springer Science+Business Media, LLC 2013

10 years has contributed to the rapid development and adoption of aCGH from the much lower resolution CGH technique. Also, early iterations of aCGH technology, using probes based on bacterial artificial chromosome or BACs, have been virtually replaced initially by cDNA then by oligonucleotide-based probes for everyday applications. These have the advantage of additional design flexibility and technical reproducibility (7). High-density BAC arrays are still useful and informative in specialized applications such as focused, high-density investigations of unique regions of the genome (8). Nowadays, aCGH chips based on oligonucleotide-based probes are the preferred platform for genome-wide copy number interrogation. Additionally, the routine use of aCGH has entered the realm of clinical molecular diagnostics and is now used as a molecular adjunct in the study of complex medical disorders. This is assisted by the diminution in the cost of this technology, increased consistency and reproducibility of its laboratory workflow, as well as improved bioinformatic pipeline for in the management of the information.

As noted above, copy number changes are cardinal events in oncogenesis, and these often involve amplifications and homozygous losses of canonical oncogenes and tumor suppressor genes, respectively. Accurate determinations of these changes in cancer have the potential to inform regarding unique cancer drivers and therefore of significant diagnostic, prognostic, as well as predictive values. One example is amplification of the *EGFR* locus in chromosome 7p12 coding for a receptor tyrosine kinase of the ErbB family. Amplification of this locus is seen in approximately 40% of high-grade gliomas or glioblastomas and constitutes the most commonly amplified genetic locus in this very aggressive form of glioma (9). Genomic-wide technology such as aCGH has the advantage of providing copy number information of the entire cancer genome.

One of the more pressing issues in clinical application of aCGH technology is the quality of template DNA, specifically, DNA extracted from cancer specimens in the form of formalin-fixed paraffin-embedded (FFPE) tissue. Recent technical improvements in the aCGH workflow have overcome the traditional barriers imposed by FFPE-derived genomic DNA in aCGH (10, 11). This carries tremendous implications for the clinical pathology laboratory since a vast majority of the pathologic specimens are processed and stored as FFPE tissues; in addition, this allows for the exploration of a vast collection of archived tissue specimens in storage.

Emerging technologies such as single-nucleotide polymorphism or SNP chips encompassing millions of probes that not only measure copy numbers, but also identify SNPs at numerous loci throughout the genome (12). Next-generation sequencing (NGS) can also provide genome-wide copy number information at base-pair resolution (13). This information is invaluable when combined with mutational data and gene expression to provide an integrative picture of the cancer genome. However, routine sampling of specimens using NGS remains costly and not practical at this moment. NANOSTRING is an emerging technology which can provide rapid and economical determination of CNAs and may be suited for routine laboratory workflow (14, 15). Currently, NANOSTRING is optimized for multiplex quantitative determinations for the abundance of specific mRNAs in FFPE tissue; however, the same technology has been adapted into a copy number variation panel for genomic DNA from frozen tissue. Ongoing improvements in the chemistry and workflow will allow for similar determination in FFPE tissue which can provide, in a 2 day workflow, genome-wide copy number information for multiple specimens. Regardless of these novel technologies, chip-based aCGH remains the most robust method for genome-wide copy number determination, and ongoing improvement to its workflow will continue to strengthen this platform.

2. Materials

2.1. For Genomic DNA Extraction from Frozen Tissue

2.1.1. Reagents

Phosphate-buffered saline (PBS): Dissolve the following in 800 mL distilled H_2O:

8 g of NaCl.

0.2 g of KCl.

1.44 g of Na_2HPO_4.

0.24 g of KH_2PO_4.

Adjust pH to 7.4. Adjust volume to 1 L with additional distilled H_2O. Sterilize by autoclaving and store at room temperature.

Cell lysis buffer: 1% Triton X-100, 5 mM $MgCl_2$, 0.32 M sucrose, 10 mM Tris–HCl, pH 8.0.

Protein digestion buffer: 25 mM EDTA, 100 mM NaCl, 10 mM Tris–HCl, pH 8.0.

SDS: 10% (w/v) sodium dodecyl sulfate (SDS). Filter the solution with a syringe-driven filter unit. Store at room temperature.

Proteinase K: Proteinase K (Fungal), Invitrogen Corp (cat# 25530-015). Dissolve 50 mg/mL in distilled water and stored in aliquots at −20 °C.

Phenol/chloroform/isoamyl alcohol (25:24:1, v/v) Invitrogen Corp (cat# 15593-031). Store at 4 °C.

10 M ammonium acetate solution. Store at 4 °C.

Ethanol absolute (200% proof). Store at room temperature.

10 mM Tris–HCl, pH 8.0. Store at room temperature.

DNA digestion: DpnII (New England BioLabs, cat# R0543S, 10,000 U/mL).

Cleanup of digested DNA: QIAquick PCR purification columns (Qiagen, cat#28104).

Purification of labeled DNA: QIAquick PCR purification columns (Qiagen, cat#28104).

Precipitation and hybridization of labeled DNA:

1. Human Cot-1 DNA (Invitrogen, 1 mg/mL stock, cat# 15279-011). Store at –20 °C.
2. 10× Blocking Agent (Agilent Technologies). To prepare the 10× Blocking Agent, add 1,350 μL of nuclease-free water to the vial containing lyophilized 10× Blocking Agent (supplied with Agilent Oligo aCGH Hybridization Kit). Leave at room temperature for 60 min, and mix on a vortex mixer to reconstitute sample before use or storage. The 10× Blocking Agent can be prepared in advance and stored at –20 °C.
3. TE buffer, pH 7.4: 1 mM Tris–HCl, 0.1 mM EDTA. Store at room temperature.
4. Agilent 2× Hi-RPM hybridization buffer.
5. Agilent SureHyb hybridization chamber.
6. Gasket slide (Agilent Technologies).
7. Hybridization oven with rotator rack (Agilent Technologies).

Labeling of genomic DNA:

1. 2.5× random priming buffer from the BioPrime® Total Genomic Labeling System (Invitrogen, cat# A18097-010): 125 mM Tris–HCl, pH 6.8, 12.5 mM $MgCl_2$, 25 mM 2-mercaptoethanol, 750 mg/mL random primers. Store at –20 °C.
2. 10× dNTP mixture: 1.2 mM dATP, 1.2 mM dGTP, 1.2 mM dCTP, and 0.6 mM dTTP dissolved in Tris-EDTA buffer, pH 8.0. Store at –20 °C.
3. 25 nanomole Cy3- and Cy5-dUTP (Amersham Biosciences, Piscataway, NJ). Store in the dark, at –20 °C.
4. 40 U/mL Klenow fragment (Large Fragment of DNA Polymerase I) from the BioPrime DNA Labeling System kit. Store at –20 °C.
5. Stop buffer: 0.5 M EDTA, pH 8.0. Store at room temperature.

Washing of microarray slides:

1. Agilent Oligo aCGH Wash Buffer 1 and Oligo aCGH Wash Buffer 2. Store at room temperature.
2. Acetonitrile (Sigma). Store at room temperature.
3. Stabilization and drying solution (Agilent Technologies).
4. Blunt-ended plastic forceps.
5. Large horizontal staining dishes.
6. Magnetic stir plate with heating.

Scanning and image processing:

1. Axon 4000B scanner (Axon Instruments Inc., Union City, CA).
2. GenePix Pro 6.0 software (Axon Instruments Inc.).
3. Agilent Feature Extraction (FE) software v10.5.

3. Methods

Array comparative genomic hybridization (aCGH) is a microarray-based method for analyzing genomic DNA to detect CNAs between samples (16). In aCGH, two genomic DNA samples are labeled with different fluorophores. The samples are hybridized to a microarray, and the ratio of the fluorescent intensities of the fluorophores is measured for each feature on the array (16–19). This ratio provides a relative measure of the difference in DNA copy number between the samples.

Cancer genomes harbor widespread structural chromosomal aberrations leading to amplification and deletion of cancer-related loci. Comparative genomic hybridization has emerged as the technology of choice for identification and characterization of DNA CNAs on a genome-wide level. This approach has been further enabled by the availability of human genome sequence and the adaptation of CGH to a microarray format. Bacterial artificial chromosome (BAC)-based arrays have proven highly effective in defining the CNAs in gliomas and other brain tumors (20, 21). In addition to the construction of a 700 kb BAC array (21), we have gained considerable experience in the use of Agilent oligonucleotide-based platform in the analysis of more than 250 human brain tumors (22) (unpublished data). Based on our experience, we feel that commercial oligonucleotide aCGH platforms are affordable and sufficiently robust to detect CNAs at a relatively high resolution. Here we describe our experience with oligonucleotide aCGH, primarily the 60-mer oligonucleotide arrays manufactured by Agilent Technologies.

aCGH is performed to determine DNA copy number changes using Agilent Human 105 K oligonucleotide microarrays (http://www.home.agilent.com/agilent/home.jspx). Genomic coordinates for this array are based on the NCBI build 36, March 2006 freeze of the assembled human genome (UCSC hg18), available through the UCSC Genome Browser. This array provides an average spatial resolution of 21.7 kb.

3.1. Genomic DNA Extraction from Frozen Tissue

Day 1

1. Let tissues thaw over dry ice.
2. Using a sterile blade, cut a piece of the tissue to be used for DNA extraction.

3. Chop up the tissue as small as possible.
4. Put sample in Falcon tube with 1–2 mL PBS. Pipette sample up and down to homogenize tissue.
5. Suspend the sample in 10 mL PBS.
6. Make sure cells are suspended via vortex, flicking, etc.
7. Spin at 500 × *g* for 10 min, aspirate supernatant.
8. Add 10 mL cell lysis buffer, vortex the vials, and centrifuge at 1,500–2,000 rpm for 10 min, aspirate supernatant.

   Add protein digestion buffer, depending on size of pellet, 1–5 mL, add proteinase K to a final concentration of 500 μg/mL and SDS to a final concentration of 0.1%.
9. Vortex at level 6–7 to mix.
10. Incubate vials in 55 °C shaking water bath overnight.

Day 2

1. In the hood, add equal amounts of Phenol/chloroform/isoamyl alcohol (PCI). Phenol aids in precipitating the digested protein. Note: PCI has a top layer of buffer. Phenol is extremely corrosive. Make sure to use a glass pipette to aspirate the phenol and not the buffer.
2. Centrifuge the tubes at 1,500–2,000 rpm for 30 min in a microfuge.
3. Transfer the top aqueous phase, without disturbing the interphase, to new tubes.
4. Add 1/10 volume of 10 M Ammonium Acetate and 2× volume of 100% EtOH. This aids in precipitation of genomic DNA.
5. Shake the mixture by inverting tubes up and down. A DNA precipitate should become visible. Transfer the high molecular weight DNA precipitate to a new Eppendorf tube.
6. Spin down tubes at high speed for approx 5 min and discard supernatant. Resuspend DNA pellet in TE (10 mM Tris–HCl pH 8.0, 1 mM EDTA) thoroughly.
7. High molecular weight genomic DNA should be stored in 4 °C.

3.2. Genomic DNA Extraction from Formalin-Fixed Paraffin-Embedded Tissues

Day 3

1. Cut 5–7 μm sections from the block onto slides. Less than 10 μm thickness is preferred. The number of sections required differs depending on the size of the tissue (in general 3–5 slides are sufficient). Marking the area of your interest on the back of the slide is necessary, determined by looking at H&E-stained slides (see Note 1).
2. Add 500 μL of xylene in a 2 mL clean Eppendorf tube. Using a sterile surgical blade, scrape tissues into the Eppendorf tube (see Note 2). Add another 500 μL of xylene to the tube.

3. Vortex and incubate 5–10 min at room temperature to allow paraffin to be dissolved.
4. Centrifuge 10 min at 16,100 ×g at 4 °C. Transfer tubes carefully, holding the top of the tube, not the bottom (see Note 3).
5. Carefully remove 900 μL of xylene by pipette. Use clean tips.
6. Add 900 μL of xylene and vortex well. Repeat steps 3–5 (see Note 4).
7. Add 1 mL of 100% EtOH (room temp) and vortex well. This step removes the xylene. Leave 5–10 min at room temperature.
8. Centrifuge 5 min at 14,000 rpm at 4 °C. Carefully remove as much EtOH as possible without disturbing pellet (see Note 5).
9. Add 1 mL of 100% EtOH. Repeat steps 7 and 8.
10. Repeat step 9.
11. Add 1.5 mL PBS. Make sure cells are suspended by vortexing at high speed.
12. Spin for 10 min at 14,000 at 4 °C (this allows for the cleaning of cells, and put proteins and other junk in solution).
13. Carefully remove as much PBS as possible without disturbing pellet (an additional 10–15 min spin may be necessary for the tissues to pellet).
14. Add 500 μL of protein digestion buffer, 500 μg/mL of Proteinase K, and SDS to a final concentration of 0.1%.
15. Vortex. Incubate in 55 °C shaking water bath for three nights (see Note 6).

Day 4

16. Vortex and spin mixtures down. In the hood, add equal amounts of Phenol/chloroform/isoamyl alcohol to each sample.
17. Centrifuge the tubes at 14,000 rpm for 30 min at 4 °C.
18. Transfer the top aqueous phase using a wide-mouth pipette (cut the pointy end off with a razor blade), without disturbing the interphase, to new tubes.
19. Add 1/10 volume of 10 M ammonium acetate, 40 μg of glycogen regardless of volume, and 2.5× volume of 100% EtOH (see Note 7).
20. Shake the mixture by inverting tubes several times (see Note 8).
21. Spin the solution at room temperature for 20 min at 14,000 rpm, discard supernatant.
22. Resuspend DNA in TE (10 mM Tris–HCl pH 8.0, 1 mM EDTA). The amount depends on the size of the DNA pellet. Store in −20 °C (see Note 9).

3.3. Preparation of Samples for Hybridization

1. Please see DNA extraction protocol for fresh-frozen and FFPE DNA above.
2. For array hybridizations, 5 μg each of tumor and normal genomic DNA were fragmented by DpnII restriction digest before labeling. After purification with QIAquick PCR purification columns, digested DNA was visualized on 1% agarose gel. Digested genomic DNAs should have $A_{260/280}$ ratios of 1.8 to 2.0.
3. For each labeling reaction, 1 μg each of tumor and normal digested DNA were used. Add 1 μg of test DNA to a 0.5 mL tube for Cy3 labeling and 1 μg of control DNA to a 0.5 mL tubes for Cy5 labeling. To each tube, add distilled water to bring the total volume to 21 μL and 20 μL of 2.5× random priming reaction buffer mix. Then, denature DNA and random primer mix by boiling for 5 min and place on ice for 5 min.
4. Incubate denatured samples with 5 μL of 10× dNTP mix, 3 μL of Cy3- or Cy5- dUTP, and 1 μL of Klenow fragment for 2 h at 37 °C in a thermal cycler. Terminate reactions by adding 0.5 M EDTA.
5. Purify labeled DNA samples using QIAquick PCR purification columns. Elute in 50 μL of elution buffer provided in the kit.
6. The degree of labeling or specific activity of each dye is measured by NanoDrop reading. First use 1.5 μL of TE to blank the instrument. Then use 1.5 μL of purified labeled genomic DNA for quantitation. Measure the absorbance at A_{260} nm (DNA), A_{550} nm (cy3), and A_{650} nm (cy5). A successful hybridization requires a minimum Cy3 incorporation of 0.5 pmol/μL and Cy5 incorporation of 0.3 pmol/μL.
7. In a 1.5 mL Eppendorf tube, combine Cy3- and Cy5-labeled DNA, 50 μL of human Cot-1 DNA, 15 μL of 3 M NaOAc, and 415 μL of 100% EtOH. Vortex briefly and spin at 14,000 rpm for 30 min.
8. Discard supernatant, briefly dry pellet at room temperature.
9. Add 25 μL of 10× blocking solution, 125 μL of 2× high-RPM hybridization buffer, and 100 μL of dH_2O. Vortex until the pellet is completely resuspended and the probe solution appears clear.
10. Incubate probe mixture at 95 °C for 5 min, then cool at 37 °C for 30 min. Use aluminum foil to protect tubes from light.
11. Load a clean gasket slide into the Agilent SureHyb chamber base with the gasket label facing up and aligned with the rectangular section of the chamber base. Ensure that the gasket slide is flushed with the chamber base. Slowly dispense 250 μL (for 2× microarray) of hybridization sample mixture onto the gasket well.

Table 1
Wash conditions.

Wash conditions

	Dish	Wash buffer	Temperature	Time
Disassembly	#1	Oligo aCGH Wash Buffer 1	Room temperature	
1st wash	#2	Oligo aCGH Wash Buffer 1	Room temperature	5 min
2nd wash	#3	Oligo aCGH Wash Buffer 2	37 °C	1 min
Acetonitrile wash	#4	Acetonitrile	Room temperature	10 s
3rd wash	#5	Stabilization and drying solution	Room temperature	30 s

12. Put a microarray slide "active side" down onto the gasket slide, so the numeric barcode side is facing up, and the "Agilent"-labeled barcode is facing down. Make sure that the sandwich pair is properly aligned.
13. Put the SureHyb chamber cover onto the sandwiched slides, and slide the clamp assembly onto both pieces. Hand-tighten the clamp firmly onto the chamber. Put assembled slide chamber in the rotator rack in a hybridization oven set to 65 °C. Set hybridization rotator to rotate at 20 rpm. Hybridize for 40 h.

3.4. Preparation and Washing of Microarray Slides (see Table 1 for Wash Conditions)

Always use fresh Oligo aCGH Wash Buffer 1 and Oligo aCGH Wash Buffer 2 for each wash group (up to five slides/batch). The temperature of Oligo aCGH Wash Buffer 2 must be at 37 °C for optimal performance. Add the volume of buffer required to a disposable plastic bottle and warm overnight in an incubator set to 37 °C. Put a slide-staining dish into a 1.5 L glass dish three-fourths filled with distilled water and warm to 37 °C by storing overnight in an incubator.

1. Fill staining dish #1 with Oligo aCGH Wash Buffer 1 at room temperature.
2. Put a slide rack into staining dish #2. Add a magnetic stir bar. Fill staining dish #2 with enough Oligo aCGH Wash Buffer 1 at room temperature to cover the slide rack. Put the dish on a magnetic stir plate.
3. Put the pre-warmed 1.5 L glass dish filled with water and containing staining dish #3 on a magnetic stir plate with heating element. Fill the staining dish #3 approximately three-fourths full with Oligo aCGH Wash Buffer 2 (warmed to 37 °C). Add a magnetic stir bar. Turn on the heating element to maintain temperature of Oligo aCGH Wash Buffer 2 at 37 °C.
4. Remove one hybridization chamber from incubator, and resume rotation of the others. Record whether bubbles formed during hybridization and if all bubbles are rotating freely.

5. Prepare the hybridization chamber disassembly.
   (a) Put the hybridization chamber assembly on a flat surface, and loosen the thumbscrew, turning counterclockwise.
   (b) Slide off the clamp assembly, and remove the chamber cover.
   (c) With gloved fingers, remove the array-gasket sandwich from the chamber base by lifting one end and then grasping in the middle of the long sides. Keep the microarray slide numeric barcode facing up, and quickly transfer the sandwich to staining dish #1.
   (d) Without letting go of the slides, submerge the array-gasket sandwich into staining dish #1 containing Oligo aCGH Wash Buffer 1.
6. With the sandwich completely submerged in Oligo aCGH Wash Buffer 1, pry the sandwich open from the barcode end using the blunt ends of the forceps between the slides, and then gently twist the forceps to separate the slides. Let the gasket slide drop to the bottom of the staining dish. Remove the microarray slide, and quickly put into slide rack in the staining dish #2 containing Oligo aCGH Wash Buffer 1 at room temperature. Minimize exposure of the slide to air. Touch only the barcode portion of the microarray slide or its edges.
7. Repeat step 4 through step 6 for the additional slides in the batch. A maximum of five microarray slides are washed at one time in order to facilitate uniform washing.
8. When all slides in the batch are put into the slide rack in staining dish #2, stir using setting 4 for 5 min to ensure good but not vigorous mixing.
9. Transfer slide rack to staining dish #3 containing Oligo aCGH Wash Buffer 2 at 37 °C, and stir using setting 4 for 1 min.
10. Slowly remove the slide rack trying to minimize droplets on the slides. It should take 5–10 s to remove the slide rack.
11. Remove the slide rack from Oligo aCGH Wash Buffer 2 and tilt the rack slightly and blot the wash buffer using paper towel to minimize carry-over. Quickly transfer the slide rack to staining dish #4 containing acetonitrile, and stir using setting 4 for 10 s.
12. Transfer slide rack to slide-staining dish #5 filled with stabilization and drying solution, and stir using setting 4 for 30 s.
13. Slowly remove the slide rack to minimize droplets on the slides.
14. Scan slides immediately to minimize the bleaching of signal intensities. For temporary storage use plastic slide box wrapped in aluminum foil in dark.

3.5. Scanning of Microarray and Image Processing

Immediately after washing slides are scanned in Axon 4000B scanner using GenePix pro 4.0. Microarray images are analyzed, and data points were generated using the Feature Extraction software (version 9.1, Agilent Technologies) with linear normalization (protocol-v4_91). Data are subsequently imported into CGH Analytics software (version 3.4.40, Agilent Technologies). Detection of gains and losses are based on the z-score algorithm (threshold 2.5) and visual inspection of the $\log_2$ ratios. In general, $\log_2$ ratios ≥ 0.4 in at least five consecutive probes are considered as a reliable copy number alteration (CNA). Genes with $\log_2$ ratios greater than 2 are considered as highly amplified.

3.6. Data Analysis Methods and Interpretation

DNA CNA is identified through dynamic thresholding of segmented aCGH data. Circular binary segmentation (CBS) is used to segment each hybridization into regions of common mean (23). In the next step, the median absolute deviation (MAD) across all segments is then obtained for each hybridization. Probes assigned to segments with mean value greater than a scaled MAD are identified as gain. Likewise, probes corresponding to segments with mean value less than a scaled MAD are identified as loss. A default MAD scaling factor of 1.11 is utilized for both gains and losses (24). Following segmentation and classification, data are further reduced, without compromising the continuity and breakpoints, to facilitate downstream analyses (25). To identify minimal regions of common alteration across all hybridizations, the Genomic Identification of Significant Targets in Cancer (GISTIC) approach is utilized (26).

4. Discussion

Copy number alterations or CNAs are common events in brain tumors. A recent systematic survey of glioblastoma (GBM) has confirmed the presence of recurrent amplifications of genomic loci containing the *EGFR, CDK4, PDGFRA*, *MDM2*, and *MDM4* genes. Several other loci were also identified to be recurrently amplified but at lower frequency (27). Additionally, homozygous deletions of loci containing *CDKN2A/B* and *PTEN* are also frequently detected in glioblastoma. Gliomas progress through the malignant transformation of low-grade gliomas (WHO grade II) to GBM (WHO grade IV), the most biologically aggressive form of glioma (28, 29). Molecular stratification of GBM based on age of onset has divided these histologically distinguishable tumors into primary and secondary GBM (30). Primary GBM, which predominantly appears in older individuals and without a preceding low-grade growth phase, is strongly associated with amplifications of

EGFR and *MDM2* and deletion of chromosome 10q. These changes, especially the first two, are often not seen in secondary GBM found predominantly in younger patients. Amplification of the *EGFR* locus and resultant EGFR protein overexpression is found in one-third of GBM and carries significant functional consequences (31). Molecular diagnostic tests for *EGFR* amplification are important adjuncts in the diagnosis of glioma and might provide crucial predictive information for response to targeted therapy (32). Systematic analysis of CNAs in meningioma, an extra-axial brain tumor arising from meningothelial cells, has identified gain of chromosome arm 1q as a poor clinical prognosticator (22, 33). Lastly, recurrent losses of genomic regions containing the *CDKN2A* and *PTEN* loci are identified in sporadic chordomas using oligonucleotide-based aCGH chips (the same type that is discussed in the methodology session). Interestingly, many of these tumors do not have amplification of the *brachyury* locus in chromosome 6q (34). It appears recurrent CNAs are present and function as drivers of different types of brain tumors, and aCGH technology affords a well-validated and robust method for genome-wide interrogation of copy number status in these tumors. Recent technical improvements have increased the versatility of this constantly evolving technology to better investigate copy number changes in brain tumors (21).

5. Notes

Note 1. Sections should be placed on slides to be scraped. Do not put sections into Eppendorf tubes directly. Mincing of tissues by scraping often leads to better dissolution in extraction buffer.

Note 2. Rinsing the blade off into the tube with the 1 mL of xylene will help to get all pieces of tissue into the tube. Change blades for different samples, and work in an isolated area to avoid contamination of other DNA samples.

Note 3. Touching the bottom could cause the xylene to heat and cause current, stirring the tissue.

Note 4. For very old or large samples, you may repeat step 6, since it is very important to remove paraffin completely for better dissolution of tissues in extraction buffer.

Note 5. This helps remove all traces of paraffin.

Note 6. During incubation, make sure to vortex and replenish proteinase K in the morning and afternoons to facilitate DNA extraction.

Note 7. This aids in precipitation of genomic DNA.

Note 8. The solution may look cloudy at this stage.

Note 9. Since the yield for FFPE samples are so small, instead of loading 5 μL of sample for the gel, only load 1 μL of sample. A decent gel picture can still be taken. Also, a digest is not required for FFPE samples because the DNA is already degraded due to age and the fixing process. For quantification, aim to get 150–250 ng/μL. If the yield is higher, dilute the sample and quantitate again.

References

1. Iafrate AJ, Feuk L, Rivera MN et al (2004) Detection of large-scale variation in the human genome. Nat Genet 36(9):949–951. doi:10.1038/ng1416 ng1416 [pii]
2. Lee C, Iafrate AJ, Brothman AR (2007) Copy number variations and clinical cytogenetic diagnosis of constitutional disorders. Nat Genet 39(7 Suppl):S48–S54. doi:ng2092 [pii] 10.1038/ng2092
3. Sanders SJ, Ercan-Sencicek AG, Hus V et al (2011) Multiple recurrent de novo CNVs, including duplications of the 7q11.23 williams syndrome region, are strongly associated with autism. Neuron 70(5):863–885. doi:doi:S0896-6273(11)00374-6 [pii] 10.1016/j.neuron.2011.05.002
4. Stefansson H, Rujescu D, Cichon S et al (2008) Large recurrent microdeletions associated with schizophrenia. Nature 455(7210):232–236. doi:nature07229 [pii] 10.1038/nature07229
5. Weiss LA, Shen Y, Korn JM et al (2008) Association between microdeletion and microduplication at 16p11.2 and autism. N Engl J Med 358(7):667–675. doi:NEJMoa075974 [pii] 10.1056/NEJMoa075974
6. Beroukhim R, Mermel CH, Porter D et al (2010) The landscape of somatic copy-number alteration across human cancers. Nature 463(7283):899–905. doi:nature08822 [pii] 10.1038/nature08822
7. Ou Z, Stankiewicz P, Xia Z et al (2011) Observation and prediction of recurrent human translocations mediated by NAHR between nonhomologous chromosomes. Genome Res 21(1):33–46. doi:gr.111609.110 [pii] 10.1101/gr.111609.110
8. Basinko A, Audebert-Bellanger S, Douet-Guilbert N et al (2011) Subtelomeric monosomy 11q and trisomy 16q in siblings and an unrelated child: molecular characterization of two der(11) t(11;16). Am J Med Genet A 155(9):2281–2287. doi:10.1002/ajmg.a.34162
9. Nicholas MK, Lukas RV, Jafri NF et al (2006) Epidermal growth factor receptor—mediated signal transduction in the development and therapy of gliomas. Clin Cancer Res 12(24):7261–7270. doi:12/24/7261 [pii] 10.1158/1078-0432.CCR-06-0874
10. Hostetter G, Kim SY, Savage S et al (2010) Random DNA fragmentation allows detection of single-copy, single-exon alterations of copy number by oligonucleotide array CGH in clinical FFPE samples. Nucleic Acids Res 38(2):e9. doi:10.1093/nar/gkp881
11. Panzeri E, Conconi D, Antolini L et al (2011) Chromosomal aberrations in bladder cancer: fresh versus formalin fixed paraffin embedded tissue and targeted FISH versus wide microarray-based CGH analysis. PLoS One 6(9):e24237. doi:10.1371/journal.pone.0024237
12. Dutt A, Beroukhim R (2007) Single nucleotide polymorphism array analysis of cancer. Curr Opin Oncol 19(1):43–49
13. Meyerson M, Gabriel S, Getz G (2010) Advances in understanding cancer genomes through second-generation sequencing. Nat Rev Genet 11(10):685–696. doi:nrg2841 [pii] 10.1038/nrg2841
14. Geiss GK, Bumgarner RE, Birditt B et al (2008) Direct multiplexed measurement of gene expression with color-coded probe pairs. Nat Biotechnol 26(3):317–325. doi:10.1038/nbt1385
15. Payton JE, Grieselhuber NR, Chang LW et al (2009) High throughput digital quantification of mRNA abundance in primary human acute myeloid leukemia samples. J Clin Invest 119(6):1714–1726. doi:10.1172/JCI38248
16. Pinkel D, Segraves R, Sudar D et al (1998) High resolution analysis of DNA copy number variation using comparative genomic hybridization to microarrays. Nat Genet 20(2):207–211. doi:10.1038/2524
17. Snijders AM, Nowak N, Segraves R et al (2001) Assembly of microarrays for genome-wide measurement of DNA copy number. Nat Genet 29(3):263–264. doi:10.1038/ng754
18. Cai WW, Mao JH, Chow CW et al (2002) Genome-wide detection of chromosomal imbalances in tumors using bac microarrays. Nat Biotechnol 20(4):393–396. doi:10.1038/nbt0402-393

19. Beheshti B, Braude I, Marrano P et al (2003) Chromosomal localization of DNA amplifications in neuroblastoma tumors using cDNA microarray comparative genomic hybridization. Neoplasia 5(1):53–62
20. Nigro JM, Misra A, Zhang L et al (2005) Integrated array-comparative genomic hybridization and expression array profiles identify clinically relevant molecular subtypes of glioblastoma. Cancer Res 65(5):1678–1686. doi:10.1158/0008-5472.CAN-04-2921
21. Mohapatra G, Engler DA, Starbuck KD et al (2011) Genome-wide comparison of paired fresh frozen and formalin-fixed paraffin-embedded gliomas by custom BAC and oligonucleotide array comparative genomic hybridization: facilitating analysis of archival gliomas. Acta Neuropathol 121(4):529–543. doi:10.1007/s00401-010-0773-z
22. Gabeau-Lacet D, Engler D, Gupta S et al (2009) Genomic profiling of atypical meningiomas associates gain of 1q with poor clinical outcome. J Neuropathol Exp Neurol 68(10):1155–1165. doi:10.1097/NEN.0b013e3181ba3952 00005072-200910000-00011 [pii]
23. Olshen AB, Venkatraman ES, Lucito R et al (2004) Circular binary segmentation for the analysis of array-based DNA copy number data. Biostatistics 5(4):557–572. doi:10.1093/biostatistics/kxh008
24. Korkola JE, Heck S, Olshen AB et al (2008) In vivo differentiation and genomic evolution in adult male germ cell tumors. Genes Chromosomes Cancer 47(1):43–55. doi:10.1002/gcc.20504
25. Van Wieringen WN, Van De Wiel MA, Ylstra B (2007) Normalized, segmented or called aCGH data? Cancer Inform 3:321–327
26. Beroukhim R, Getz G, Nghiemphu L et al (2007) Assessing the significance of chromosomal aberrations in cancer: methodology and application to glioma. Proc Natl Acad Sci U S A 104(50):20007–20012. doi:10.1073/pnas.0710052104
27. Cancer Genome Atlas Research Network (2008) Comprehensive genomic characterization defines human glioblastoma genes and core pathways. Nature 455(7216):1061–1068. doi:nature07385 [pii] 10.1038/nature07385
28. Louis DN, Ohgaki H, Wiestler OD et al (eds) (2007) Who classification of tumours of the central nervous system. IARC, Lyon
29. Wen PY, Kesari S (2008) Malignant gliomas in adults. N Engl J Med 359(5):492–507. doi:359/5/492 [pii] 10.1056/NEJMra0708126
30. Ohgaki H, Kleihues P (2007) Genetic pathways to primary and secondary glioblastoma. Am J Pathol 170(5):1445–1453
31. Kesari S, Ramakrishna N, Sauvageot C et al (2005) Targeted molecular therapy of malignant gliomas. Curr Neurol Neurosci Rep 5(3):186–197
32. Yip S, Iafrate AJ, Louis DN (2008) Molecular diagnostic testing in malignant gliomas: a practical update on predictive markers. J Neuropathol Exp Neurol 67(1):1–15. doi:10.1097/nen.0b013e31815f65fb
33. Jansen M, Mohapatra G, Betensky RA et al (2011) Gain of chromosome arm 1q in atypical meningioma correlates with shorter progression-free survival. Neuropathol Appl Neurobiol. doi:10.1111/j.1365-2990.2011.01222.x
34. Le LP, Nielsen GP, Rosenberg AE et al (2011) Recurrent chromosomal copy number alterations in sporadic chordomas. PLoS One 6(5):e18846. doi:10.1371/journal.pone.0018846

Chapter 21

Finding Common Regions of Alteration in Copy Number Data

Oscar M. Rueda, Ramon Diaz-Uriarte, and Carlos Caldas

Abstract

In this chapter, we review some recent methods designed for detecting recurrent copy number regions, that is, genomic regions that show evidence of being altered in a set of samples. We analyze Affymetrix SNP6 data from 87 Her2-type breast tumors from a recent study using three different methods, showing different definitions and features of common regions: studying heterogeneity in copy number profiles, refining candidates for driver oncogenes, and consolidating broad amplifications.

Key words: aCGH, SNP, Copy number alterations, Copy number variation, Minimal common regions

1. Introduction

Cancer is a heterogeneous disease. The study of genomic profiles of individual copy number alterations (CNAs) is only the first step in the search of oncogenes or tumor suppressors. Most of the aberrations are passengers, and the discovery of driver alterations often involves finding those that are present in a significant number of samples (1). This is what has been called in the literature "common regions of alteration" or "recurrent alterations." However, the main problem in the identification of recurrent CNA regions (rCNA regions) is the definition itself. Rueda and Diaz-Uriarte (2) defined an rCNA region as "a set of contiguous probes (a region) that, as a group, shows a high enough probability (or evidence) of being altered (e.g., gained) in at least some samples or arrays." This definition is open in the sense of how we decide how big must be the set of probes, the strength of the evidence in the alteration, and the number of probes or length of the region. In the same review (2), we presented several different scenarios that fit the above definition of "recurrent region." For instance, sometimes

Diponkar Banerjee and Sohrab P. Shah (eds.), *Array Comparative Genomic Hybridization: Protocols and Applications*, Methods in Molecular Biology vol. 973, DOI 10.1007/978-1-62703-281-0_21, © Springer Science+Business Media, LLC 2013

the alteration is only present in a subset of the samples, while in other situations, the same region is gained in some samples and lost in others, as it is often the case for germline copy number aberrations (copy number variations, CNVs). Several regions may overlap with different frequencies of alteration in the boundaries of driver cancer genes, while others can show different copy number patterns between samples but the same pattern within samples, thus can be summarized with the copy number of just one probe. Many algorithms have been proposed; all of them designed to detect alterations under one or several of these scenarios. In this chapter, we will not review all methods available (we refer the reader to the aforementioned review (2) or (3)). Instead, we will mention some recently developed methods, and we will select some methods for different practical situations and offer some guidance through the analysis.

To introduce the problems and methods, we will start with a simple, naive approach. We could just compute the frequency of alteration (gain and loss) for each probe and merge consecutive probes with a frequency larger than a given threshold. The problem with this approach is that it does not lead to a common region, because the set of samples that shows alteration in a given probe doesn't have to be the same set with alterations in the next probe. Nevertheless, this method can be implemented with almost any software, even with a spreadsheet. We only have to create a column with the sum of the samples that present an alteration for each probe and select those with a sum larger than a given threshold. We will not discuss this approach any further in the sections that follow.

2. Methods

There are two possible starting points in the computation of common regions of alteration (CRAs): first, when we have raw data (not segmented) and we want to do the segmentation and the computation of common regions in one step using the same method, and second, when we have already segmented the data using one of the many available algorithms for that task. We will cover both cases, and we will also present a web application that allows for both types of input data, waviCGH.

2.1. Input: Raw Data

Here, we assume that, whatever it is aCGH data with two colors or SNP data with just one, the data have been normalized so it is in the form of $\log_2$-ratios (0 is supposed to be neutral copy number).

2.1.1. RJaCGH

RJaCGH (4, 5) is a Bayesian nonhomogeneous hidden Markov model (NH-HMM) for the analysis of copy number data. Instead of selecting a priori the number of hidden states and assuming that each hidden state is a copy number, RJaCGH fits simultaneously several HMMs with different number of states and assign a posterior probability to each of them using reversible jump (RJ). The states are classified a posteriori into states of loss, neutral copy number, or gain using a window parameter (see Subheading 2.2.1). The two algorithms available for computing common regions, pREC-A, and pREC-S (6) work with probabilities of alteration. They are designed to detect regions with a joint probability of alteration larger than a given threshold over all samples or over a subset of samples, respectively. This can be very useful in situations when there is a lot of heterogeneity present and important alterations are expected to be present in only a small fraction of the samples. RJaCGH is available as an R package.

2.1.2. Other Recently Developed Methods

CNAnova (7) fits a linear model to a window of probes on all samples. The individual t-statistics for each sample detect copy number alterations in each array, and an overall F-statistic contrasts that the region is commonly altered. Then it uses a procedure based on mean shift to obtain the boundaries of the regions and finally computes the significance of each region correcting for multiple testing. It is also available as an R package `http://www.compbio.group.cam.ac.uk/software.html`.

CMDS (8) uses the correlation blocks between samples that share a common alteration. It computes the normalized Pearson correlation for each pair of probes. As it only needs to look at the correlations for neighbor probes, it performs a diagonal transformation computing for each probe the correlation with their *b* adjacent probes that speeds up the computations. It finally uses a *t*-test to obtain significant probes. R and C code are available at `https://dsgweb.wustl.edu/qunyuan/software/cmds/`.

These algorithms are designed to detect regions that are altered, on average, in all samples.

2.2. Input: Segmented/ Called Data

In this section, we suppose that we have segmented our data using one of the many methods available. In that case, and depending on the output of the algorithm, we can have segmented means (continuous data that represent regions that share the same copy number, but no information about which ones are gains/losses) or called data (copy number calls, or categories such as neutral, lost, or gained for each of the probes and samples). Since some CRA algorithms use continuous input and others discrete data, we will make some comments first about how to obtain called data from segmented data. Next (starting on Subheading 2.2.2), we discuss several methods that find recurrent regions using the called data.

2.2.1. Calling Alterations

We assume that our input data is in the following format (note that this is the standard DNAcopy (9, 10) output, one of the most used segmentation methods):

```
Sample Chrom Start End Number.probes Segmented.Mean
```

where each row is a region of the same copy number for a given sample.

Threshold-Based Approach

It was the first method applied to aCGH analysis, and it is based on setting individual thresholds for each sample based on its variability. If we call $\bar{y}$ the mean, σ_y the standard deviation of the raw data (median and IQR are valid choices too, or using the statistics from segmented data instead of the raw data), and m_t the segmented mean of the region t, we classify each region t as a region of gain or loss according to these equations:

$$t / m_t \geq \bar{y} + k_G \sigma_y \Rightarrow \text{Gain}$$

$$t / m_t \leq \bar{y} + k_L \sigma_y \Rightarrow \text{Loss}$$

We have to define the thresholds k_G and k_L for gains and losses. We recommend using nonsymmetrical thresholds, because the expected loss of one copy is $\log_2(1/2) = -1$, while the expected gain of one copy is $\log_2(3/2) = 0.58$, although the actual values will depend on the particular platform used. Plateau plots in DNAcopy (9) can be very useful for setting these numbers. In a recent study with Affy SNP6 (11), we have used values of 2 for gains and −2.5 for losses. Additional adjustments can be done to correct for cellularity (the proportion of non-tumoral cells in the sample) using different thresholds depending on that proportion in each sample.

Classification of Hidden States

There are many HMM-based algorithms for copy number segmentation. Some of them assign a priori a copy number to each hidden state, but sometimes this is not optimal. RJaCGH (12) includes a method that can also be used on an arbitrary HMM to classify each hidden state into loss, neutral, or gain. It is based on a normal reference μ_N (the expected log-ratio for a normal copy number; that would be 0 in most cases), and a window for losses and for gains with widths W_L, W_G (similar to the thresholds k_L and k_G). We use the mean (μ_i) and the variance (σ^2_i) of each hidden state to classify each hidden state following the algorithm in Fig. 1.

The same adjustments for cellularity can be done using different values for the windows in each array.

Other Methods

Other calling methods take segmented data as input and classify each segment into copy number states. CGHcall (13), available as a Bioconductor package, uses a mixture of normals to classify each segment into 3 (loss, neutral, or gain) or 4 (amplification) states. FastCall (14), part of the TASSO R package, uses a similar methodology but is significantly faster.

Algorithm 1: RJaCGH state labelling in HMM models

Input: normal reference: μ_N.
Input: window: W_L, W_G.
foreach *state i* **do**
  Compute $NI = (\mu_N - W_L, \mu_N + W_G)$;
  $P(i = Loss) = P(N_{(\mu_i, \sigma_i^2)} \leq NI_1)$;
  $P(i = Gain) = P(N_{(\mu_i, \sigma_i^2)} \geq NI_2)$;
  $P(i = Normal) = 1 - P(i = Loss) - P(i = Gain)$;
  i is classified as $arg.max\{P(i = Loss), P(i = Gain), P(i = Normal)\}$
end
Output: Each hidden state classified as neutral, lost or gained

Fig. 1. RJaCGH algorithm for labeling hidden states.

2.2.2. GISTIC

GISTIC (15, 16) is a method for recurrent CNA regions whose main objective is to differentiate between driver and passenger aberrations. It uses a G-score based on the frequency of the copy number alterations and the average amplitude in log-ratio. Then it uses a permutation test to add significance to those regions and finally incorporates biological assumptions to define "peak regions." The most recent version is GISTIC2.0 (17). It starts with segmented data (in continuous form) and follows this pipeline:

- Deconstruction of segmented profiles into somatic CNAs (sCNAs). The main goal of this stage is to model the background rate of SCNAs. In this new version, it uses an algorithm that separates arm-level and focal SCNAs by length instead of by using high-amplitude thresholds.
- Probabilistic scoring of SCNAs. The aim is to compute the probability that the observed SCNAs occurred by chance alone. The functional form for the log-probability distribution of observing a SCNA with a given length and amplitude (in copy numbers, not in log copy number, as the previous version) has been derived based on a large set of samples from two Affymetrix platforms. To avoid biases in other platforms, the method only uses a certain range of the data where probes don't saturate. Then, it uses a permutation test to compute *p*-values. For deletions, the authors propose a procedure named GeneGISTIC that scores genes instead of probes.
- Identification of independent significantly altered regions. This is the so-called peel-off algorithm, designed to capture independent peaks. It has been revised, being an improvement over the old version (15).
- Localizing target genes for each significantly altered region. Instead of selecting the minimal common region (see Rouveirol (18)), they employ at this stage an algorithm called RegBounder, that models the random fluctuation of the G-scores to capture driver genes.

Related Methods

GISTIC has proved to be a very popular method used in many studies (see (17) for references), and several other improvements have been proposed. Etemadmoghadam et al. (19) present a method for comparing common regions between groups of samples called Δ-GISTIC. The idea is to apply GISTIC independently to the two groups and compute for each region of interest the difference between G-scores (ΔG). Then, the labels of the two groups are permuted and this statistic is computed for each permutation to generate the distribution under the null hypothesis. This approach has some practical problems, such as when some of the regions do not appear in certain permutations. A major difficulty with using this approach is that the authors do not provide code.

JISTIC (20) uses a variant for the peeling-off procedure called limited peel-off that decomposes the G-score into the part corresponding to the peak and another that would include an independent contribution. This helps to distinguish between "genuine multiple peaks and a single peak within noisy fluctuations." The code is platform independent as a JAVA application and is available at `http://www.c2b2.columbia.edu/danapeer-lab/html/software.html`.

DINAMIC (21) uses a permutation scheme based on cyclic shifts of the copy number profiles. It takes a probe *k* and considers the sequence of probes starting from *k* (merging the last probe and the first). The index *k* can be different for each sample. For each of these permutations they consider a statistic based on the maximum value of the sum of gains/losses for each probe. Then they use a peeling procedure specifically designed for these shifts to identify peaks. R code is available in `http://www.bios.unc.edu/research/genomic_software/DINAMIC/`.

GAIA (22) is also similar in spirit to GISTIC, but it works only with the calls and not with the log-ratios and uses homogeneous peel-off, based on within-sample homogeneity. They define a homogeneous region as a sequence of probes with the same profile within samples (but not necessarily between samples). Their peel-off algorithm can expand the boundaries of the region if two contiguous probes are homogeneous.

These authors show in simulations and real datasets improvements over the old GISTIC, but they show no comparisons with version 2.0 of GISTIC (comparisons between GISTIC 2.0 and the other methods are missing in (17) too).

2.3. waviCGH

waviCGH (23) is a web application for the analysis and comparison of genomic copy number profiles. It has a complete pipeline with the following steps:

1. Input data: it accepts log-ratios or called copy number data.
2. Previsualization: for log-ratios or segmented data in the case of called input.

3. Background correction, normalization, and preprocessing: for log-ratios. waviCGH can update the genomic coordinates to a new assembly, median subtract the data, average duplicated probes, and impute missing data using several methods.
4. Segmentation: for log-ratios, it provides several parallelized methods for segmentation and calling of alterations.
5. Common regions: there are two methods available. SuperSORI is a fast method that returns regions shared by at least two individuals (called smallest regions of imbalance). It has a couple of parameters for filtering segments smaller than a given number of probes and to remove gaps smaller than a given number of probes between two regions. The other method ("permutations" in the web interface) is in fact a method for finding common probes; not regions, and it is based on permutations.
6. Cytogenetic browser: waviCGH allows us to inspect the results using a simple interface with links to Ensembl. It also displays a summary ideogram.

3. Data Analysis and Interpretation

The data we will use comes from the Metabric study (11), a collection of 997 Affymetrix SNP6 breast tumor samples (the data is publicly available; see (11) for details). We will only work with the 87 Her2-type subset.

3.1. Exploring Heterogeneity with RJaCGH

In the Metabric study (11), 997 breast tumors were classified into five intrinsic subtypes (Basal, Her2, Luminal A, Luminal B, and normal-like) based in their expression profiles using the PAM50 classifier (24). We know that these groups don't reflect all the heterogeneity present in breast tumors. Here, we will analyze the 17 q-arm of the 87 Her2 samples to explore that variability in copy number in the ERBB2 oncogene region.

RJaCGH package can be installed from R using the following command:

```
> install.packages("RJaCGH")
```

We have the data in a file named "17qHer2.txt" that has a column for chromosome, a column for probe position and the rest of the columns for the log-ratios for each sample. This R code would read the data and fit the HMM model:

```
> library(RJaCGH)
> X <- read.table('17qHer2.txt', header=TRUE,
sep='\t')
> Chrom <- X$Chrom
```

```
> Pos <- X$Pos
> y <- X[,3:89)
>jp <- list(sigma.tau.mu=rep(0.05, 6), sigma.
tau.sigma.2=rep(0.03, 6), sigma.tau.
beta=rep(0.07, 6), tau.split.mu=0.1, tau.
split.beta=0.1)
> set.seed(3423)
> fit <- RJaCGH(y=y, Chrom=Chrom, Pos=Pos, jump.
parameters=jp, k.max=6, burnin=2500, TOT=1000)
```

This code reads the data and splits the genomic information from the log-ratios. Then, the call to `RJaCGH()` runs the segmentation for all the arrays. It has several parameters: the object `jp` is a list of parameters for the reversible jump (see (4) for details). If it is not provided, the function will make a search for a set of appropriate values. `k.max` is the maximum number of hidden states to fit and `burnin` and `TOT` are the number of burn-in and definite samples to draw from the distribution (equivalent to iterations of the algorithm). Depending on the computational power available, these numbers should be increased. Note that the function will create a temporary folder that must be kept if we plan to do further work with the object.

Sometimes it is necessary to adjust the thresholds for calling alterations, as we have seen in the previous section. Inspecting the distribution of the posterior means of hidden states in each array, we can see if the states classified as "gain" or "loss" have sensible log-ratio medians. If not, we can correct for that using the `relabelStates()` function and the parameter `window`. Increasing its default value, 1, will decrease the number of alterations called (see help file for details).

Finally, we run pREC-S to find regions of at least 0.5 probability of gain shared by at least 15 arrays:

```
mCR <- pREC_S(fit, p=0.5, freq.array=15,
alteration="Gain")
```

Figure 2. shows all regions found by the algorithm. The top panel shows the q arm of chromosome 17, where clearly the most important region is located in the ERBB2 locus. There are other interesting areas, but some of them are CNVs that are inherited alterations not related to the disease. The bottom panel shows the zoomed ERBB2 locus, where we can see that the minimal common region only contains ERBB2, as is well known for Her2-type breast cancer tumors.

One thing that comes to mind when looking at Fig. 2. is that not all samples have amplification in ERBB2. We can use the output of pREC-S to cluster (using Ward's hierarchical method) the samples based on the number of common probes in the recurrent regions obtained. If we label the samples based on the expression of ERBB2, we will be able to detect if the Her2-type group is homogeneous:

```
> plot(mCR, stats=FALSE, cex.axis=0.5, array.
labels=Clinical$Her2.Expr, method="ward")
```

Figure 3. shows part of the output of the above call to `plot`. In that dendrogram, we can see that the samples that cluster

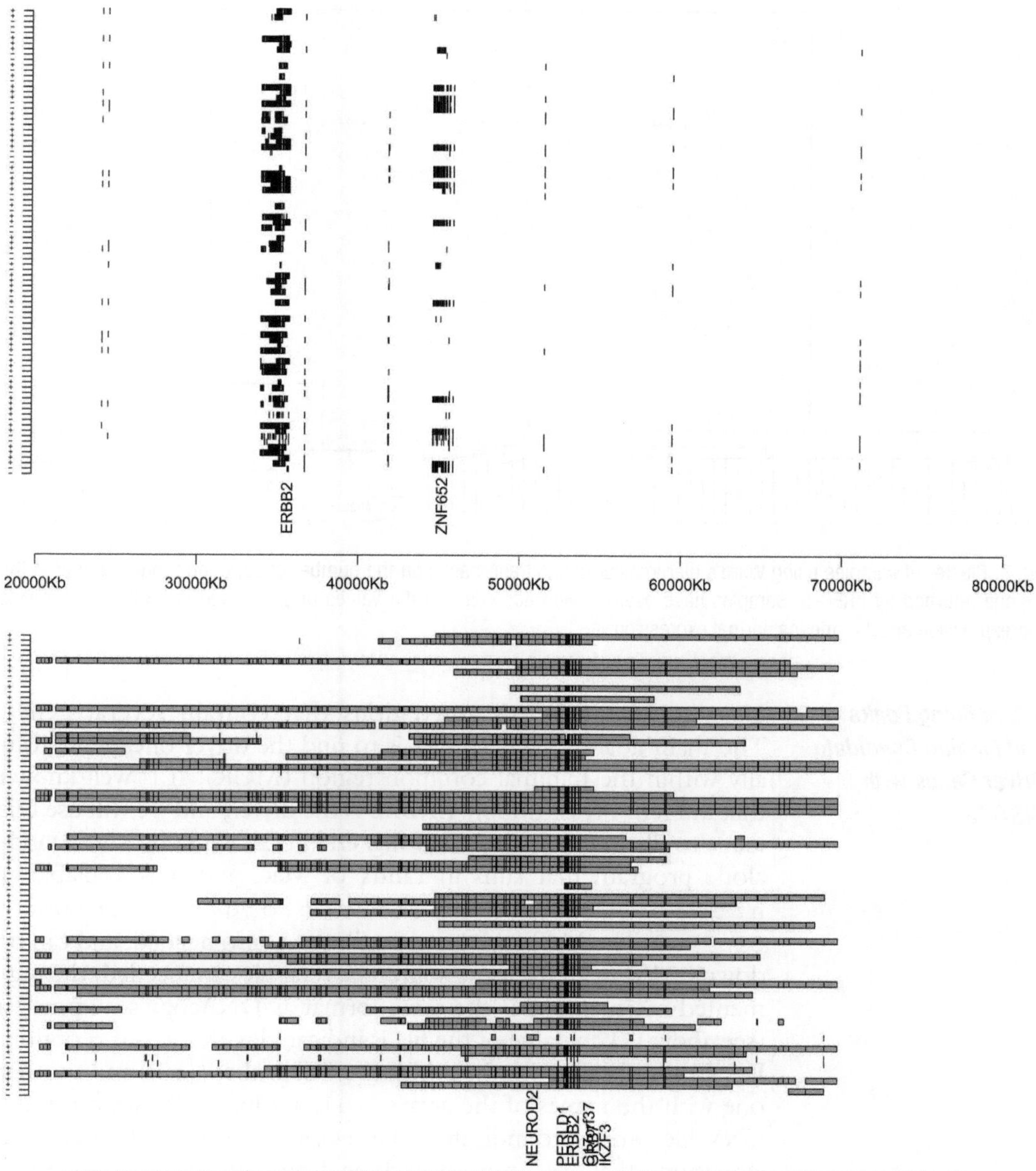

Fig. 2. Top panel shows common regions of gain in q arm of chromosome 17 with a probability of alteration larger than 0.5 and shared by at least 15 samples as returned by pREC-S. Bottom panel zooms in the region of ERBB2. The samples have been labeled according to their values of expression in ERBB2: "+" means overexpression and "−" means normal expression.

together with respect to copy number are essentially the same samples that show overexpression of Her2. As we have seen, the ERBB2 amplicon is the most important in this arm, therefore not all the samples in the Her2 subgroup show both amplification and overexpression of this gene (see (11) for a proper identification of this subgroup).

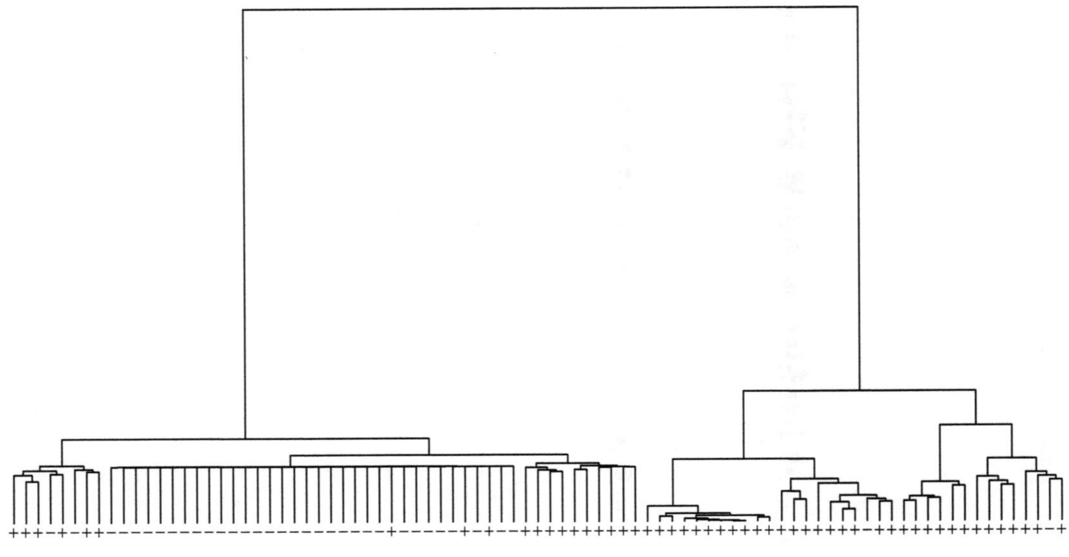

Fig. 3. Cluster of samples using Ward's hierarchical method and based on the number of common probes shared in the regions returned by pREC-S. Samples have been labeled according to the values of expression in ERBB2: "+" means overexpression and "−" means normal expression.

3.2. Refining Peaks and Finding Candidate Driver Genes with GISTIC2

Usually, amplifications are regions that contain several genes. The main goal of many studies is to find the driver oncogene, usually within the minimal common region (MCR). It is well known that in Her2 breast tumors ERBB2 is the driver, and we will use this locus to illustrate an analysis using GISTIC2. GISTIC2 is a standalone program that runs in Linux or Mac, and it is available at `http://www.broadinstitute.org/cancer/pub/GISTIC2`. The MATLAB runtime library is also needed, and can be downloaded from the same source. Several files are needed: the segmented means file uses the same format as DNAcopy's output (9) (see above). The name of the file is indicated with the `-seg` option. We also need a file with the positions of all probes (`-mk`) and another one with the names of the arrays (`-alf`). Optionally, we can use a CNV file (`-cnv`) to indicate some regions of probes with germline alterations that we want to exclude from our analysis. We must choose a directory to save the results (`-b`) and specify the reference genome file (`-refgene`, included with the installation). There are many parameters for the function call, but the most important are the thresholds for gains and losses (`-ta`, `-td`) . They must have the same value for every array. We think that it would be a better idea to use individual thresholds for each sample to take into account differences in noise or cellularity values. We could scale every array with its mean (or median) and standard deviation (or IQR) to obtain the same effect, but unfortunately this procedure can change drastically the range of the log-ratios and make the distributional assumptions of the method implausible (personal communication with the authors). Instead, we will use for this example a value of 0.25 for calling gains and a value of −0.3 (note that it must be specified in the

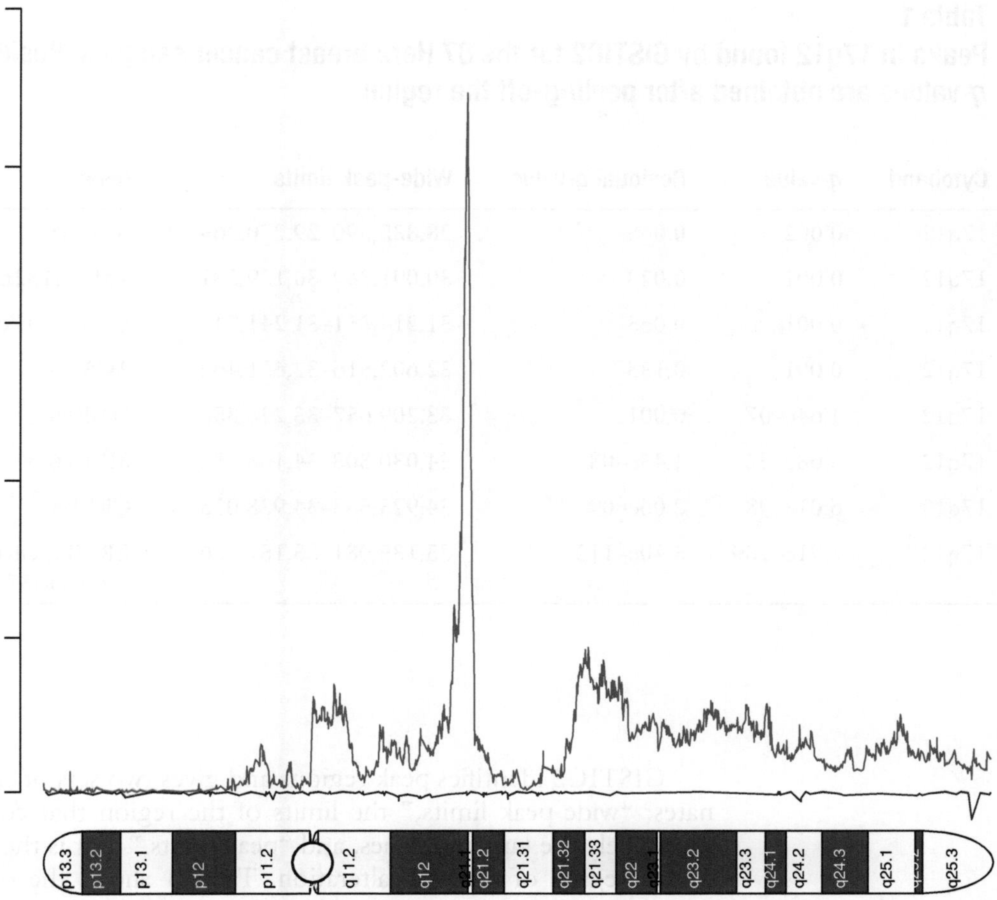

Fig. 4. G-scores for the regions in chromosome 17 for the 87 Her2 breast cancer samples. For amplifications, the plot shows the G-score and for deletions, -G-score.

function call in absolute value) for calling losses. We will also set the smallest size in probes of the regions considered as 4 (`-js 4`), we will choose to perform gene analysis for deletions (`-genegistic 1`), set the confidence level to 0.90 (`-copnf 0.90`), verbose output (`-v 30`), and exclude samples with a number of segments larger than 10,000 (`-maxseg 10000`).

The call, after setting the right environment variables (see the example file included with the distribution), is the following:

```
./gp_gistic2_from_seg -b $basedir -seg $segfile
-mk $markersfile -refgene $refgenefile -alf $alf
-cnv $CNV -ta 0.25 -td 0.3 -js 4 -genegistic 1
-conf 0.90 -v 30 -maxseg 10000
```

GISTIC2 produces several files for output. For lack of space, we do not provide details; the reader can refer to the manual for a complete explanation. Here, we will focus on the G-score, a combination of the frequency of alteration and the copy number. Figure 4. plots the values of this statistic for gains/amplification (positive axis) and losses/deletions (negative axis). We can see a high peak in ERBB2 locus, in q12.

Table 1
Peaks in 17q12 found by GISTIC2 for the 87 Her2 breast cancer samples. Residual *q*-values are obtained after peeling-off the region

Cytoband	*q*-value	Residual *q*-value	Wide-peak limits	Genes
17q12	0.002	0.068	28,835,590–29,270,269	ACCN1
17q12	0.001	0.023	30,091,349–30,139,331	TMEM132E
17q12	0.001	0.065	31,214,351–31,241,245	CCL5, C17orf66
17q12	0.001	0.133	32,602,516–32,651,462	ACACA
17q12	4.64e−07	0.001	33,209,657–33,228,356	HNF1B
17q12	3.08e−22	4.43e−08	34,030,803–34,108,152	MLLT6
17q12	6.01e−98	3.06e−09	34,925,553–34,938,025	CRKRS
17q12	7.91e−139	3.70e−112	35,135,081–35,157,736	ERBB2, GRB7, C17orf37

GISTIC2 identifies peak regions and gives two sets of coordinates: "wide-peak limits," the limits of the region that contains most likely the targeted genes, and "peak limits," that is the limits of the region of maximal alteration. Table 1 shows the regions found in arm 17q12 with their wide-peak limits and the genes they target. As expected, ERBB2 region is the most significant one although it also includes GRB7 and C17orf37. Refining this region with the peak limits does not remove any of the three genes.

3.3. Exploring Common Regions with waviCGH

We are going to run chromosome 17 for the 87 Her2 samples using called data based on DNAcopy (9) and the thresholds described in Subheading 2 (for more details, see (11)). waviCGH requires a zipped file with a column named `ProbeNames` with the names of the probes, a column named `Chr` with the chromosomes, a column named `Position` with the positions, and a column for each sample with the call. We then select SuperSORI with the default values and run the analysis. Figure 5 shows on the left panel a screen capture of the output of the tool. There are two small regions that contain ERBB2, both wider than the one found by GISTIC2. The reason is that, with the default parameters, SuperSORI merges nearby regions removing gaps between them (in this case, no more than 3 probes close). On the right panel, we can see the samples that contribute to each region.

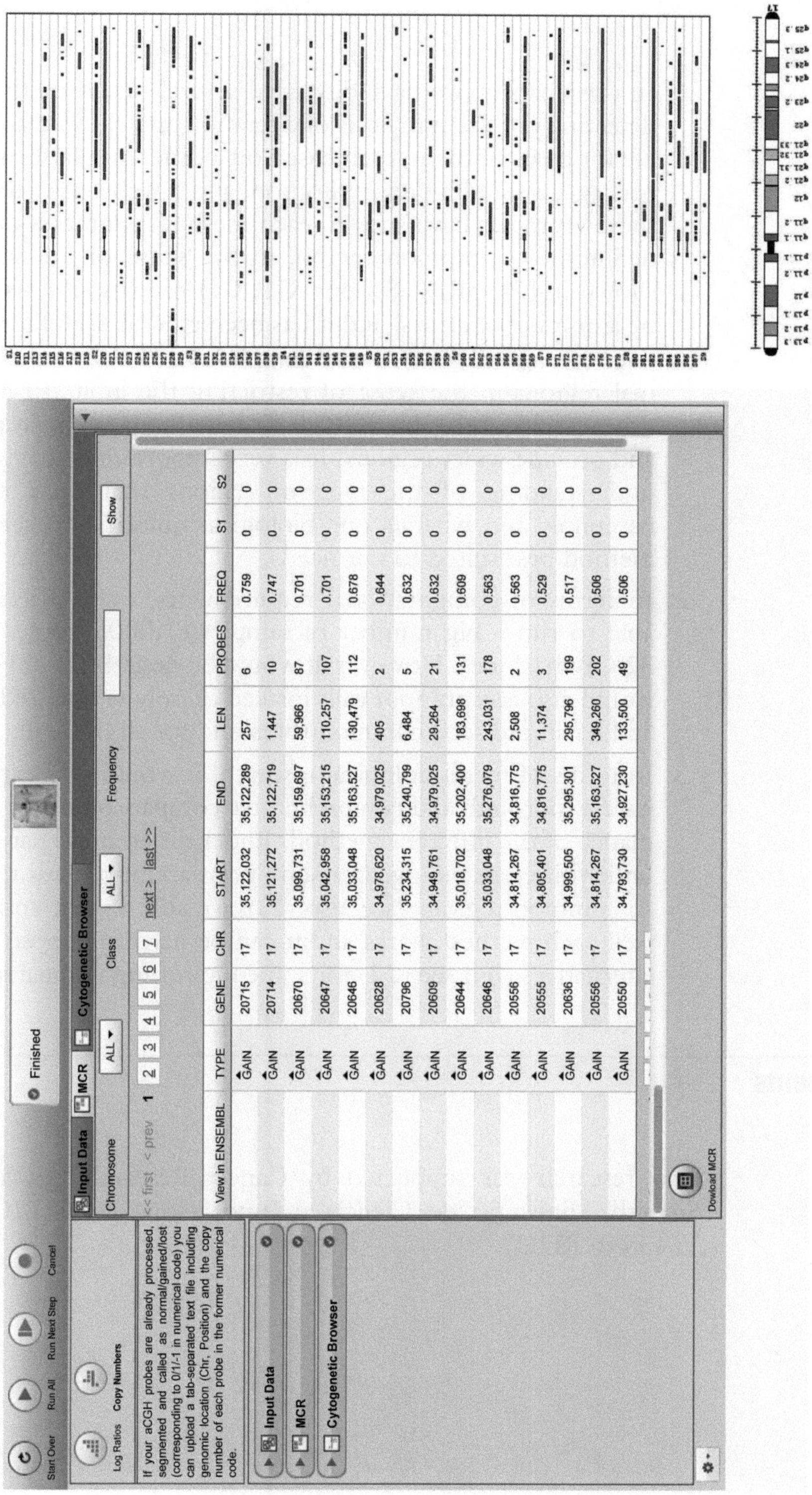

Fig. 5. SuperSORI output. Left panel shows an excerpt of the regions found by the method. Right panel shows an ideogram and the composition of each region.

4. Notes: Potential Pitfalls and Troubleshooting

1. In this chapter, we have discussed several recent methods for detecting common regions of alteration and showed an example for three of them, each based on a different platform (an R package, a stand-alone application, and a web-server tool). As these examples have shown, each method assumes a different definition of common region, in the sense of the scenarios shown in (2). pREC-S from RJaCGH returns regions shared by a number of arrays, and it has been proved useful for studying heterogeneity in a set of samples. GISTIC2 returns minimal regions in the sense of restricting the limits to find the driver genes, and SuperSORI is designed to clean noisy profiles and provide wider regions. It is very important to have a clear understanding of the type of regions that each method computes and what is our biological question to select the method best suited to our needs.
2. In terms of computational requirements, only GISTIC was able to run a big number of samples (1,000 samples with a 32 GB machine). However, it would be desirable that GISTIC would also accept input calls instead of only segmented means and having to set a common threshold for all of them.
3. In terms of methodological research, it would be desirable to have a thorough and comprehensive comparison of the performance of the different methods under different scenarios, and under different biological assumptions about process that lead to common regions (see also (2)); we still do not know how well each method works and it can be hard to choose among alternatives (but see (22) for some interesting comparisons).

Acknowledgments

This research was supported by Cancer Research UK and the Spanish Ministerio de Ciencia e Innovación (grant BIO2009-12458 to RDU).

References

1. Haber DA, Settleman J (2007) Cancer: drivers and passengers. Nature 446:145–146
2. Rueda OM, Diaz-Uriarte R (2010) Finding recurrent copy number alteration regions: a review of methods. Curr Bioinform 5(1):1–17
3. Shah SP (2008) Computational methods for identification of recurrent copy number alteration patterns by array CGH. Cytogenet Genome Res 123(1–4):343–351
4. Rueda OM, Diaz-Uriarte R (2007) Flexible and accurate detection of genomic copy-number changes from aCGH. PLoS Comput Biol 3(6):e122
5. Rueda OM, Diaz-Uriarte R (2009) RJaCGH: Bayesian analysis of aCGH arrays for detecting copy number changes and recurrent regions. Bioinformatics 25:1959–1960
6. Rueda OM, Diaz-Uriarte R (2009) Detection of recurrent copy number alterations in the genome: taking among-subject heterogeneity seriously. BMC Bioinformatics 10(308)
7. Ivakhno S, Tavaré S (2010) CNAnova: a new approach for finding recurrent copy number abnormalities in cancer SNP microarray data. Bioinformatics 26(11):1395–1402
8. Zhang Q et al (2010) CMDS: a population-based method for identifying recurrent dna copy number aberrations in cancer from high-resolution data. Bioinformatics 26(4):464–469
9. Seshan VE, Olshen A (2011) DNAcopy: DNA copy number data analysis. R package version 1.26.0
10. Olshen AB et al (2004) Circular binary segmentation for the analysis of array-based DNA copy number data. Biostatistics 5:557–572
11. Curtis C et al (2012) The genomic and transcriptomic architecture of 2,000 breast tumours reveals novel subgroups. Nature 486(7403):346–352. doi:10.1038/nature10983, Published online 18 April 2012
12. Rueda OM, Rueda C, Diaz-Uriarte R (2011) A Bayesian HMM with random effects and an unknown number of states for DNA copy number analysis. J Stat Comput Simul doi:{10.1080/00949655.2011.609818}
13. van de Wiel MA et al (2007) CGHcall: calling aberrations for array CGH tumor profiles. Bioinformatics 23(7):892–894
14. Benelli M et al (2010) A very fast and accurate method for calling aberrations in array-CGH data. Biostatistics 11(3):515–518
15. Beroukhim R et al (2007) Assessing the significance of chromosomal aberrations in cancer: methodology and application to glioma. Proc Natl Acad Sci USA 104(50): 20007–20012
16. Weir B et al (2007) Characterizing the cancer genome in lung adenocarcinoma. Nature 450(7171):893–898
17. Mermel CH et al (2011) Gistic2.0 facilitates sensitive and confident localization of the targets of focal somatic copy-number alteration in human cancers. Genome Biol 12(R41)
18. Rouveirol C et al (2006) Computation of recurrent minimal genomic alterations from array-CGH data. Bioinformatics 22:2066–2073
19. Etemadmoghadam D et al (2009) Integrated genome-wide DNA copy number and expression analysis identifies distinct mechanisms of primary chemoresistance in ovarian carcinomas. Clin Cancer Res 15(4):1417–1427
20. Sanchez-Garcia F et al (2010) JISTIC: identification of significant targets in cancer. BMC Bioinformatics 14(189)
21. Walter V, Nobel AB, Wright FA (2011) DINAMIC: a method to identify recurrent DNA copy number aberrations in tumors. Bioinformatics 27(5):678–685
22. Morganella S, Pagnotta S, Ceccarelli M (2011) Finding recurrent copy number alterations preserving within-sample homogeneity. Bioinformatics 27(21):2949–2956
23. Carro A et al (2010) waviCGH: a web application for the analysis and visualization of genomic copy number alterations. Nucleic Acids Res 38(suppl 2):W182–W187
24. Parker JS et al (2009) Supervised risk predictor of breast cancer based on intrinsic subtypes. J Clin Oncol 27(8):1160–1167

Chapter 22

Distinguishing Somatic and Germline Copy Number Events in Cancer Patient DNA Hybridized to Whole-Genome SNP Genotyping Arrays

Gavin Ha and Sohrab Shah

Abstract

Chromosomal aneuploidy and segmental copy number changes are common genomic aberrations in cancer. Copy number alterations (CNAs) arise from deletions, insertions, or duplications resulting in chromosomal aberrations and aneuploidy. Genomes of normal cells also exhibit variable copy number called germline copy number variants (CNVs). CNVs in the general population tend to confound interpretation of predictions when attempting to extract relevant driver somatic events in cancer. In large studies of CNAs in cancer patients, it becomes necessary to accurately identify and separate CNAs and CNVs so as to prioritize candidate tumor suppressors and oncogenes. We have developed a probabilistic approach, HMM-Dosage, for segmenting and distinguishing CNAs and CNVs as separate, discrete events in cancer SNP genotyping array data. We outline the steps and computer code for the analysis of whole-genome cancer DNA hybridized to SNP genotyping arrays, focusing on distinguishing somatic CNA and germline CNVs, and describe the combined approach of HMM-Dosage for probabilistic inference and classification of somatic and germline copy number changes.

Key words: Copy number alterations, Germline copy number variants, Probabilistic, HMM-Dosage, Cancer SNP genotyping array data, High-density SNP genotyping arrays, Affymetrix Genome-Wide Human SNP6.0, Illumina Human1M BeadChip

1. Introduction

Human cancer contains a complex combination of aberrations in the genome (1). Chromosomal aneuploidy and segmental copy number changes are among the most common genomic aberrations in cancer (2). CNAs, which can range in size from 1 kb to entire chromosomes, arise from events such as deletions, insertions, and duplications. These events can have pathogenic implications if regions containing oncogenes and tumor suppressor

Diponkar Banerjee and Sohrab P. Shah (eds.), *Array Comparative Genomic Hybridization: Protocols and Applications*, Methods in Molecular Biology vol. 973, DOI 10.1007/978-1-62703-281-0_22,

genes are affected. Large-scale studies profiling multiple types of genomic aberrations, including copy number and loss of heteozygosity (LOH), in cancer genomes have recently emerged. The Cancer Genome Atlas (TCGA) analyzed copy number and LOH (among many other genomic alterations) for 206 glioblastoma (3) and 489 high-grade serous ovarian adenocarcinomas (4) hybridized to single nucleotide polymorphism (SNP) genotyping arrays. In other studies, 746 cell-lines (5) and 2,520 primary tissues (6) across a range of cancer types were analyzed for copy number and LOH profiles. Moreover, the inclusion of somatic copy number events in the analysis of tumor samples, such as breast cancer, can aid in refining existing and discovering novel subtype classifications (7, 8).

Genomes of normal cells also exhibit, natural polymorphic regions with variable copy number called germline copy number variants (CNVs) (9–13). CNVs are often prevalent near segmental duplications (14) and contribute to global variations and phenotypic differences in healthy individuals. Substantial effort has been made in sequencing large cohorts to comprehensively explore CNVs as part of the 1000 Genomes Project (15, 16). Results from a majority of these studies surveying CNVs have been archived in the Database of Genomic Variants (DGV, http://www.projects.tcag.ca/variation/) (17). While germline CNVs in the population may be associated with disease (18–21), in the case of cancer, these events tend to confound interpretation of predictions when attempting to extract relevant driver somatic events. Analogous to germline SNPs in somatic point mutation analysis, CNVs are present as a source of genetic variation whose role in tumor pathophysiology is undetermined. As more studies involve ever-increasing cohort sizes of cancer patients, it becomes necessary to routinely use methods that are capable of accurately identifying and systematically separating CNA and CNV events so as to prioritize candidate tumor suppressors and oncogenes.

1.1. Detecting Copy Number Changes in Cancer Patient DNA Hybridized to SNP Genotyping Arrays

High-density SNP genotyping arrays are effective high-throughput assays for analyzing genome-wide copy number in humans. This technology is well established and cost-effective for large cohort cancer studies and analysis for recurrent copy number events.

Affymetrix Genome-Wide Human SNP6.0 and Illumina Human1M BeadChip are popular platforms used for genotyping, copy number, and LOH profiling. A single array, consisting of more than 1 million SNP and CNV probes, measures hybridization intensity as a surrogate for the amount of DNA at each loci. Analysis of SNP arrays primarily use two quantitative metrics derived from the allelic intensities. The first is the *log ratio*, which is computed as the ratio between the total (sum of A + B alleles) intensity of the tumor sample and the total intensity of the match normal or reference sample at a particular locus. The log ratio data at each SNP and CNV probe is used for performing total copy number prediction while allele-specific or parent-specific copy number uses only the set of SNP probes. In the absence of matched normal tissue, a pooled *reference* sample can be used. Typically, the reference is generated by taking the median probe intensity across a pool of healthy samples such as from the HapMap project (22). While the reference sample can help reduce the amount of population CNVs from the tumor data, using matched normal samples will be best for removing patient-specific CNVs. The second metric is B-allele frequency (BAF; usually derived from minor allele identification in population studies), which is the proportion of B-allele intensity relative to the total intensity computed solely using SNP probes. BAF naturally allows for analysis of genotypes and LOH whereby clusters of consecutive SNP probes are heterozygous with BAF of ~0.5 in the normal and becomes homozygous with BAF equal to 0 or 1 due to somatic allelic loss.

Similar to gene expression microarrays, SNP array intensities also need to be normalized to eliminate background noise. Normalization tools are platform-specific and generally categorized based on the manufacturer. For Affymetrix, *CRMAv2* (23, 24) and *ACNE* (25) perform cross-hybridization of probes between the alleles, GC content, and restriction enzyme induced fragment-length biases. For Illumina, *crlmm* (26, 27) and *tQN* (28) are tools that normalize raw and BeadStudio-processed intensities (29), respectively. In general, copy number neutrality, gain, and loss are inferred from values near, greater than, and less than zero in log scale, respectively.

Following normalization, segmentation algorithms are used to detect regions of copy number gain and loss that contiguously span a variable number of probes. Circular binary segmentation (30, 31) is a nonparametric technique for detecting significant change at breakpoints in aCGH and SNP genotyping data of a single sample. More recently, hidden Markov models (HMM) designed specifically for analyzing cancer data have been applied to segment aCGH and SNP genotyping data to handle noise and outliers by probabilistically accounting for spatial correlation (32–35) (Table 1). Furthermore, HMM-based tools are attractive due to their flexibility in modeling variable number of discrete biologically interpretable states and use statistical modeling approaches to account for polyploidy, normal contamination, and tumor heterogeneity.

1.2. Germline CNVs and Somatic CNAs in Cancer

We have previously determined the comprehensive landscapes of CNA and CNV in a cohort of 2,000 breast cancer patients whose DNA were hybridized to SNP genotyping arrays (8). We found that CNVs in tumor data have distinct statistical patterns compared

Table 1
Segmentation algorithms

Algorithm/software	Type of genomic aberration	Comments
DNAcopy (CBS) (36)	Copy number	Nonparametric. Requires post hoc step for discrete copy number classification
PICNIC (33)	Copy number and LOH	Ploidy correction, allele-specific copy number
GPHMM (35)	Copy number and LOH	Ploidy correction, GC content correction, stromal contamination
OncoSNP (34)	Copy number and LOH	Ploidy correction, GC content correction, stromal contamination, intra-tumoral heterogeneity
HMM-Dosage (8)	Copy number	Distinguish somatic CNA and germline CNV

to CNAs likely due to the polymorphic nature of germline variants and contributing normal tissue contamination in the sample. Furthermore, CNVs tend to be much shorter in length compared to CNAs based on inspection of records in Database of Genomic Variants (DGV) (17) and analysis of the breast cancer cohort. Normal contamination leads to weaker CNA event signals and shifting of log ratio amplitudes toward a diploid state. This is because CNAs are only found in the subpopulation of the sample making up the cancer cells; moreover, CNV signals are exaggerated relative to CNAs because all cells in the sample contain these germline events.

There are several strategies that can be used to separate CNAs and CNVs in cancer genome analysis. The best solution is to use an array from the corresponding matched normal DNA to compute log ratios prior to segmentation. This generates neutral ratio signals for germline variant probes, which allows for excluding CNVs inherently during somatic analysis but removes the ability to predict CNVs during segmentation.

In many instances, normal DNA from patients cannot be acquired; therefore, a reference is used (as described earlier). CNAs and CNVs can be detected and distinguished using post hoc approaches. The simplest approach is to manually "subtract" (or filter) copy number events that overlap records in a database containing germline polymorphic CNVs such as DGV. This presents a drawback; however, as there are uncertainties in region sizes due to differing resolutions from various technologies used in DGV entries. Furthermore, DGV regions collectively account for a large fraction of the genome, which can be problematic when applying the filtering approach. This can potentially exclude a large percentage of the genome from the results. To address these issues, PredictCNV (36), a regression tree classifier trained on DGV features, classifies segments of copy number changes as CNA or CNV. The distributed classifier is trained only on the lower resolution arrays in the TCGA glioblastoma data, and it is not clear if it is suitable for other cancer types or data using high-density arrays. These methods applied separately after segmentation analysis and do not probabilistically factor into improving segmentation of these events.

We have developed a probabilistic approach, HMM-Dosage, which extends ref. (32) for segmenting and distinguishing CNAs and CNVs as separate, discrete events in cancer SNP genotyping array data. This algorithm, described in ref. (8), is a combined approach that simultaneously performs segmentation and classification of CNAs and CNVs. HMM-Dosage models the different statistical patterns of CNVs and CNAs in a single, unified probabilistic framework by incorporating position-specific CNV prior knowledge.

We outline the steps and computer code for the complete analysis pipeline of whole-genome cancer DNA hybridized to SNP genotyping arrays, with the specific focus on distinguishing somatic CNA and germline CNVs. For this protocol, we will first describe the normalization of Affymetrix SNP6.0 arrays using CRMAv2. Next, we will detail the procedure for segmenting log ratio data using two different approaches in CBS and OncoSNP, and distinguishing CNAs and CNVs using the post-processing approach of the PredictCNV classifier. Finally, we will describe the combined approach of HMM-Dosage for probabilistic inference and classification of somatic and germline copy number changes.

The advantages of separating CNA and CNV predictions are twofold. First, and most importantly, filtering germline events from somatic CNAs helps prioritize regions that contain candidate driver tumor suppressors or oncogenes. Secondly, the dichotomized list of events can allow separate downstream analysis including investigating CNV-specific patterns in the cohort of cancer patients such as its cis-acting effect on gene expression.

2. Materials

2.1. Tumor Whole-Genome SNP Genotyping Array Data

1. Raw hybridization intensities of a tumor sample hybridized to an Affymetrix Genome-Wide Human SNP6.0 array. Match normal sample array or reference sample also hybridized to Affymetrix SNP6.0 (see Note 1). A precomputed reference derived from the HapMap270 dataset can be downloaded at http://www.compbio.bccrc.ca/software/hmm-dosage/.

2.2. Normalization Software Download and Installation

1. R programming software is required to run many software packages described in this protocol. R (version R-2.10.1 or higher) can be obtained at http://www.r-project.org/.
2. Allele-specific CRMAv2 normalization tool is part of the R library package *aroma.affymetrix* (v1.4). To install these packages, use these commands while in an active R session:

```
>source("http://www.aroma-project.org/hbLite.R");
>hbInstall("aroma.affymetrix");
```

3. Download required annotation data files from http://www.affymetrix.com and http://www.aroma-project.org/chipTypes/GenomeWideSNP_6 (see Note 2). Set up the directory structure as described in http://www.aroma-project.org/setup.

2.3. Copy Number Segmentation Software Download and Installation

1. CBS is part of the R package, DNAcopy, which can be installed within an R session. Method and parameter definitions are comprehensively detailed in http://www.bioconductor.org/packages/2.6/bioc/manuals/DNAcopy/man/DNAcopy.pdf.

```
>install.packages("DNAcopy");
```

2. PredictCNV is implemented in R. The files required ("*prediction_code.R*" and "*CNV_prediction_objects.RData*") to run the algorithm can be downloaded from http://www.mskcc.org/mskcc/html/72726.cfm. The authors provide a document with detailed definitions and instructions for each function. The *randomForest* R library is also required and can be downloaded here http://www.cran.r-project.org/web/packages/randomForest/index.html.
3. OncoSNP (https://sites.google.com/site/oncosnp/) is a cancer-specific copy number segmentation software that can be downloaded as compiled MATLAB executables (see Note 3).
4. HMM-Dosage (http://www.compbio.bccrc.ca/software/hmm-dosage/) is software for segmentation and prediction of copy number in tumor data (see Note 3).

In the next section, we assume that all software has been correctly installed.

3. Methods

3.1. Normalization of Array Intensities Using AS-CRMAv2

1. Within a new R session (version R-2.10.1 or higher), load the *aroma.affymetrix* library package that contains the CRMAv2 normalization tool (see Fig. 1, A1).

```
> library("aroma.affymetrix")
```

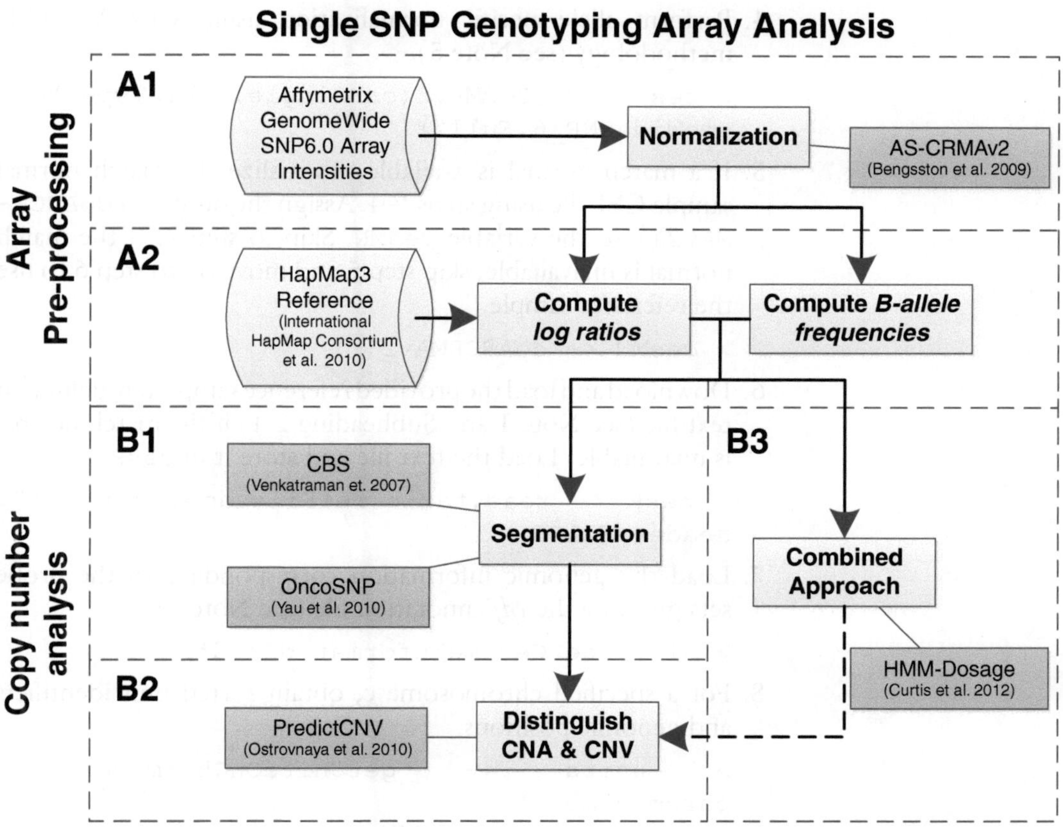

Fig. 1. Workflow diagram for single genotyping array analysis procedure of the Affymetrix Genome-Wide SNP6.0 platform. (A1) First step in the analysis consists of intensity normalization using allele-specific (AS)-CRMAv2. (A2) Computing the *log ratios* requires a reference sample that can be generated from any dataset consisting of normal (not tumor) individuals; HapMap3 dataset is described. The *B-allele frequency* (BAF) is computed as the relative fraction of the B-allele intensity with respect to the total intensity. (B1) Nonparametric (e.g., CBS) and parametric (e.g., OncoSNP) algorithms are used to segment log ratio and/or BAF data into regions of discrete copy number changes. However, most approaches do not inherently consider distinguishing germline CNV and somatic CNA events. (B2) The classification model, PredictCNV, was designed to probabilistically separate CNVs and CNAs in results generated from other segmentation algorithms. (B3) HMM-Dosage is a hidden Markov model that simultaneously accounts for CNAs and CNVs, segmenting and predicting regions of copy number changes for both somatic and germline events. As an optional post-processing step, PredictCNV can be used to filter the output of HMM-Dosage (*dashed arrow*).

2. Link the location of the chip definition file (*cdf*) for the *Affymetrix GenomeWideSNP_6* platform (see Note 2).

```
> cdf <- AffymetrixCdfFile$byChipType("GenomeWideSNP_6", tags="Full")
```

3. Load a single CEL file into an R object and assign it to a set containing only the one sample (see Note 4).

```
> celFile <- AffymetrixCelFile$fromFile(celFilename)
>csRSingle <- AffymetrixCelSet(list(celFile))
> setCdf(csRSingle, cdf=cdf)
```

4. Perform allele-specific normalization using CRMAv2 (24) methodology (see Note 5).

```
> ces <- doASCRMAv2(csRSingle, chipType="GenomeWideSNP_6,Full")
```

5. If a match normal is available, normalize the match normal sample CEL file using steps 1–4. Assign the output of `doASCRMAv2()` to the variable `cesMN`. Skip to step 7. If the match normal is unavailable, skip step 5 and move on to step 6 to use the reference sample.

```
> cesMN <- doASCRMAv2(...)
```

6. Download and load the provided reference sample tab-delimited text file (see Note 1 and Subheading 2.1) if the match normal is unavailable. Load the text file and store it in `cesR`.

```
> cesR <- read.table(refFilename, sep="\t", header=FALSE)
```

7. Load the genomic information corresponding to the probe sets given by the *cdf* annotation file (see Note 2).

```
> gi <- getGenomeInformation(cdf)
```

8. For a specified chromosome c, obtain sorted unit identifiers and genomic positions.

```
> units <- getUnitsOnChromosome(gi, chromosome=c)
> pos <- getPositions(gi, units=units)
> probes <- getUnitNames(cdf, units=units)
> N <- length(pos)
> chrIndex <- as.data.frame(rep.int(c,N))
> sI <- order(pos)
```

9. Extract the total intensities and B-allele frequencies (BAF) for the probe units of the given chromosome c (see Note 6).

```
> theta <- extractTotalAndFreqB(ce, units=units)
> # computed BAF
> BAF <- theta[,"freqB"]
> BAF[grepl("CN",probes)] <- 2 # use 2 for CN probes
```

10. Compute the raw copy number for each probe by first extracting the normalized intensities of the match normal or reference (see Notes 1 and 7, and Fig. 1, A2). Here, the given example uses the reference. One can use the matched normal instead by replacing `cesR` with `cesRMN`.

```
> # extract reference intensities
> thetaR <- cesR[cesR[,1]==j,]
```

```
> # compute raw copy number
> # assume that reference positions match exactly with the sorted positions of theta (that is, pos[sI])
> C <- 2.0*theta[sI]/thetaR[,3]
```

11. Output two sets of results for raw copy number and *log2 ratios*. Each file consists of 5 columns, sorted by positions for given chromosome c using `sI` generated in step 8. The columns are probe name, chromosome number, genomic coordinate of probe, raw copy number or log ratio, and B-allele frequency.

```
> cnDataFrameRawCN <- cbind(probes[sI],chrIndex,pos[sI],C[sI],BAF[sI])
> cnDataFrameLogR <- cbind(probes[sI],chrIndex,pos[sI],log2(C[sI]/2),BAF[sI])
> write.table(cnDataFrameRawCN, file=fc1, append=T, quote=FALSE, sep="\t", eol="\n", na="NaN", dec=".", row.names=F, col.names=F)
> write.table(cnDataFrameLogR, file=fc2, append=T, quote=FALSE, sep="\t", eol="\n", na="NaN", dec=".", row.names=F, col.names=F)
```

12. Repeat steps 8–11 for each of the 22 autosomes and 2 sex chromosomes. To do this in a single code block, use a loop around the code commands listed in steps 8–11. The results are output to the file "*rawCN-BAF_output.txt*" and "*logR-BAF_output.txt*" Fig. 2, Panel 1 shows a plot of the normalized log ratios for chromosome 10 of a breast cancer sample.

```
> fc1 <- file("rawCN-BAF_output.txt","w+") #open file handles
> fc2 <- file("logR-BAF_output.txt","w+")
> for (j in 1:24){
## Use chr numbers 1-22, X, and Y
if (j==23){ c='X' }
else if (j==24){ c='Y' }
else { c=j }
## code from steps 8-11…
}
> close(fc1) #close the file handle
> close(fc2)
```

3.2. Segmentation Analysis of Copy Number Changes: CBS

1. Within a new R session (version R-2.10.1 or higher), load the *DNAcopy* library package that contains the circular binary segmentation (CBS) algorithm (see Fig. 1, B1).

```
> library(DNAcopy)
```

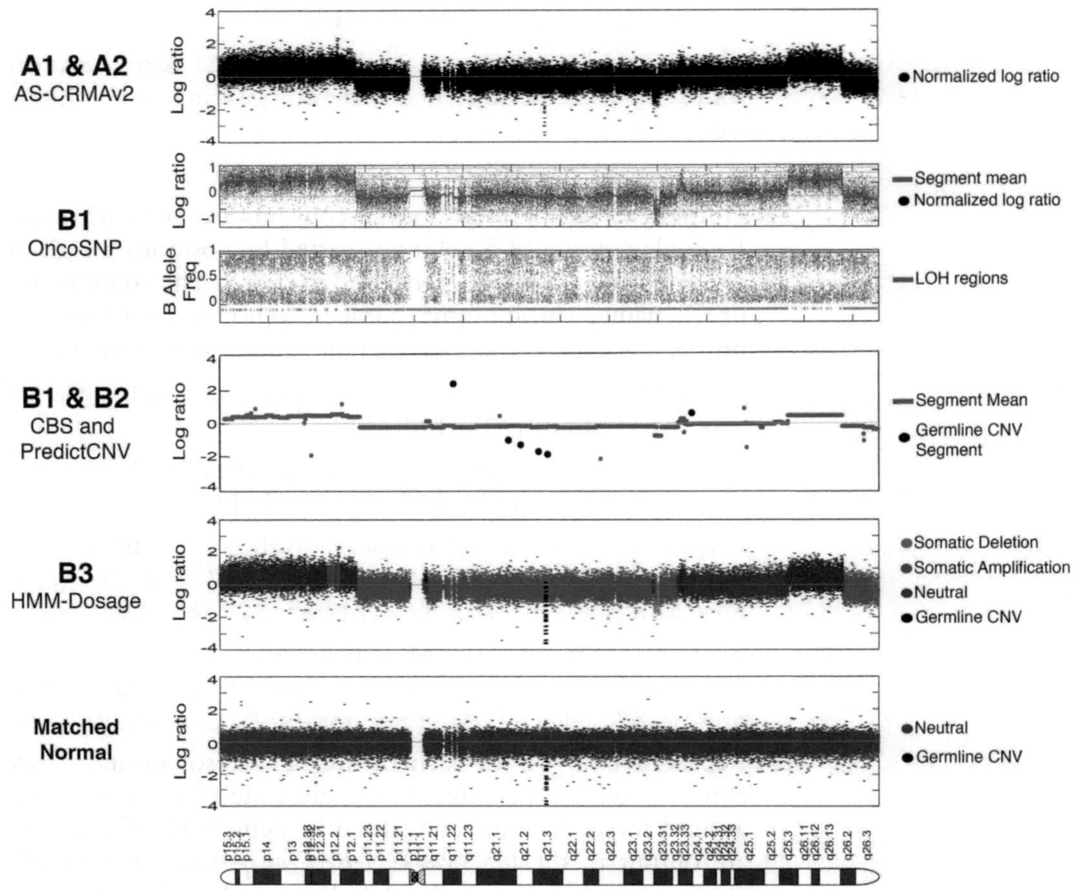

Fig. 2. Copy number results for chromosome 10 of a triple negative breast cancer sample (38) hybridized to the Affymetrix SNP6.0 platform. Panel 1 displays the results for AS-CRMAv2 (24) normalization and computation of log ratios without copy number state classification (Steps A1 and A2, see Fig. 1). Panels 2 and 3, which are plots generated directly by OncoSNP (34), show the predicted segment means and loss of heterozygosity (LOH). Panel 4 shows the predicted segments using CBS and classification of germline CNVs using PredictCNV (36) (Steps B1 and B2). Panel 5 displays copy number predictions generated by HMM-Dosage (8), which distinguishes germline CNVs and somatic CNAs (Step B3). An example of a CNV predicted by HMM-Dosage can also be found in the matched normal sample of the same patient (Panel 6).

2. Load in the log ratio data file "*logR-BAF_output.txt*" for a single sample generated from the output of AS-CRMAv2 generated in Subheading 3.1 (see Note 8).

```
> sampleRawCN <- read.table(file="logR-BAF_
output.txt", header=F, stringsAsFactors=F)
```

3. Prepare a "copy number array" object that will be passed into the main function for CBS analysis (see Note 9).

```
> sampleID = "test1"
sampleLogR <- sampleRawCN[,4]
> cna.obj <- CNA(sampleLogR,
chrom=sampleRawCN[,2], maploc=sampleRawCN[,3],
data.type="logratio", sampleid=c(sampleID))
```

4. (Optional) Smoothing can be applied to the CNA object prior to segmentation (see Note 10).

```
> cna.obj <- smooth.CNA(cna.obj)
```

5. Perform segmentation using CBS (see Note 11).

```
> segment.cna.obj <- segment(cna.obj)
```

6. Prepare the results in a new variable and save it as an image for downstream analysis (see Note 12).

```
> dataseg$data <- cbind(sampleRawCN[,2:3], sampleLogR)
> names(dataseg$data) <-
c("chrom","maploc",sampleID)
> dataseg$output <- segment.cna.obj$output
> dataseg$data[is.nan(dataseg$data[,3]),3] <-0
> save(list=dataseg, file="cbs_results.RData")
#save the variable
```

3.3. Segmentation Analysis of Copy Number Changes: OncoSNP

1. Ensure that the current operating system is Linux x64 architecture and that OncoSNP is properly installed (see author website listed in Subheading 2.3, step 2, Note 3, and Fig. 1, B1).
2. Prepare a 3-column text file that specifies the sample ID, locations to the input tumor *log ratio* and *BAF* data, and location of normal sample *log ratio/BAF* data (see Note 13). For this example, let this file be named "*singleSampleInputForOncoSNP.txt.*" The tumor *log ratio/BAF* file for this example is "*logR-BAF_output.txt*," which was generated in Subheading 3.1.
3. Download and install the GC content files for NCBI genome build 37 (hg19) into a folder named "quantisnp."

```
OncoSNP_v1.1> mkdir quantisnp; cd quantisnp;
wget ftp://ftp.stats.ox.ac.uk/pub/yau/quan-
tisnp2/download/b37.tar.gz
tar xvfz b37.tar.gz
```

4. Create an output directory for storing the results.

```
OncoSNP_v1.1> mkdir results/
```

5. Run OncoSNP from the Linux terminal command line while in the installation directory (see Note 14). Fig. 2, Panel 2–3 shows a plot of the copy number and loss of heterozygosity (LOH) predictions.

```
OncoSNP_v1.1> run_oncosnp.sh \
MATLAB_Compiler_Runtime/v714/ \
--batch-file singleSampleInputForOncoSNP.txt \
--output-dir results/ \
--gcdir quantisnp/b37/ \
```

```
--paramsfile configuration/hyperparameters-affy.
   dat \
--levelsfile configuration/levels-affy.dat \
--trainingstatesfile configuration/trainingStates.
   dat \
--tumourstatesfile configuration/tumourStates8.
   dat \
--subsample 30 --emiters 15 --headerlines 0 \
--stromal --intratumour --female --plot –fulloutput
   --headerlines 0
```

3.4. Distinguishing Somatic CNAs and Germline CNVs: PredictCNV Classifier

1. Within a new R session (version R-2.10.1 of higher), load the library and scripts necessary to run PredictCNV (see Subheading 2.3, step 2 and Fig. 1, B2). Ensure that the required files of "*prediction_code.R*" and "*CNV_prediction_objects.RData*" are in the current working directory, otherwise, specify the full path.

```
> library(randomForest)
> source('prediction_code.R')
> load('CNV_prediction_objects.RData')
```

2. Load in the segmentation data generated from Subheading 3.2 (see Note 15) using the load command in R. Ensure that the saved "*.RData*" file is in the current working directory, otherwise, specify the full path.

```
> load("cbs_results.RData")
```

3. Run the prediction software using the PredictCNV classifier (see Notes 16 and 17).

```
> newSegs <- predict.CNVs(dataseg, use.
cohort=FALSE, smoothed=TRUE, glad=FALSE,
gainloss.defined=FALSE, nmad=1)
```

4. Output the results into a file consisting of tab-delimited columns (see Note 18). Figure 2, Panel 4 shows examples of classified CNV segments.

```
> write.table(newSegs, file=outFile, col.
names=T, row.names=F, sep="\t", quote=F)
```

3.5. HMM-Dosage: A Combined Approach for Segmenting and Predicting CNAs and CNVs

1. Ensure that the current operating system is Linux x64 architecture and that HMM-Dosage is properly installed (see author website listed in Subheading 2.3, step 4, Note 3, and Fig. 1, B3).
2. Prepare the input raw copy number file by processing the file "*rawCN-BAF_output.txt*," which was generated in Subheading 3.1. In the command line, extract the columns 2–4, which contain chromosome number, genomic position, and log ratio, and substitute the chromosome X and Y symbols with 23 and 24, respectively. This is assigned to the new file "*rawCN-BAF_output_processed.txt*."

```
HMMK11_0.1.0> cut -f 2-4 rawCN-BAF_output.txt
| sed s/X/23/g | sed s/Y/24/g > rawCN-BAF_out-
put_processed.txt
```

3. Download the CNV frequency file from the website http://www.compbio.bccrc.ca/software/hmm-dosage/ (see Subheading 2.3, step 4 and Note 19). For this example, let's assume the desired frequencies are derived from the HapMap3 (22) dataset.

```
HMMK11_0.1.0> wget http://www.compbio.bccrc.
ca/wp-content/uploads/2010/10/
HapMap3CNV1258_probeFreq_AffySNP6.txt.gz;
gunzip    HapMap3CNV1258_probeFreq_AffySNP6.
   txt.gz
```

4. Create an output directory to store the results.

```
HMMK11_0.1.0> mkdir results/
```

5. Run HMM-Dosage from the Linux command line (see Note 20). Figure 2, Panel 5 shows an example plot of a predicted CNV, distinguished from somatic events.

```
HMMK11_0.1.0> cd bin/;
./run_hmmK11LogR.sh \
../MATLAB_Component_Runtime/v77/ \
../rawCN-BAF_output_processed.txt \
../HapMap3CNV1258_probeFreq_AffySNP6.txt \
../test/initialParams.mat \
../results/hmmdosage_results_cna.txt \
../results/hmmdosage_results_segs.txt \
../results/hmmdosage_results.mat \
-1
```

4. Notes

1. When a matched normal is unavailable, the reference sample can be generated from an external dataset of normal individuals such as from HapMap270 (37) or HapMap3 (22) cohorts. The dataset is summarized by computing the median across all individuals for each probe *i*, θ_i^R . This value is used as the denominator in the log ratio. The choice of the reference dataset should take into consideration gender, which can affect analysis of chromosome X, and ethnic origin, which can affect population specific germline CNVs.
2. Ensure that the chip definition file (cdf), "*GenomeWideSNP_6, Full.cdf*," is in the location, "*/annotationData/chipTypes/GenomeWideSNP_6/*." This file contains the genomic coordinates

and probe sequences corresponding to the probe sets found on the array. Furthermore, ensure the presence of a file for each of these three extensions: *.acs*, *.ufl*, *.ugp*; these files should all correspond to the same genome build (e.g., *hg18*).

3. These MATLAB executables were originally compiled on Linux x64 architecture operating systems; thus, they can only be run on machines with the this architecture.

4. Scripting the analysis to run one sample, independently of others, allows for easy parallelization. CRMAv2 was designed as a single-sample normalization technique.

5. The function `doASCRMAv2()` can take up to a few hours. It performs allelic cross-talk calibration, base-pair normalization, probe-level summarization, and fragment-length normalization (23, 24).

6. The output to function `extractTotalandFreqB()` consists of 2 columns: (1) Total intensity and (2) Fraction of B-allele intensity (BAF) with respect to total intensity. Total intensity θ_i is computed as the sum between the normalized intensities of the major allele θ_i^A and minor allele θ_i^B for each SNP probe *i*, and the measured intensity of each CNV probes. BAF is computed as $\theta_i^B / \theta_i^A + \theta_i^B$. CNV probes are non-polymorphic and will only contain one intensity value for each probe; hence, BAF values will not be computed for these.

7. Raw copy number is computed as the ratio between the normalized tumor intensity and normalized reference (or match normal) intensity of probe *i* given by equation: $C_i = 2(\theta_i / \theta_i^R)$. The factor of 2 is simply for interpretability of a neutral ratio being "two" copies. When computing raw copy number, you must ensure that the positions in the reference file correspond exactly with the positions in the tumor sample.

8. This demonstration of CBS only involves segmentation of total copy number; hence, the BAF data (5th column) in the input data file (output from AS-CRMAv2) is not used. It is possible to segment the BAF data to find segments of allelic imbalance for LOH analysis. For this example, the log ratio data is used; therefore, the file "*logR-BAF_output.txt*" is used.

9. The `Copy Number Array` object requires several arguments. `chrom` specifies the chromosome for each probe; this is given by the 2nd column in the input raw copy number file. `maploc` specifies the genomic coordinate of the probes; this is given by the 3rd column. `data.type` is set to "log ratio." `sampleid` is user specified.

10. Smoothing is applied to the `Copy Number Array` object prior to segmentation. Default settings for this function using sliding window of 5 data points for detecting outliers with the 3rd value being the data point of interest. Smoothing of an

outlier involves shifting the data point to 2 standard deviations from the median of the sliding window.

11. The `segment()` function does *not* generate the discrete copy number call; it only performs segmentation. The $output list of the return object for the `segment()` function consists of 6 columns: (1) sample ID, (2) chromosome, (3) genomic start coordinate of segment, (4) genomic end coordinate of segment, (5) number of markers for segment, and (6) average log ratio across all probes in the segment.
12. Processing of the `segment()` function results into the `dataseg` variable as shown is necessary for running PredictCNV in a downstream analysis step.
13. OncoSNP version 1.1 accepts a 3-column file that specifies the sample IDs, locations of tumor sample *log ratio/BAF* data files, and locations of normal sample *log ratio/BAF* data files (if available). For the purposes of parallelization, a single sample is specified: the input tumor *log ratio* and *BAF* data is the file "*logR-BAF_output.txt*." Note that "*logR-BAF_output.txt*" is already in the correct format.
14. The full details of the parameter arguments for OncoSNP version 1.1 are described in the author website listed in Subheading 2.3, step 3. `--batch-file` specifies the file created in Subheading 3.3, step 2. `--gcdir` specifies the directory of the GC content files discussed in Subheading 3.3, step 3; this is used for correcting local GC content bias during total copy number analysis of *log ratios*. `--paramsfile` contains the software settings and hyperparameter of the model. `--levelsfile` contains the mean (log ratio) for each distribution corresponding to a copy number state. `--trainingstatesfile` contains the settings used for model training via expectation maximization. `--tumourstatesfile` specifies the list of copy number and allelic imbalance states used in the model. `--subsample` specifies the proportion of randomly sampled data points to use for model training. `--emiters` specifies number of iterations for model training. `--stromal` is a flag that, when used, will inform the model to account for normal contamination, which by default, will be estimated as between 0 and 90% (at intervals of 10%). `--intratumour` is a flag that will inform the model to estimate intra-tumoral heterogeneity. Use `--female` for including chromosome X in the analysis whereas `--male` includes chromosome Y. `--fulloutput` flag indicates for the software to also output the probe-level results in addition to the segments. The description of output files and plots are provided in the author website listed in Subheading 2.3, step 3.

15. This example will only demonstrate the application of the PredictCNV classifier using CBS segmentation output generated in Subheading 3.2. The classifier can use segmentation results generated from any method as described in ref. (36). We leave it to the reader to experiment with segments produced from other algorithms such as OncoSNP or HMM-Dosage. To be compatible with PredictCNV, the outputs from these algorithms will need to be adjusted to the acceptable format.

16. PredictCNV is a supervised approach that uses a random forest decision tree classification model. The model in the distributed code was trained using TCGA glioblastoma data hybridized to Aglient 244k arrays (3). This algorithm is designed to produce more accurate results when a cohort of tumor patients is used. Therefore, the segments from each patient can be concatenated into a single object for input into the classifier. For this guide, we use the single sample analysis approach to allow for convenient parallelization of large number of samples.

17. The default input parameters to the `predict.CNVs()` function are used with the exception of `use.cohort=FALSE`. The `gain.loss.defined` parameter specifies whether a column with header name "*state*" and values of {"*Loss*," "*Normal*," "*Gain*"} is included in the object `dataseg$output`. OncoSNP and HMM-Dosage provide discrete copy number predictions that can be used to populate the "*state*" column after proper formatting (see Note 15). These results can be analyzed with `predict.CNVs()` while setting `gain.loss.defined=TRUE`. For CBS, the copy number prediction is not provided; thus, `gain.loss.defined=FALSE` and `nmad` parameters are used. `nmad` specifies the "threshold for the number of median absolute deviations of the array residuals" for calling "*Gain*" or "*Loss*" (see Documentation provided by the authors, Subheading 2.3, step 2).

18. The output of the `predict.CNVs()` the same as `dataset$output` except with an additional column, "*predict.CNV*," where TRUE denotes the segment as a predicted CNV and FALSE denotes a predicted CNA segment.

19. The CNV frequency file consists of probe-level frequencies of the germline copy number gains and losses across a normal (healthy) dataset such as HapMap (37). The frequency file contains 6 columns, each corresponding to the CNV frequencies of homozygous deletion, hemizygous deletion, gain, amplification, and high-level amplification. Inclusion of these frequencies, which is part of a key feature in HMM-Dosage, provides prior knowledge of a particular locus being a germline CNV event in the population. This allows HMM-Dosage to simultaneously segment and predict copy number, returning calls from an increased state-space that includes both somatic CNAs and germline CNVs.

20. The full details of the parameter arguments for HMM-Dosage (version 0.1.0) are described in the author website listed in Subheading 2.3, step 4. The first argument is the `infile`, which in the example was generated at Subheading 3.5, step 2, contains 3 columns for chromosome number, genomic coordinate of probe, and raw copy number (see Note 7). `freqfile` specifies the file downloaded as described in Subheading 3.5, step 3 (see Note 19). `paramSetFile` contains the software settings, hyperparameter, means (*log ratio*) for each distribution corresponding to the copy number state, and settings used for model training via expectation maximization. `outfile` specifies the probe-level output file which will contain 5 columns for chromosome number, start position of probe, end position of probe, log ratio, and state predictions. Somatic CNA states are correspond to 1 (homozygous deletion), 2 (hemizygous deletion), 4 (gain), 5 (amplification), and 6 (high-level amplification). For germline CNVs, the same corresponding discrete state definitions are assigned to 7–11, respectively. State 3 is the neutral genotype. `paramfile` specifies the output file that stores the converged training parameters for the model. This information is useful for advanced users who wish to inspect parameter changes throughout expectation maximization estimation. The file is in MATLAB binary format, thus, MATLAB installation is required. `chr` use -1 for this argument.

References

1. Stratton MR, Campbell PJ, Futreal PA (2009) The cancer genome. Nature 458:719–724
2. Negrini S, Gorgoulis VG, Halazonetis TD (2010) Genomic Instability—an evolving hallmark of cancer. Nat Rev Mol Cell Biol 11:220–228
3. Cancer Genome Atlas Research Network (2008) Comprehensive genomic characterization defines human glioblastoma genes and core pathways. Nature 455:1061–1068
4. Cancer Genome Atlas Research Network (2011) Integrated genomic analyses of ovarian carcinoma. Nature 474:609–615
5. Bignell GR, Greenman CD, Davies H et al (2010) Signatures of mutation and selection in the cancer genome. Nature 463:893–898
6. Beroukhim R, Mermel CH, Porter D et al (2010) The landscape of somatic copy-number alteration across human cancers. Nature 463:899–905
7. Chin SF, Teschendorff AE, Marioni JC et al (2007) High-resolution aCGH and expression profiling identifies a novel genomic subtype of ER negative breast cancer. Genome Biol 8:R215
8. Curtis C, Shah SP, Chin S et al (2012) The genomic and transcriptomic architecture of 2,000 breast tumours reveals novel subgroups. Nature 486:346–352
9. Sebat J, Lakshmi B, Troge J et al (2004) Large-scale copy number polymorphism in the human genome. Science 305:525–528
10. Tuzun E, Sharp AJ, Bailey JA et al (2005) Fine-scale structural variation of the human genome. Nat Genet 37:727–732
11. Redon R, Ishikawa S, Fitch KR et al (2006) Global variation in copy number in the human genome. Nature 444:444–454
12. Kidd JM, Cooper GM, Donahue WF et al (2008) Mapping and sequencing of structural variation from eight human genomes. Nature 453:56–64
13. Conrad DF, Pinto D, Redon R et al (2010) Origins and functional impact of copy number variation in the human genome. Nature 464:704–712

14. Sharp AJ, Locke DP, McGrath SD et al (2005) Segmental duplications and copy-number variation in the human genome. Am J Hum Genet 77:78–88
15. 1000 Genomes Project Consortium, Durbin RM, Abecasis GCR et al (2010) A map of human genome variation from population-scale sequencing. Nature 467:1061–1073
16. Mills RE, Walter K, Stewart C et al (2011) Mapping copy number variation by population-scale genome sequencing. Nature 470:59–65
17. Iafrate AJ, Feuk L, Rivera MN et al (2004) Detection of large-scale variation in the human genome. Nat Genet 36:949–951
18. Friedman JM, Baross A, Delaney AD et al (2006) Oligonucleotide microarray analysis of genomic imbalance in children with mental retardation. Am J Hum Genet 79:500–513
19. Sebat J, Lakshmi B, Malhotra D et al (2007) Strong association of de novo copy number mutations with autism. Science 316:445–449
20. Lee C, Iafrate AJ, Brothman AR (2007) Copy number variations and clinical cytogenetic diagnosis of constitutional disorders. Nat Genet 39:S48–S54
21. Sharp AJ, Mefford HC, Li K et al (2008) A recurrent 15q13.3 microdeletion syndrome associated with mental retardation and seizures. Nat Genet 40:322–328
22. The International HapMap 3 Consortium and Principal investigators and Altshuler, David M and Gibbs, Richard A and Peltonen, Leena and Project coordination leaders and Altshuler, David M and Gibbs, Richard A and Peltonen, Leena and Dermitzakis, Emmanouil and Manuscript writing group and Schaffner, Stephen F and Yu, Fuli and Peltonen, Leena and Dermitzakis, Emmanouil and Bonnen, Penelope E and Altshuler, David M and Gibbs, Richard A and Genotyping, QC, de Bakker, Co-Leader, Paul IW et al (2010) Integrating common and rare genetic variation in diverse human populations. Nature 467:52–58
23. Bengtsson H, Irizarry R, Carvalho B et al (2008) Estimation and assessment of raw copy numbers at the single locus level. Bioinformatics 24:759–767
24. Bengtsson H, Wirapati P, Speed TP (2009) A single-array preprocessing method for estimating full-resolution raw copy numbers from all affymetrixgenotypingarraysincludingGenomeWideSNP 5 & 6. Bioinformatics 25:2149–2156
25. Ortiz-Estevez M, Bengtsson H, Rubio A (2010) ACNE: a summarization method to estimate allele-specific copy numbers for Affymetrix SNP arrays. Bioinformatics 26:1827–1833
26. Scharpf RB, Ruczinski I, Carvalho B et al (2011) A multilevel model to address batch effects in copy number estimation using SNP arrays. Biostatistics 12:33–50
27. Ritchie ME, Carvalho BS, Hetrick KN et al (2009) R/Bioconductor software for Illumina's Infinium whole-genome genotyping BeadChips. Bioinformatics 25:2621–2623
28. Staaf J, Vallon-Christersson J, Lindgren D et al (2008) Normalization of Illumina Infinium whole-genome SNP data improves copy number estimates and allelic intensity ratios. BMC Bioinformatics 9:409–409
29. Peiffer DA, Le JM, Steemers FJ et al (2006) High-resolution genomic profiling of chromosomal aberrations using Infinium whole-genome genotyping. Genome Res 16:1136–1148
30. Venkatraman ES, Olshen AB (2007) A faster circular binary segmentation algorithm for the analysis of array CGH data. Bioinformatics 23:657–663
31. Olshen AB, Venkatraman ES, Lucito R et al (2004) Circular binary segmentation for the analysis of array-based DNA copy number data. Biostatistics 5:557–572
32. Shah SP, Xuan X, DeLeeuw RJ et al (2006) Integrating copy number polymorphisms into array CGH analysis using a robust HMM. Bioinformatics 22:e431–e439
33. Greenman CD, Bignell G, Butler A et al (2010) PICNIC: an algorithm to predict absolute allelic copy number variation with microarray cancer data. Biostatistics 11:164–175
34. Yau C, Mouradov D, Jorissen RN et al (2010) A statistical approach for detecting genomic aberrations in heterogeneous tumor samples from single nucleotide polymorphism genotyping data. Genome Biol 11:R92
35. Li A, Liu Z, Lezon-Geyda K et al (2011) GPHMM: an integrated hidden Markov model for identification of copy number alteration and loss of heterozygosity in complex tumor samples using whole genome SNP arrays. Nucleic Acids Res 39:4928–4941
36. Ostrovnaya I, Nanjangud G, Olshen AB (2010) A classification model for distinguishing copy number variants from cancer-related alterations. BMC Bioinformatics 11:297–297
37. International HapMap Consortium, Frazer KA, Ballinger DG et al (2007) A second generation human haplotype map of over 3.1 million SNPs. Nature 449:851–861
38. Shah SP, Roth A, Goya R et al (2012) The clonal and mutational evolution spectrum of primary triple-negative breast cancers. Nature 486:395–399

Index

Diponkar Banerjee and Sohrab P. Shah (eds.), *Array Comparative Genomic Hybridization: Protocols and Applications*, Methods in Molecular Biology vol. 973, DOI 10.1007/978-1-62703-281-0, © Springer Science+Business Media, LLC 2013

D

E

F

G

N

O

P

Q

R

U

V

W

X

Z

MIX
Papier aus verantwortungsvollen Quellen
Paper from responsible sources
FSC® C105338

If you have any concerns about our products,
you can contact us on
ProductSafety@springernature.com

In case Publisher is established outside the EU,
the EU authorized representative is:
Springer Nature Customer Service Center GmbH
Europaplatz 3, 69115 Heidelberg, Germany

Printed by Libri Plureos GmbH
in Hamburg, Germany